Philosophical Ecology

Roberto Marchesini

Philosophical Ecology

Anthropocentrism as an Epistemological Obstacle

palgrave
macmillan

Roberto Marchesini
School of Human-Animal Interaction and Centre for the Study of Posthumanist Philosophy
Bologna, Italy

ISBN 978-3-032-04986-5 ISBN 978-3-032-04987-2 (eBook)
https://doi.org/10.1007/978-3-032-04987-2

This Palgrave Macmillan imprint is published by the registered company Springer Nature Switzerland AG.
The registered company address is: Gewerbestrasse 11, 6330 Cham, Switzerland

Acknowledgments

I would like to express my sincere gratitude to Sarah De Sanctis for her valuable work on the translation, and to Sabrina Golfetto for her crucial editorial contribution.

Contents

About the Author

Roberto Marchesini (1959) is Director of School of Human-Animal Interaction and of the Centre for the Study of Posthumanist Philosophy, both based in Bologna, Italy. He has collaborated with the World Phenomenology Institute in Boston and has published several essays in *Analecta Husserliana*. His main publications include: *Over the Human. Post-humanism and the Concept of Animal Epiphany* (Springer, 2017); *The Philosophical Ethology of Roberto Marchesini*, a collection of essays edited by Jeffrey Bussolini, Brett Buchanan, and Matthew Chrulew (Routledge, 2017); *Dialogo Ergo Sum. My Pathway into Posthumanities* (University of Virginia Press, 2018); *Beyond Anthropocentrism* (Mimesis International, 2018); "Philosophical Ethology and Animal Subjectivity," in *Animality in Contemporary Italian Philosophy*, edited by Felice Cimatti and Carlo Salzani (Palgrave Macmillan, 2020); *The Virus Paradigm* (Cambridge University Press, 2021); with Marco Celentano, *Critical Ethology and Post-Anthropocentric Ethics* (Springer, 2021); *The Creative Animal: How Every Animal Builds Its Own Existence* (Palgrave Macmillan, 2022); *Technophysiology, or How Technology Modifies the Self* (Cambridge Scholars Publishing, 2023); and *Posthumanist Manifesto* (Rowman & Littlefield, 2024).

CHAPTER 1

Introduction

We are currently facing an ecological crisis—that much is now abundantly clear. A vast body of literature discusses what could be done to address it, and which human practices are to blame. But this book takes a different approach. Rather than focusing on solutions or symptoms, I aim to examine the underlying mindset that drives these practices—in other words, the deep-seated way of thinking that has led us here. My answer lies in anthropocentrism (not to be confused with the anthropocentric perspective). This is a philosophical ideology, rooted primarily in humanism, that asserts both the distinctiveness and superiority of the human over all other life forms. In what follows, I argue that before any meaningful change can occur, we must first let go of this intuitive, inherited worldview and embrace ecological thinking.

Ecology, as a descriptive science, has shown that no living form exists in isolation: every biological process relies on, and thus implicitly presupposes, the presence of a counterpart that enables it to unfold (Margulis 1998; Gilbert and Epel 2015). Through this perspective, which views the biosphere as a kind of super-organism with its own unitary physiology, ecology has laid the foundations for rethinking the predicates, constitutive relationships, symbiotic dependencies, ontopoietic hybridizations, as well as the temporal dimension of entities and events (Lovelock 1979). In other words, ecology is not only a framework for describing biological

R. Marchesini, *Philosophical Ecology*,
https://doi.org/10.1007/978-3-032-04987-2_1

relationships among living beings, but a way of understanding and cognizing entities in transcendental terms (Iofrida 2019; Iovino 2018).

This approach thus extends beyond description; it implies *an ontological and epistemological revolution* that can no longer be postponed. In this sense, it is essential to speak of *philosophical ecology* (Amoroso 2023), which is not to be confused with ecological fundamentalism or an attempt to turn description into prescription. Rather, ecological thinking encourages a new mindset, one that can be applied across different contexts (Brenner 2018; deLaplante et al. 2011; Keller and Golley 2000). In this book, I examine what I see as the core principles of ecological thinking and show how they can drive far-reaching philosophical change (Bateson 1972). Additionally, I explore how our way of conceiving reality has limited our ability to fully understand the current ecological crisis and the urgent actions it calls for.

This book thus aims to examine the contributions of ecological thought, which have gradually developed into distinct currents—such as environmentalism, deep ecology, the ethics of responsibility, critiques of anthropocentrism, antispeciesism, the Gaia Hypothesis, affective ecologies, and the phenomenology of life. However, rather than merely summarizing these perspectives, my goal is to uncover a unified framework for philosophical ecology as a discipline. This book emphasizes a twofold necessity: that of adopting a broad philosophical perspective to establish an ethics of human-biosphere relationships that goes beyond purely deontological, external principles; and that of redefining ontology, which currently remains essentialist and lacks a relational approach (Deleuze 1994). This goes beyond simply cataloging dependencies among living beings or coevolutionary processes (Attfield 2018); instead, it requires a deeper exploration of life as an expression of both fundamental and contextual relationships.

To advance the field of philosophical ecology, the first crucial step is to clarify various terms and approaches and relate them to one another to identify convergences and overlaps; this will help build a conceptual reference map. As mentioned, this book also seeks to provide guiding principles for shifting from the essentialist perspective that still dominates philosophy today toward a relational and emergent viewpoint (McKeown 2014). Concepts such as *heteronomy*—the predisposition to be shaped by external influences—and *hybridization*—the tendency to internalize otherness—complement ideas like *symbiosis* (Odum and Barrett 2005) and *sympoiesis* (Haraway 2016) within an eco-ontological framework. In this view, the predicates manifested by an entity are understood as arising from its relational constitution.

Among the founding principles of ecology, it is thus important to stress the premise of the interconnection of living organisms, which understands life itself, in its ontological dimension, as a relational process unfolding within specific spheres of proximity rather than as a condition of accidental interaction. I consider it necessary to clarify this point because, in ecology, one does not speak in vague terms of relations among living beings, but rather of specific positions of exchange, linked to the differentiation of environments and the particular phylogenetic trajectories of species—trajectories that must be treated as singular events (Gould 1990). Ecological niches define gradients of influence and dependence, not a generic situation where "everything is connected to everything else." For this reason, I prefer to speak of an *eco-ontology* rather than of an undifferentiated relational ontology, even though my debt to that approach is evident.

In this book, I address the question of connections starting from the debate between those who emphasize the primacy of internal relations—those that organize an entity and confer certain predicates to it—and those who, by contrast, assign greater importance to the being's external relations (Frigerio 2023). I put forward three basic criticisms against these approaches: first, they fall back into a dualistic, antinomic logic, a product of the non-ecological understanding which conceives ontology as a process of disjunction (Plomwood 2002); second, they rest on a generic and occasional conception of the connections among living beings, which is precisely what the ecological approach calls into question (Attfield 2018); and finally, they fail to take into account that, in an evolutionary conception of life, external relations are incorporated, and internal ones are reflected in the niche (Lewontin 1982).

In ecological thought, it is crucial to speak of *hetero-referentiality*: specific connections that generate conditions of interdependence which i) do not remain confined to the phenomenological level but also involve the evolutionary processes of the entity (Margulis 1998) and ii) are based on a selectivity of connection that makes certain relations necessary or highly probable (Bateson 1972). In other words, philosophical ecology cannot ignore the *meta-principles of ecology* and must be able to grasp the peculiarities of the phenomenology of life. In this sense, I believe it is important that such a vision draw on significant contributions from new feminist materialism—particularly concepts such as *transcorporeality* (Alaimo

2016), *contamination* (Tsing 2015), *vibrancy* (Bennett 2010), and *intra-action* (Barad 2007).

An ecological perspective must begin with a conception of the range of virtualities of life, which should neither be explained through an essentialist framework, nor thought of as an empty receptacle—as if the entity were a fluid awaiting an imposed form. Rather, every being actualizes itself in its relations with the external world (Deleuze 1994). The essay presented here focuses above all on a critique of the disjunctive and dualistic thinking characteristic of Western culture. At the same time, it acknowledges its debt to Indigenous epistemologies, which place interconnection at the center. These ways of thinking are valuable not only for ecology, but also for reconnecting different domains of knowledge and overcoming rigid oppositions (Chatterjee et al. 2022). In this light, ecological thinking must reach beyond the study of interactions among organisms alone (Guattari 2000). We should also remember Aldo Leopold's key idea of a land ethic. His essay *A Sand County Almanac* (1949) is the seminal work that sparked all the following reflections in the environmental movement. An ecological ethic must be able to foster open dialogue among the many diverse voices and stances in today's debate, building on Leopold's fundamental legacy and making each contribution part of the ongoing development of environmental ethics (Callicott 1989).

This book presents itself as a journey designed to challenge the main epistemological obstacles to the development of a philosophical ecology—one capable of bringing together, within a coherent framework, the shifts in the foundations of Western humanist culture brought about by Darwinian evolutionism, ecology, the rise of bioethics, and the major conceptual transformations introduced by twentieth-century continental philosophy. It is therefore a multi-stage path, in which each theme, addressed in a single chapter, prepares the way for the next. It was first necessary to reconnect the ecological crisis to the metamorphosis of the anthropological question, given that the century we have just left behind has put into question—if not entirely dismantled—the serene and self-referential vision embodied by the Vitruvian Man. Only then does it become possible to examine the question of the individual and its existential ambivalence within ecological issues, no longer understood as external to it but as intimately involved with it.

Grasping this condition of being part of a whole leads back to the meta-principles of ecology, using an organismic reading of the biosphere as a framework for a critique of anthropocentrism and human disjunction.

Connected to this are reflections on the phenomenology of life and on the recognition of agency in the living—dimensions denied by neo-Darwinian reductionism but now regaining ground thanks to the development of the Extended Synthesis (Pigliucci and Müller 2010). Only in this way is it possible to address the question of *ecological* ethics, capable of going beyond environmental ethics because it is grounded in philosophical premises consistent with the meta-principles of ecology. The theoretical aspect of this vision is therefore treated at the conclusion of the book's journey.

To begin this path of reflection, I focus on the need to broaden the ecological approach beyond the purely descriptive and scientific analysis of the relations between living organisms and their environment. Indeed, this is necessary in order to avoid falling into both the biologicist trap and the equally sterile trap of transcendence. The goal, rather, is to acknowledge the meta-principles of ecology—such as heteronomy, symbiosis, hybridization, and sympoiesis—that can serve as useful tools of philosophical meta-analysis for overcoming the aporias of dualistic and disjunctive frameworks. In developing a philosophical ecology attentive to current debates in the international cultural landscape, I also consider the contributions of contemporary new materialism and the critique of essentialist thought advanced by authors such as Gilles Deleuze and Félix Guattari, who laid the groundwork for a reconsideration of ontology in an ontopoietic sense. Equally urgent is the need to question the principle of purity, in favor of a hybrid conception of being—an element of the changes in contemporary thought with which philosophical ecology must necessarily engage (Braidotti 2013).

The overcoming of an anthropocentric stance in both ontology and epistemology also leads us to re-evaluate the position of non-human entities, no longer understood as mere resources but as integral parts of the human dimension. This also includes the field of eco-critical studies, which explores the relationship between literature and the biosphere. What changes, above all, is the ontological perspective, which highlights the impossibility of understanding human predicates within a self-enclosed framework. The anthropo-poietic aspect—human becoming across the different stages of ontogenesis—thus emerges as a central issue in our relationship with the biosphere.

In the second chapter, I examine the ecological crisis, in its various dimensions and in the ways it is typically presented: (i) as a threat to human survival, for example in the metaphor whereby "our house is on fire"; (ii) as the depletion of resources necessary for human well-being; (iii) as the loss of natural beauty, often approached from a conservationist perspective. Philosophical ecology, however, aims to show that the environmental crisis is not simply a matter of poor resource management, but rather *the result of a flawed way of thinking*. The major environmental crises humanity faces today—climate change, biodiversity loss, disruption of homeostatic cycles, deforestation, and pandemic risk—demonstrate that this is not merely an alteration of our habitat, but a crisis of our anthropological and cultural paradigm.

Indeed, the ecological crisis compels humans to take a hard look at themselves, confronting both the vulnerabilities inherent in their nature and the limits of their finitude, which, rather than acting as constraints, can become catalysts for existential awareness and openness to the world. This crisis is a product of modernity—a cultural paradigm that sought to separate humanity's destiny from that of the biosphere. It stems from a divisive conception of humanity, one that has devalued the non-human and, in parallel, denied the possibility of a shared destiny. Philosophy, of course, cannot ignore the existential condition of the individual; on the contrary, it must be able to place it at the center of its reflection. Yet it is precisely in overcoming individual self-referentiality that the goal of a paradigmatic change must be sought. An ecological conception of ontology understands individuation not as an a priori principle that finds meaning by turning inward, but as a process realized through relationships with human and non-human others.

In this sense, philosophical ecology emphasizes the subject's capacity not only to safeguard their own space of survival, but also their relational niche—that is, conviviality. Once the foundation of belonging and social bonds, the convivial dimension is now fractured, with serious consequences for psychological well-being and developmental processes, particularly for younger generations increasingly channeled by capitalist society into a solipsistic drift. The cultural attitudes responsible for the ecological crisis—hedonism, consumerism, egocentrism, narcissism—as direct expressions of a particular social model, are precisely what fosters the pervasive sense of loneliness and dissatisfaction in the individual (Lasch 1979). In the third chapter, I therefore explore the relationship between the collapse of conviviality and the disruption of the environmental context, alongside the rising feelings of frustration, distress, and depression (Roszak 2001; Danon 2020). The individual, in other words, is not spared from the collapse of ecological equilibria, but ends up at the center of one of the greatest existential crises of the contemporary era.

Chapter 4 considers life on Earth in terms of geological evolution and the dynamic balance among different living forms. It is often claimed that our planet presents optimal conditions for sustaining life; however, this assessment overlooks the fact that these conditions *were brought about by living organisms* over roughly four billion years. In this context, one cannot ignore James Lovelock's Gaia Hypothesis, which was among the first to highlight the relationship between the evolution of life and planetary terraforming. Biodiversity, with its richness of forms and relationships, represents the only guarantee for maintaining these parameters within acceptable margins of fluctuation and compatible timescales of change (Cain et al. 2011). It must also be remembered that the dynamic behavior of planetary conditions is tied to a state of threshold feedback: below a certain gradient, homeostatic mechanisms buffer disturbances and maintain the system in a given climax; above that threshold, autocatalytic mechanisms pave the way for catastrophic events. Life on Earth has already experienced catastrophic phases that caused mass extinctions and radical transformations of living forms. But while in the past these extinctions were driven by external factors—such as asteroid impacts or intense volcanic activity—today humanity itself has become the cause of a new mass extinction.

The fifth chapter examines anthropocentrism as an epistemological obstacle to the development of philosophical ecology and to a full understanding of the ongoing ecological crisis. Anthropocentrism is analyzed in two main forms: (i) the anthropocentric perspective, understood as the human tendency to interpret the world through a self-centered lens, which leads to the transformation of the biosphere into mere backdrop or to the anthropomorphization of otherness; and (ii) paradigmatic anthropocentrism, which represents a full-fledged anthropocentric ideology that reached its apex in the humanist revolution. The anthropocentric perspective is often portrayed as an insurmountable constraint for humans. In fact—as I show in this chapter—just as egocentrism can be mitigated through relationships with others and the practice of empathy, the human horizon *can* be broadened through interaction with non-humans. This process has been historically facilitated by the development of scientific thought (Bachelard 1984) and technological advances, which have highlighted the numerous distortions inherent in human reasoning (Tversky and Kahneman 1974). As for anthropocentrism as an ideology, the humanist framework that places humans at the center—as measure, subsumption of the world, and as an autonomous entity separated from the biosphere—it is now clearer than ever that it has to be abandoned.

The sixth chapter considers the characteristics of the living, not from a descriptive point of view—focused on the adaptive and ecological traits of different species—but from the perspective of an ontological inquiry into life in its existential and creative dimension (Bergson 1911). Addressing the phenomenology of life means concentrating not so much on the morphological, physiological, and ethological structures of organisms, but rather on what underlies the adaptive and interactive processes that distinguish them. In other words, it requires reflecting on the ontological foundations that drive life to adapt by relating with its environment. It is not, then, a matter of describing forms, functions, expressions, and interactions among organisms—the task of descriptive ecology and evolutionary biology—but of bringing to light the metapredicates of life that guide the adoption of such predicative traits. What comes into focus is life's intrinsically relational condition, its intentionality, its evolution based on dependence on external factors, the dissipative nature of inherited information, and its development through interaction with other living beings. The phenomenology of life rests on a close dialogue between processes of autopoiesis—for instance, the organism's active participation in its relations—and processes of heteropoiesis, that is, openness to external influence in the shaping of the being.

Chapter 7 examines the concept of the agency of living beings. The protagonism of life in shaping the environment forces us to reconsider the image of the biosphere as mere backdrop or stage for humanity, conceived as the sole actor in earthly affairs, and to inaugurate a new vision of transcendence within immanence, drawing on the thought of Baruch Spinoza. To understand the development of the biosphere, it is necessary to abandon the misperception of passivity long attributed to living organisms—except for humans, traditionally considered the only subject endowed with self-determination. From the conception of the animal as an automaton, a Neo-Cartesian expression of North American behaviorism and Central European psychoenergetics, to the Neo-Darwinian view that stripped living beings of any authorship over their phylogenetic trajectory—while framing a gene-centric reading of organic constitution—twentieth-century biology grounded the development of biodiversity in the stochasticity of accidental mutation and the constraints imposed by the environment (Monod 1971). Contemporary evolutionary theories, while remaining faithful to the basic Darwinian framework, show that living beings are in fact far from passive in their relationship with the environment. Organisms are now recognized as active agents, capable of influencing their own phylogenetic trajectory—not according to a predetermined plan, but through their agency, understood as the capacity to define their own life direction. Concepts such as niche construction, epigenetic inheritance, and polyphenism are now central to the new evolutionary synthesis.

The eighth chapter explores the various paradigms of environmental ethics, offering a historical overview of early reflections, particularly from the 1960s onward, up to contemporary proposals in ecological ethics and environmental bioethics. Over the past fifty years, approaches to moral reflection on human conduct toward the biosphere have undergone profound changes, demonstrating an ever-growing interest among scholars and the general public (Danowski and de Castro 2016). First, I illustrate the main moral philosophy frameworks based on their anthropocentric or biocentric assumptions: from the distinctly anthropocentric approach of frontier ethic to the ecocentric perspective of holistic or deep ecology. To understand the positions of different authors, it was also necessary to examine the principal ethical theories they adopt: utilitarianism, natural law theory, contractualism, virtue ethics, the ethics of sympathy, and the capabilities approach. Among the issues I address, what stands out is the relationship with non-human others. The chapter concludes by emphasizing the need for an ecological ethics grounded in the recognition of relation and diversity, in which the interests of others are acknowledged through an empathetic approach.

Drawing on all that was discussed thus far, Chap. 9 delves into the core theoretical analysis of philosophical ecology, aiming to show how its meta-principles can be applied to construct a new framework for reflection. To analyze the foundations of philosophical ecology, I begin with the philosophy of relations and the debate between internal and external relations, which first started with Bertrand Russell in the early twentieth century. Today, discussions of relations increasingly recognize the importance of not separating an entity from its connections and from its internal systemic organization (Ladyman and Ross 2007). My own proposal for a philosophical ecology, while embracing certain considerations from the philosophy of relations, departs from it in asserting that the relations among living beings cannot be evaluated in the same way as those among non-living entities. When dealing with a biological organism, a relation cannot be considered outside an evolutionary framework and a specific ecological placement; connections cannot be regarded as merely occasional or accidental. For this reason, such relations must be approached from a perspective that highlights the peculiarities of living phenomena and assigns the connective matrix a diachronic dimension, in which the time factor cannot be reduced to a mere constant: exchange processes are already implied in the constitution of the entity (*heteronomy*) and have consequences for its ontopoiesis (*heterogenesis*). That is why I speak of *eco-ontology*: an ontology that not only embraces the relational conception but also draws on the meta-principles of ecology.

Finally, the conclusions I propose aim to serve as a stimulus for opening a new field of research on philosophical ecology, based on fundamental principles capable of presupposing and recognizing the peculiarities of the biosphere (Attfield 2018). My intention was to create a close connection between areas of knowledge that continue to proceed along parallel tracks, particularly between the human sciences and the life sciences, because I believe that within this dualism lies the greatest epistemological obstacle to understanding the ongoing crisis. My contribution thus primarily lies in creating problematic connections—where each acquisition is never an end in itself, but constitutes a matter with which other disciplines must engage—in order to bring forth a space of theoretical possibility that I call *eco-ontology*. This is what occurs in the evolutionary dynamics of the biosphere, and this is what ecology itself shows, as every structural organization of an entity is in itself able to refer to something else. The final chapter also analyzes the various domains of application of philosophical ecology and the shifts in approach it calls for in disciplines such as anthropology and pedagogy, with the aim of opening a new field of research. Philosophical ecology seeks to serve as an interpretative lens capable of highlighting aspects often overlooked or misunderstood by the humanist approach. In particular, it emphasizes how anthropocentrism constitutes first and foremost an *epistemological* obstacle, which not only prevents a full understanding of the ecological crisis, but also hinders a deep comprehension of the intrinsically relational nature of human beings and their ontopoietic vocation.

To sum up, I believe the ecological crisis cannot simply be treated as a technical problem to be solved with new ethical appeals or fear-based strategies. Instead, *what's needed is a fundamental shift in the way humans think*. Traditionally, philosophy has primarily focused on the human dimension, often viewing it as separate from the biosphere. The human being has been conceived as an autonomous entity, defined *iuxta propria principia*, with the biosphere often seen as a backdrop for humanity to exist against, if not an antithesis from which humans ought to emancipate themselves. This disjunctive perspective has given rise to anthropocentrism, not only in ethical terms but also as a general philosophical stance. To construct a new integrative vision, it is therefore essential to develop *a different ontological conception* that regards the predicates of beings not as expressions of their pure essence but as manifestations of their *relationships with life as a whole.*

To this end, this book examines the broader aspects of the ecological crisis, which encompasses not only climate change but also other interconnected events with potentially synergistic and emergent effects. It argues that our failure to grasp the severity of the problem stems from our difficulty to adopt ecological thinking. To truly address the current crisis, in fact, we must abandon the ontological and epistemological frameworks that still position humanity in opposition to the biosphere. This requires an overall reassessment of the ontology of the living, moving beyond the outdated view of the biosphere as a mere background or space to inhabit. Only through such a profound philosophical shift can we develop an ethics that deeply internalizes the imperative to think ecologically, embedding it within the human perspective itself.

References

Alaimo, Stacy. 2016. *Exposed: Environmental Politics and Pleasured in Posthuman Times.* Minneapolis: University Minnesota Press.

Amoroso, Prisca. 2023. *Ecologia.* Roma: Castelvecchi.

Attfield, Robin. 2018. *Environmental Ethics.* Oxford: Oxford University Press.

Bachelard, Gaston. 1984. *The New Scientific Spirit.* Boston: Beacon Press.

Barad, Karen. 2007. *Meeting the Universe Halfway: Quantum Physics and the Entanglement of Matter and Meaning.* Durham: Duke University Press.

Bateson, Gregory. 1972. *Steps to an Ecology of Mind: Collected Essays in Anthropology, Psychiatry, Evolution, and Epistemology.* Chicago: University of Chicago Press.

Bennett, Jennet. 2010. *Vibrant Matter. A Political Ecology of Things.* Durham: Duke University Press.

Bergson, Henri. 1911. *Creative Evolution.* New York: Henry Holt and Company.

Braidotti, Rosi. 2013. *The Posthuman.* Cambridge: Polity Press.

Brenner, Joseph E. 2018. The Philosophy of Ecology and Sustainability: New Logical and Informational Dimensions. *Philosophies* 3 (2): 16. https://doi.org/10.3390/philosophies3020016.

Cain, Michael L., William D. Bowman, and Sally D. Hacker. 2011. *Ecology.* Sunderland: Sinauer Associates.

Callicott, J. Baiard. 1989. Animal Liberation: A Triangular Affair. In *In Defense of the Land Ethic: Essays in Environmental Philosophy*, 15–38. Albany: State University of New York Press.

Chatterjee, Uday, et al. 2022. *Indigenous People and Nature: Insights for Social, Ecological, and Technological Sustainability*. Amsterdam: Elsevier.

Danon, Marcella. 2020. *Ecopsicologia. Come sviluppare una nuova consapevolezza ecologica*. Sansepolcro: Aboca Museum Edizioni.

Danowski, Deborah, and Eduardo Viveiros de Castro. 2016. *The Ends of the World*. Cambridge: Polity Press.

deLaplante, Kevin, Bryson Brown, and Kent A. Peacock, eds. 2011. *Philosophy of Ecology*. Waltham: North-Holland.

Deleuze, Gilles. 1994. *Difference and Repetition*. New York: Columbia University Press.

Frigerio, Christian. 2023. *Ricomporre un Cosmo in Frammenti: Filosofia, Ecologia e Relazioni*. Sesto San Giovanni: Mimesis.

Gilbert, Scott, and David Epel. 2015. *Ecological Developmental Biology: The Environmental Regulation of Development, Health and Evolution*. Oxford: Oxford University Press.

Gould, Stephen Jay. 1990. *Wonderful Life*. New York: W.W Norton & Company.

Guattari, Félix. 2000. *The Three Ecologies*. London/New York: Continuum.

Haraway, Donna Jeanne. 2016. *A Cyborg Manifesto*. Minneapolis: University of Minnesota Press.

Iofrida, Manlio. 2019. *Per un paradigma del corpo: una rifondazione filosofica dell'ecologia*. Macerata: Quodlibet.

Iovino, Serenella. 2018. *Filosofie dell'ambiente*. Roma: Carocci.

Keller, David R., and Frank B. Golley, eds. 2000. *The Philosophy of Ecology: From Science to Synthesis*. Athens: University of Georgia Press.

Ladyman, James, and Don Ross. 2007. *Every Thing Must Go: Metaphysics Naturalized*. Oxford: Oxford University Press.

Lasch, Christopher. 1979. *The Culture of Narcissism*. New York: W.W. Norton & Company.

Leopold, Aldo. 1949. *A Sand County Almanac and Sketches Here and there*. New York: Oxford University Press.

Lewontin, Richard C. 1982. Organism and Environment. In *Learning, Development, and Culture*, ed. Henry C. C. Plotkin, 151–170. New York: Wiley.

Lovelock, James E. 1979. *Gaia: A New Look at Life on Earth*. Oxford: Oxford University Press.

Margulis, Lynn. 1998. *Symbiotic Planet: A New Look at Evolution*. New York: Basic Books.

McKeown, Greg. 2014. *Essentialism: The Disciplined Pursuit of Less*. London: Virgin Books.

Monod, Jacques. 1971. *Chance and Necessity: An Essay on the Natural Philosophy of Modern Biology*. New York: Alfred A. Knopf.
Odum, Eugen P., and Gary W. Barrett. 2005. *Fundamentals of Ecology*. Belmont: Thomson Brooks-Cole.
Pigliucci, Massimo, and Gerd B. Müller. 2010. *Evolution. The Extended Synthesis*. Cambridge: MIT Press.
Plomwood, Val. 2002. *Environmental Culture: The Ecological Crisis of Reason*. London: Psychology Press.
Roszak, Theodore. 2001. *The Voice of the Earth: An Exploration of Ecopsychology*. Grand Rapids: Phanes Press.
Tsing, Anna Lowenhaupt. 2015. *The Mushroom at the End of the World: On the Possibility of Life in Capitalist Ruins*. Princeton: Princeton University Press.
Tversky, Amos, and Daniel Kahneman. 1974. Judgement Under Uncertainty: Heuristics and Biases. *Science* 185:1124–1131.

CHAPTER 2

An Anthropological Crisis

2.1 Premise

Over the past two decades, the effects of a severe ecological crisis on Earth have become increasingly apparent. Only a few negationists still reject or downplay this urgency, despite the crystal clear evidence presented by the scientific community (WMO 2025). Beyond the well-known issue of climate change, the twenty-first century presents us with more complex, articulated, and pressing problems compared to the previous one. These challenges go beyond traditional concerns about the limits of development or resource deficits, involving a set of degenerative processes that threaten the resilience of our biosphere. The world appears to be teetering on the brink of disruption due to the convergence of several different problems, endangering the equilibrium mechanisms governing life as we know it (Jamieson 2014; Servigne et al. 2021).

The primary cause, though likely not the only one, of this dramatic situation is human action. Several anthropic activities have a devastating impact on the biosphere's capacity to absorb their effects. Critical issues include waste accumulation (Meijer et al. 2021), land overexploitation, ecosystem disruption, rampant deforestation (Hecht and Cockburn 2011), excessive greenhouse gas emissions (Lesch et al. 2023), resource depletion (Raveendranathan 2018), interference with recycling processes, soil erosion (Morgan 2005), biodiversity loss, and disruption of vital natural cycles (Archer 2010). The human impact on the planet, already

R. Marchesini, *Philosophical Ecology*,
https://doi.org/10.1007/978-3-032-04987-2_2

exacerbated by the demographic boom of the twentieth century, is made worse by the energy-intensive lifestyle of a growing population segment. This is coupled with an increase in pollutants and poorly biodegradable effluents. We stand on the brink of a point of no return due to the irreversibility and acceleration of certain processes.

That being said, labeling humans as a disease (Lovelock 2006), a cancer (Sebastiani 2017), or a highly infectious virus (Farnham 2020) for the planet, despite being an evocative metaphor and aligning with certain environmentalist views, is ultimately misleading. In fact, responsibility for this critical situation cannot be attributed generically to all humans, as there are populations living in acceptable balance with the ecological turnover and others that do not (Diamond 2011). Only a specific fraction, not even a majority, of the global population bears blame for this situation (Moore 2015). For instance, carbon dioxide emissions and other greenhouse gases predominantly affect certain economies, while deforestation is driven by the extensive consumption of animal-based food in Western countries (Safran Foer 2010). Therefore, the urgency of these crises can be traced back to a very specific social and economic model.

In particular, it can be traced back to forms of unrestrained capitalism, often labeled as *turbo-capitalism*—a self-referential development model fixated on profit at the expense of the well-being of populations (Bevilacqua 2011). Of course, there are also virtuous examples of social economy that coexist with this globalized state of hyper-exploitation, with deep-rooted traditions even in Europe (Rozzoni and Limata 2022). Unfortunately, however, the profit-based model, born out of the industrial revolution, has seen unprecedented growth in recent decades. The issue is primarily due to an ideological trend that has gained momentum amid the decline of previous grand narratives. This perspective perceives the entire biosphere as a resource entirely at humanity's disposal; the turbo-capitalist approach maximizes the exploitation of every resource—human and material alike—to generate profit. This vision sustains market demand for goods by perpetuating a consumerist lifestyle and exploiting natural resources, transforming them into low-value entities.

Now, the capitalist model has thrived in many societies due to certain inherent tendencies in human nature as well as cultural influences that have accentuated its value (Shepard 1973; Wilson 1978; Moffett 2019). It is therefore crucial to reflect on some of humanity's innate tendencies—such as the use of tools, the strong competitive drive, and the adaptive generalism in foraging behavior that makes humans naturally prone to shortsightedly exploiting available resources in the here and now. On top of this, certain egocentric and narcissistic traits are nurtured and reinforced by the market itself, fostering the belief that happiness is attained through accumulative behavior, if not outright exploitation of the external world, and promoting a self-centered lifestyle (Benasayag and Schmit 2003; Han 2009). Similarly, competitive, conformist, and social motivations are often encouraged by contemporary culture, undoubtedly contributing to the promotion of individualistic attitudes centered around the possession of goods (Marchesini 2023).

Therefore, the current ecological crisis cannot be solely attributed to the economic aspect or the impact of human presence on the planet. This is more than a problem of resources, even though it undeniably affects what we call consumer goods (Rushkoff 2013; Beck 1997). Rather, it is a pervasive attitude that needs to be addressed on various fronts, emphasizing the interconnectedness between humans and other living forms. The crisis cannot be solely attributed to a disruption of survival resources; it primarily concerns humanity's ability to comprehend the relationships binding it to the biosphere (Dark 2019). In other words, the fundamental problem lies in our very thought processes, the cultural foundations shaping our social and economic models, and the technological development that, instead of prioritizing eco-compatibility, widens the gap between operational capacities and the ability to manage such potential. For all these reasons, what we are witnessing is first and foremost an anthropological crisis.

2.2 Human Nature and the Ecological Crisis

Talking about human nature is always complex. The fear is that any definition of human nature could limit humanity's ontopoietic space, presenting it as a constraint or even an inescapable fate that humans cannot avoid, and potentially serving as a justification. This aversion to acknowledging a well-defined human nature in terms of innate tendencies and cognitive traits becomes even more pronounced when discussing certain problematic behaviors. A clear example of this is the critical reception of books such as *On Aggression* by Konrad Lorenz (1966) and *Human Ethology* by Irenäus Eibl-Eibesfeldt (2007). However, this fear is unjustified. Interpreting human relationships in terms of conflict and domination does not deny the existence of empathetic, convivial, and cooperative tendencies in humans. Still, acknowledging the high level of aggression in our species should serve as a warning, a call for attention, not something to be repressed out of fear of justifying it. When Lorenz, in his essay *Civilized Man's Eight Deadly Sins* (1974), links these tendencies to their impact on technological society—showing that a missile operator or even a bomber pilot is incapable of moderating their aggression due to the lack of direct contact with others—he illustrates how culture is not the overcoming of nature (nature being somehow "other"), but rather a direction and modification of the operational field of nature itself.

The idea that culture somehow overshadows human nature, or that human nature is so insignificant it can be disregarded—an outlook that has prevailed in Western thought from the Sophists to the neo-humanism of philosophical anthropology—skillfully avoids, in my view, a question that cannot be overlooked: "Is there something in human nature that predisposes us to ecologically problematic behaviors?" Again, this is not about adopting a fatalistic—let alone justifying—stance, nor an inclination toward extinctionism (humanity is a cancer from which the biosphere must heal), but rather about finding a solution to this problem. The issue, I believe, does not lie in questioning the phylogenetic definition of humanity—a condition of adaptive specificity that, after Charles Darwin (1809–1882), particularly in works like *The Descent of Man and Selection in Relation to Sex* (1871) and *The Expression of Emotions in Man and Animals* (1872), cannot be denied—but rather in reflecting on the very definition of nature and its ontogenetic plasticity. Today, we need an emancipation of the concept of nature: a rethinking of biological phenomena not in terms of determinism but as spaces of possibility. As we will see when discussing the phenomenology of life and animal agency, it is necessary to move beyond determinism to understand that life operates within fields of virtuality—fields that, while defined by specific perimeters, are far from lacking an internal constitutive freedom. This means that acknowledging human nature as such does not preclude self-determination; instead, it points to tendency-driven traits that then produce specific outcomes.

The problem of human nature has traditionally been addressed through two opposing hypotheses: (1) the hypothesis of human incompleteness, advanced by prominent figures in philosophical anthropology (Scheler 2009; Gehlen 1988), which suggests that our species lacks adaptive specializations; and (2) the hypothesis of the dominance of innate traits, central to sociobiology (Wilson 1978), which posits that every human behavior stems directly—i.e., is a direct expression—of phylogenetic characteristics. In my essay *Post-Human* (2002), I proposed a third possibility, which can be summarized in four key points:

1. The innate traits of the human being predispose our species to a marked relational tendency toward the external world;
2. Precisely because of this interactive disposition, humans tend to engage in broad-spectrum relational behaviors, which can manifest as either conflict or conviviality, and also to develop conformist tendencies;
3. This relational inclination also translates into mimetic behaviors, leading humans to incorporate non-human qualities into their behavioral repertoire;
4. As a result, human expression is never pure—that is, it is never simply the manifestation of phylogenetic traits—but rather a hybrid product that, in some respects, goes beyond the human.

Therefore, I consider the sociobiological claim—that the human behavioral repertoire derives directly from phylogenetic traits—to be incorrect, even though I acknowledge that these traits play a fundamental role as catalysts for hybridization. Likewise, the hypothesis of philosophical anthropology is equally flawed when it emphasizes human incompleteness: culture is not about filling an empty vessel or compensating for a deficiency, but about organizing a redundant system—a human nature that is rich and highly specialized. The notion of deficiency has no empirical support: the human being is by no means biologically deficient, nor lacking in specialization. A quick look at the anatomical modifications associated with bipedalism—such as foot morphology, or the shape of the spine and pelvis—is enough to disprove that idea.

When we abandon an essentialist and determinist view of human nature and adopt a relational perspective, we realize that it is mistaken to frame cultural evolution as a mere outgrowth of phylogenetic traits. At the same time, we must also acknowledge that without these underlying "copulative engines," the adoption of behavioral styles different from those dictated by phylogeny would not be possible. This relational predisposition should lead us to a completely different understanding of human nature—one shaped by an *eco-ontological* vision, in which a strong connective tendency toward the living environment and non-human otherness becomes central. As a consequence, we can no longer regard the fruits of humanity as the product of solipsistic genius, but rather as the outcome of this relational creativity. This awareness compels me to assert that the human being is inextricably bound to the biosphere, not only for biological reasons but also ontologically: the human simply does not exist outside of its relationship with the non-human. If this claim holds any plausibility, we must ask ourselves why there is so little attention paid to the biosphere. I believe the answer lies not only in cultural traditions that placed humanity on a pedestal, separating it from the rest of nature, but also in certain features of human nature that enable it to rise above other living beings and assert itself through tool use. In other words, it is this very interactivity—the eco-ontological nature of humans—that increases the risk of impacting the biosphere. Interactivity, in fact, does not generate only conviviality, but also appropriation and conflict, to which I would add the tendency to modify the living environment and to redefine one's self-perception.

Among the most prominent traits of humans, in fact, there is certainly the tendency to profoundly modify the environment in which they live—an aspect observable in all human populations and traceable in both archaeological and paleontological records. Environmental transformation is relatively present in all living forms, a phenomenon referred to as *niche construction* (Odling-Smee 1988). But in the human case, there is something more. In ethology, the term *ecosystem engineers* (Jones et al. 1994) refers to species particularly inclined to transform their environments according to a specific plan—examples include beavers, corals, and ants—and our species can be considered the most skilled in this regard (Schilthuizen 2018). This aptitude may be attributed to two behavioral traits of humans: the use of fire (Goudsblom 1992) and stone (Sahlins 1972). According to Peter Sloterdijk, in the essay *Not Saved. Essays after Heidegger* (2016), the human being descends from stone, such that human becoming unfolds under the protection of stone technology, since it is through the use of stones for throwing, striking, and cutting that the principle of *techne* first comes into play. In the following chapters, we will see how technology—far from being a mere tool in human hands—inaugurates, as I argued in the essay *Technophysiology or How Technology Modifies the Self* (2023), a true physiological and, consequently, psychological transformation of the human being. Moreover, as Konrad Lorenz noted in *The Waning of Humaneness* (1987) and Hans Jonas in *The Imperative of Responsibility* (1984), the development of technology, by expanding the range of human intervention, creates a kind of imbalance between operational potential and management capacity.

The use of fire, for its part, represents one of the most fascinating topics in prehistoric research, while also touching on the urgent contemporary issue of ecology—considering that every year, 4 million square kilometers of forest are destroyed by fires worldwide. In 2019 alone, 12 million hectares burned in the Amazon, 27,000 hectares in the Congo Basin, 8 million hectares in the Arctic, and 328,000 hectares in Indonesia. Many of these fires are human-caused, so it is reasonable to ask whether, beyond specific economic motivations, there may also be a human predisposition to use fire as a means of expansion. We know that around 2 million years ago, drought conditions in East Africa—the cradle of the *Homo* genus—likely encouraged the spread of wildfires. Early humans may have benefited from these events by finding charred animal carcasses and plants made more edible by fire. In the article *The Pyrophilic Primate Hypothesis*, Christopher Parker et al. (2016) propose that our lineage may possess an inherent tendency toward pyrophilia as an adaptive response to environmental change. In short, prehistoric hominins learned to exploit fire as a resource, as flames made it easier to procure food. The authors note that even today, Matu women in Australia use fire for hunting. This suggests that early humans may have first scoured the landscape for charred remains after wildfires, and later learned to control fire and deliberately start fires—beginning with the emergence of *Homo erectus* (Chazan 2017). Moreover, we now know that chimpanzees in the savannas of Senegal not only remain calm near wildfires, but are also capable of tracking their movement while staying safely behind the flames.

On the other hand, this tendency to modify the environment—which today represents one of the most impactful aspects of human activity—should be understood in light of what I said about the relational characteristics of the human being. It is not in contradiction with them; rather, it stems from the human capacity to learn from everything in their surroundings and to find solutions and opportunities even in problematic situations. The reflective capacity and self-control made possible by the expansion of the prefrontal cortex are a direct consequence of this Promethean recursion: on one hand, it facilitates the use of fire and stone; on the other, it acts as a selective pressure on these very qualities. Relationship and modification are two tendencies that generate a self-reinforcing, autocatalytic loop. Alongside the lithic and pyrophilic poiesis through which humans, from the dawn of their existence, charted the course of the hominization process, we must also include other tendencies intrinsic to human nature—such as a strong mimetic inclination and a distinct social dimension. These not only contributed to the transformation of the environment but also spurred a reconfiguration of the body itself, through practices such as wearing animal furs, coloring the skin, and inserting foreign elements into the skin like quills or discs. These practices expanded the behavioral repertoire and reflected a creative, poietic vision of one's own identity.

Hunting also represents one of the most significant characteristics of human nature, although the inclination toward this activity is comparable to that of the chimpanzee (*Pan troglodytes*), and is therefore due to a convergence between nature and culture. In *On Human Nature* (1978), Edward O. Wilson argued that hunting literally shaped human culture from its origins, giving rise to practices of cooperation, sharing, and even conviviality. In particular, in his essay *The Origins of Creativity* (2017), the father of sociobiology goes so far as to claim that hunting constituted the origin of many forms of human creativity, such as the theatrical art of storytelling around the fire or the tendency to depict hunting scenes in cave paintings. It is also undeniable that it is when humans began to hunt Pleistocene megafauna, such as mastodons, that they gained access not only to large quantities of meat but also to materials useful for crafting, such as furs and ivory. During a period dated between 50,000 and 35,000 years ago, humans also forged an alliance with the wolf, which would lead to the domestication of the dog—a fundamental development for the full emergence of *Homo sapiens* in this era. The relationships of cooperation and the imitation of wolves' hunting techniques are described in detail by Barry O. Lopez in his book *Of Wolves and Men* (1978). Undoubtedly, the presence of dogs in human societies from around 30,000 years ago gave rise to an alliance with far-reaching ecological consequences. While the Pleistocene was marked by climatic fluctuations with a devastating impact on megafauna, it is, in my view, no minor detail that this extinction closely mirrors the human expansion: in Africa from around 1.7 million years ago, coinciding with the emergence of the first *Homo* species; in Eurasia from about 100,000 years ago, particularly after the arrival of *sapiens* between 60,000 and 12,000 years ago; in Australia 40,000 years ago, again corresponding to the settlement of *sapiens*; and in North America roughly 11,000 years ago—each time with a striking synchronicity between human arrival and faunal disappearance.

Hunting is perhaps the most exemplary expression of this recursive evolution of nature and culture, but there are other natural-cultural expressions that should lead us to conceive of human ontology as an emergent phenomenon of inclusive relationships developed on multiple levels. Broadly speaking, one can affirm that the cultural dimension characterizing human beings across all ethnicities—and affecting the way humans perceive themselves as entities reshaped by culture (*anthropopoiesis*), and not solely by parental learning or direct experience—stems from some peculiar traits of our species. These qualities manifest in the use of objects as tools and in the transformation of one's environment, but they also foster a deeper and more internalized relationship with the surrounding world: humans learn from everything they observe and absorb expressive styles and behavioral patterns found in other species. This *trans-specific learning*—a genuine expansion of social learning—enables humans to learn from wolves, bears, birds, and all animals with which they come into contact, thus transforming the human dimension into a kind of receptacle for models found in nature. It is a sort of expressive and stylistic dressing that echoes through every human expression: from music to dance, from the use of costumes to rites of passage, from cosmetics to technological practices, and from artistic expressions to the design of everyday tools such as clips, vials, or other utensils. This ability to adopt models found in the animal or plant worlds—this heightened permeability of the human being to external influences—is due to a range of motivational factors, such as the mimetic tendency, as well as to specific cognitive endowments, including the extensive development of the prefrontal cortex and the presence of neural areas specialized in mirroring processes (Gallese and Umiltà 2006).

Anthropopoiesis, that is, the cultural construction of the human, is therefore not an autarchic process as suggested by its common interpretations (i) as compensation for a lack of natural attributes (Gehlen 1988); (ii) as a direct expression of innate characteristics (Wilson 1978); or (iii) as the result of closure and separation (Sloterdijk 2013). Instead, it is a full opening to the world, making humans inherently porous to external influences. The relationship between humans and the world around them has a transformative effect that goes beyond simple contemplation or imitation. It brings about a sense of awe, wonder, and enthusiasm, experienced through two forms: possession (feeling the otherness within oneself) and ecstasy (projecting oneself into the otherness). I refer to this as the *epiphanic effect* (Marchesini 2017), which encompasses various experiences, such as shamanic practices (Eliade 2004).

In this sense, the epiphany is the moment when, observing the aerial acrobatics of birds, the human being is swept into their flight and suddenly realizes: it is possible to fly. Therefore, it is not merely about imitation or learning from birds the techniques of flight—that comes later—but the unveiling of a new existential dimension. The epiphany is an awed participation in flight, the revelation of flight, and ultimately, an invitation to take flight. It marks a magical threshold, where nature appears not as a resource or backdrop, but as an inspiring Muse. In this moment, the human being steps beyond their strictly phylogenetic expression: through possession or ecstasy, the boundary between self and other dissolves, giving rise to a condition that is more-than-human.

Many examples demonstrate this process of hybridization, one of the most notable being dance. Multidisciplinary observations by anthropologists and ethologists (Armstrong 1947; Chauvin 1973; Celli 1989) have revealed, for instance, the striking similarities between the rituals of the crowned crane (*Balearica regulorum*) and the dances of the Watussi; between the Andean Jivaro dances and the Andean cock-of-the-rock (*Rupicola peruvianus*); between the dances of the Mexican Tarahumara and the turkey (genus *Meleagris*); between the Australian aborigines and the emu (*Dromaius novaehollandiae*); and even between Bavarian dances, such as the Schuhplattler, and the sexual display of the black grouse (*Lyrurus tetrix*). These correspondences are remarkable, not only in terms of choreography but also in the vocalizations, tongue clicks, and rhythms used. Also fascinating are the dances of the Bushmen, which feature iterative elements that replicate the movements, postures, and mimetic reproduction of various animals found in the Kalahari. As ethologist Rémy Chauvin put it, "As birds dance, so do men" (1968, 208).

In these dances, humans use their bodies to represent the movements and anatomies of animals, for example, using their arms to imitate bird wings or an elephant's trunk. More generally, we can say that no aspect of human culture fails to reveal this inspiration from otherness: music echoes the harmonics and phrasing of many birds; many martial arts and yoga poses derive from animal morphemes; technology has often drawn from natural solutions (biomimicry); cosmetics and other body-modifying practices, such as tattoos or the insertion of objects, often aim to mimic certain animals; fashion, such as costumes and headgear, often imitates animal forms; mythology, comics, alien descriptions, and even constellations often are based on zoological references.

The permeability of humans to external influences, facilitated by remarkable cognitive plasticity and capacity, has made humanity an extraordinarily volatile entity—not due to a lack, but because of a vital exuberance that fills them with frenzy. This exuberance is fueled by an unparalleled desire and an equally intense emotional sensitivity. This inherently passionate nature of humans drives them to identify with everything around them, becoming involved and influenced by it. Therefore, we cannot think of humanity as a mere trait enclosed within the biological shell of *Homo sapiens*: it is the result of the relational propensity of human nature. It is precisely the phylogenetic characteristics of humans, what we call human nature, that make humans a shifting entity, always moving toward new existential dimensions. Thus, we should not deny human nature to avoid determinism, because it is within our phylogenetic heritage that lie the attributes of our flights of fancy into the beyond-human and ontopoietic freedom. On the other hand, this deeply relational and desiring nature also has critical points when it comes to finding balance with the biosphere, precisely due to the human tendency to expand indiscriminately, to appropriate everything around them, and to consume not only the world's resources but also spaces, time, and other lives along their path.

2.3 Culture's Influence in the Ecological Crisis

The devaluation of the biosphere cannot be fully understood without considering the cultural, philosophical, aesthetic, epistemological, and ethical factors that have shaped the Western world (Thomas 1983). This section argues that addressing the ecological crisis requires more than just economic solutions or calls for eco-friendly development models. These discussions are unquestionably important, as protecting the biosphere certainly demands a reimagined economy and new technologies that are less energy-intensive and polluting—as well as technologies capable of finding innovative solutions to the crisis (Benjaminsen and Svarstad 2021). However, I want to emphasize that such analyses alone are not sufficient. We must first ask ourselves *why* we have become so careless about the biosphere, why we feel disconnected from nature, and why this detachment has deepened particularly in the Western world (Descola 2005). Ultimately, we need to explore the cultural foundations of our value system and emotional relationship with nature.

This devaluation of nature has deep roots in Western thought, the birthplace of several cultural legacies marked by anthropocentrism and a separation from the biosphere—issues that we must now confront. The sense of estrangement that humans feel toward the biosphere is not just a condition inherent to humanity, but rather a consequence of Western culture's longstanding leitmotiv. This feeling of being other than the natural world must be understood as the result of two factors: the first, more general one, arises from the human detachment from direct engagement with natural conditions (Plessner 2019); the second, more specific, comes from the process of anthropopoiesis, which emphasizes the growing cultural dimension in human life (Remotti 1996). As noted, this detachment is partly a product of innate human characteristics—such as our ability to act as ecosystem engineers and our openness to external influences—shaped by inherent motivational traits and specific cognitive abilities. As the environment around us gradually transformed, starting in the Paleolithic and accelerating during the Neolithic revolution, and as non-human models were integrated into human lifestyle and self-perception, humanity increasingly found itself distanced from its innate heritage.

As a second aspect, we must recognize that from the development of the first great civilizations—Mesopotamia around 5000 BCE, Egypt along the Nile around 3900 BCE, India in the Indus Valley around 3500 BCE, and China in the Yellow and Yangtze river valleys around 3000 BCE—the influence of cultural traditions and cultural belonging increasingly shaped how humans perceive themselves (Quigley 1979). With the advent of cultures organized around customs, rituals, values, and societal structures, humans were molded by specific cultural traditions. In this sense, we must especially consider the rise of the Indo-European culture, which developed in the Eurasian valleys through trade promoted by the domestication of the horse (Diamond 1998), and in the Mediterranean basin through Phoenician and Minoan traders, as the first seeds of what would become Western culture.

Now I believe it is incorrect to blame the alienation from nature solely on Western culture, as this is a more global process that, to varying degrees, affected the entire Euro-Asian area. However, it is clear that the West, in particular, has witnessed the rise of a set of movements aimed at emancipation from nature and anthropocentrism. This can be seen (i) in the emergence of philosophy in the fifth century BCE with Socrates and the sophists, where the focus shifted decisively and exclusively to humanity; (ii) in the rise of Latin culture, based on *humana dignitas* and civic virtues, achieved through education; (iii) with the medieval belief in human superiority and uniqueness over other living beings, based on the concept of *imago dei*; (iv) with the advent of Humanism in the fifteenth century, which initiated the Modern Age, placing humans at the center of the universe; and (v) with the Scientific Revolution and technological development, leading to the Enlightenment and the Industrial Revolution.

Again, I do not intend to attribute the current environmental devastation exclusively to the West. However, I do want to highlight how certain cultural legacies perpetuate principles that make it difficult for humans to feel like an integral part of the ecological community. The problem of cultural anthropocentrism undoubtedly presents ambiguities, which I will explore in a later chapter. As Tim Ingold points out in *The Perception of the Environment* (2000), the risk of adopting an external perspective—common in certain environmentalist expressions—can paradoxically exacerbate the notion of separation between humans and the biosphere. In contrast, Ingold argues, a local perspective, which is inevitably anthropocentric, encourages us to imagine ourselves as part of the world. I agree with Ingold's insight, but my critique lies in conflating the local perspective, which is inevitably centered on humans, with the paradigmatic anthropocentrism typical of Western culture. A similar argument is made by Bruno Latour in *Politics of Nature* (2004), where he stresses that we cannot combat the ecological crisis by separating the "chamber of nature" from the "chamber of politics" and sheltering nature from humanity. I completely agree with this statement; the point is not to propose new forms of separation, even those aimed at protection, but to address the fracture created by anthropocentrism in the first place.

Ingold's and Latour's considerations lead us to think that it is impossible to talk about defending the biosphere without challenging the paradigm of separation that has been deeply ingrained in Western culture. This is not an easy process, because the habit of division and opposition is far more embedded in our way of thinking than we might realize. As a result, it's easy to fall back into a separative mindset, even when attempting to view the dynamics of the biosphere from an external, protectionist perspective. Ethical anthropocentrism cannot be overcome without an adequate discussion of ontological anthropocentrism—specifically, the conception of humans as distinct, separate entities that claim to view the world from the outside. To use Ingold's words, humans.

> presume a world already constituted, through the action of natural forces, which then becomes the object of human interest and concern. But it is not a world of which humans themselves are conceived to be a part. To them, it is rather presented as a spectacle. They may observe it, reconstruct it, protect it, tamper with it or destroy it, but they do not dwell in it. (Ingold 2000, 215)

Therefore, those same cultural factors that have fostered the sense of separation between humans and nature, on one hand, represent epistemological obstacles to understanding the ontological aspect of the ongoing anthropological crisis; on the other hand, become sources of disorientation and, consequently, contribute to an increased sense of anthropological despair.

To begin a preliminary discussion on the disjunctive paradigm that characterizes Western culture, I will start with its most explicit expressions. I believe there are eight key points that can help clarify the sense of separation that human beings experience in relation to nature:

1. The disjunctive view of the biosphere and the perception of it as a distant, external habitat (Lovelock 1979; Latour 2019);
2. The conception of nature in Western thought, often framed in an oppositional dichotomy with reason (Leopardi 2013; Lucretius 2020; Horkheimer and Adorno 2002);
3. The idea of emancipation from nature, seen either as the starting point of human history (Kant 2007a) or as a moment of lost authenticity (Rousseau 1991; Nietzsche 2014);
4. The concept of alienation from nature, interpreted either as a fundamental human condition (Hailwood 2015; Reed 1989) or as a cultural problem (Zerzan 1994; Diamond 2017);
5. The humanist vision of humanity's divergent and ascending destiny, separating humans from the earthly and biological dimension (Pico della Mirandola 1996; Manetti 1975);
6. The ontological distinction between humans and the biosphere, particularly the opposition between *res cogitans* and *res extensa* (Descartes 2008; Heidegger 1995);
7. The idea of the human as a being without nature, ranging from humanist thought to theories of human incompleteness (Scheler 1961; Gehlen 1988);
8. The autarkic and autopoietic vision of humanity, including conceptions of *techne* as the human exoskeleton (Harari 2011; Sloterdijk 2001).

I will now look at each of these points in detail.

1. *The disjunctive view of the biosphere*

The sense of ontological disjunction and the absence of a shared destiny between humanity and the biosphere is the first major obstacle to our ability to fully grasp the severity of the ongoing ecological crisis. This crisis does not merely concern "the environment"—as if it were a house on fire—but rather the entire living organism of which we are an integral part. Too often, the issue is reduced to purely technical or managerial aspects. While these are certainly important, they are far from sufficient to capture the full scope of the problem. Only by acknowledging this underlying mindset can we understand the indifference that unfortunately still characterizes part of public opinion toward an emergency that should, by all accounts, be treated as a top priority (Godelnik 2021; Andersen 2020). This lack of attention may result from operational shortsightedness—an inability to act before the situation becomes catastrophic. Yet I believe there is a deeper root: humanity's sense of estrangement from nature. This separation is, of course, an illusion, as our existence depends entirely on the balance of the biosphere (Loreau 2023). Still, to truly understand the reasons behind our neglect, we must confront this perceptual bias (Lesch et al. 2023; Ganapathy 2021).

Ecological problems undoubtedly stem from specific political and developmental choices. However, the critique of our social and economic model must be accompanied by a deeper reconsideration of how humans think—and, more importantly, how they think of themselves in relation to the ontological and anthropological foundations of existence. The image of a self-referential human being, cut off from the natural landscape, implies not only a distinct ontological domain that no longer communicates with the biosphere, but also a divergent destiny. In the collective unconscious, the divorce has already been finalized, producing a sense of estrangement from nature that permeates the human psyche (Latour 2019). Though we may formally express concern about the destruction of the biosphere, we remain emotionally disengaged—our attention drawn elsewhere (Giraldo and Toro 2020).

This separation is rooted in a problem of self-perception, shaped by the human tendency to interpret the world either through identification—by projecting oneself onto external entities—or through disjunction, by constructing oppositional, binary frameworks (Art 2022; Godfrey-Smith 2020). While many cultures have favored the projective approach, Western culture has largely privileged the disjunctive one. This process of estrangement is not recent. It has asserted itself gradually through an ongoing alternation between two phases: (i) extended periods of transition, in which humans increasingly perceived themselves as distant from nature and other animals; and (ii) critical turning points, marked by major cultural revolutions that further deepened the rupture.

The myth of progress and the ensuing Industrial Revolution marked another decisive stage in the long history of disjunction. This era gave rise to the cult of machinery and technology, along with imperialism and colonialism, reinforcing a view of nature as nothing more than a resource to be exploited for profit (Sennett 2011; Kempf 2008). What followed was a prolonged period of cultural adjustment, continuing well into the twentieth century, exemplified by the decline of rural life in the face of accelerating urbanization. In my view, the digital revolution then introduced a further rupture in our relationship with nature (Rosa 2010). Unlike analog technologies, which maintained a network of relationships between humans and their environments, digital technologies are immersive and often lead to a withdrawal from the physical world (Balbi 2024). One clear example is the rise of social media, which in recent decades has notably affected younger generations, weakening their connection with the natural world (Rosen 2022). All meaningful relationships are based on an emotional bond—and our bond with nature is no exception. This emotional connection, essential to nurturing care and preventing indifference toward the destruction of the biosphere, is being steadily eroded by the increasing distance we feel from the natural world. While values may be rooted in rational awareness, the affective dimension plays a crucial role in sustaining them—and this affective dimension is now being lost.

2. *Nature vs Reason*

The perception of a stark dichotomy between nature and reason emerged clearly from the earliest expressions of post-Sophistic philosophy. As Nietzsche put it, we are all children of Socrates and Plato: in their thought, the natural dimension—the Dionysian—is treated as problematic, something to be controlled and subdued in the name of rationality (Nietzsche 2014). In Plato's allegory of the chariot, nature is symbolized by the unruly black horse, which must be restrained and tamed to the point of distortion through the force of reason. In Socrates' words we find a clear appeal to rational control: a call to dominate passions and instincts, to curb our wild, untamed freedom through knowledge and intellect. This dualistic view—of a human being who must suppress an inner animal always lurking in the depths of the unconscious—has left a lasting imprint on Western culture. Think of Robert Louis Stevenson's *The Strange Case of Dr. Jekyll and Mr. Hyde*, Joseph Conrad's *Heart of Darkness*, or William Golding's *Lord of the Flies*. Even Francisco Goya's etching *El sueño de la razón produce monstruos* illustrates this, with the "monsters" of irrationality depicted as animals emerging when reason sleeps. Such images reflect how deeply the reason–nature dualism has shaped the Western worldview. So much so that in *Dialectic of Enlightenment* (2002), Theodor Adorno and Max Horkheimer describe the development of reason itself as humanity's primary tool for adapting to a hostile and threatening natural environment. This widespread idea—that the natural world is a cradle or nurturing niche for other animals, yet a looming, Goliath-like threat to humans—only reinforces the disjunction between reason and nature.

A slightly different example is found in *De Rerum Natura*, where the poet Titus Lucretius Carus invokes Venus—an implicit representation of nature—to assist him in composing his work. This hymn already reveals not only the doctrine he will expound but also the way nature is conceived: as *voluptas*, as generative power, as a creative drive and cosmic force capable of making the Earth bloom. In Lucretius' vision, nature exists neither *for* humans nor *against* them; it is neither a loving mother nor a cruel stepmother, since Epicurean philosophy allows no place for teleology or purpose. For Lucretius, the problem facing humanity is not nature, but the illusions we construct ourselves: fear of death, of the gods, superstition—in a word, ignorance of the true nature of things. In this sense, the human being must appeal to knowledge and reason in order to live a serene life.

This call to reason as a distinctive trait of humanity also appears in Blaise Pascal's *Pensées* (2023). Pascal compares man to a fragile reed swayed by the wind, whose strength and dignity lie in his capacity to think. As he points out, it is not necessary for the entire universe to arm itself to destroy the human being—a drop of water is enough to kill him. Yet even if the universe were to crush him, man would still be nobler than that which destroys him (fragment 374). Human nobility does not lie in nature, but in reason—hence Pascal's call to "think well."

The sense of distance—and, in a way, the perception of a dualism between human reason and nature—is already present in the myth of the two Titans, Epimetheus and Prometheus, as recounted by Hesiod, Aeschylus, and Plato in the *Protagoras*. Epimetheus is the distributor of embodied performative traits, while Prometheus is called upon to compensate for the human being's lack of bodily functions by gifting fire and *techne*. The story of the two Titans, allies of Zeus during the Titanomachy, is well known. Prometheus—whose name means "the one who thinks ahead" or "the one who sees forward"—is portrayed as clever, reflective, and endowed with foresight. In contrast, Epimetheus—"the one who thinks after," or who is incapable of seeing beyond the here and now—is marked by impulsiveness and a lack of foresight.

Already in this mythical sketch, we see the emergence of a dualism: on one side, the Promethean human, oriented toward the future, constantly becoming, a creator of worlds; on the other, the Epimethean animal, trapped in the immediacy of the present, driven by instinct, and—borrowing from Martin Heidegger—poor in world. This myth of double genealogy, which originally separates the lineage of humans from that of animals (as told in various forms by Hesiod, Aeschylus, and especially Plato in the *Protagoras*), is grounded in the recognition the Olympian gods grant the Titans for their loyalty.

Following the victory, Athena entrusts Epimetheus with a limited set of qualities to distribute wisely among mortal creatures. But Epimetheus, in the myth, gives away all the traits to animals and forgets the human being—leaving it the only living creature without qualities.

It is this forgetfulness that Prometheus must make up for. He gives humanity fire and *techne*—the ability to shape matter and create tools. In other words, while animals embody their traits thanks to Epimetheus' gift, humans externalize theirs into instruments, allowing them the greatest possible freedom in self-definition. From the very beginning, then, Epimetheus' gift is shown to be ambivalent: on the one hand, it enables animals to meet the challenges of their environment; on the other, it confines them to their natural place. Humans, by contrast—because they were forgotten by the impulsive Titan—are liberated from natural constraints and granted self-determination through Prometheus' legacy.

This foundational myth, which profoundly influenced Western culture—especially during the Renaissance through its humanist revival—reveals two important aspects of how we have come to view the relationship between human and animal. On the one hand, humans recognize and even admire the capabilities of animals, attributing them to a divine gift; on the other, they perceive themselves as naked and helpless when comparing themselves to them. According to this myth, it is precisely the lack of natural endowments and Epimetheus' forgetfulness that prompts

Prometheus' compensatory act, making humanity the artisan of its own existence within a separate dimension—one that stands in opposition to the natural realm typical of non-human others.

This dualism inevitably gives rise to two fundamentally opposing visions of nature. On one hand, nature is seen as a nurturing mother, caring for the human being—her intervention often takes on providential overtones. In this view, however, humans must ultimately emancipate themselves from her to achieve full maturity. On the other hand, nature is cast as a cruel stepmother, the source of all human suffering, inherently hostile to humankind. These two visions often coexist within the same author. In the early works of the Italian poet Giacomo Leopardi, for instance, nature is portrayed as benevolent—dispensing beauty and love, embodying harmony, and offering solace in times of sorrow. But in his later writings, Leopardi presents nature as a hostile and indifferent force. For the poet, however, the remedy does not lie in knowledge, nor—much less—in the progress promised by reason. On the contrary, it is precisely reason that makes humans aware of their unhappiness.

3. *The idea of emancipation from nature*

The view of nature as a maternal womb from which humans must emancipate themselves through reason in order to take control of their own destiny and achieve self-determination is the central argument in Immanuel Kant's *Conjectural Beginning of Human History* (2007b). In this work, Kant explicitly challenges Johann Gottfried Herder, who, in his *Ideas for the Philosophy of the History of Mankind* (2024), had emphasized the importance of returning to nature—meaning the original state of humanity's history—and of abandoning the vices of civilization, which he considered a form of degeneration compared to the harmonious life once lived in accordance with nature. Kant firmly rejects this idyllic vision, arguing that the state of nature is far from being the true human condition. For him, human history begins with a break from nature, initiated by reason. By freeing itself from nature's womb, reason enabled humans to detach from their natural state, learning instead to dominate and exploit it. Human society, then, is the result of this emancipation: a gradual process of civilization that lifts individuals out of the perpetual conflict that characterizes the state of nature.

Kant thus sees the journey toward autonomy as a progressive detachment from nature and a growing ownership of human life itself. He outlines four key stages in this development: (i) the selection of food; (ii) the regulation of sexuality; (iii) the ability to conceive of the future; (iv) the use of tools. In the same essay, Kant explicitly describes emancipation from nature's womb as both an honorable and perilous turning point for the human being:

> [It] drives him out from the harmless and secure condition of childcare, from out of a garden, as it were, that took care of him without any effort from him and thrust him into the wide world, where so many worries, troubles, and unknown ills awaited him. (Kant 2006, 29)

Kant's reflection then examines whether this emancipation was ultimately beneficial or harmful for humans, framing the discussion around two models of life: (i) the natural model, understood as an individual existence free from social obligations; (ii) the model governed by reason, which for Kant marks the beginning of human history and is founded on the principle of human community—a condition radically different from the former, as it is centered on social life, that is, a dimension beyond the individual. For Kant, emancipation from nature represents undeniable progress because it allows human beings to become conscious of their rational essence and fully independent from nature. At the same time, he acknowledges that while the transition to civil life may constitute progress for humanity as a whole, it is not necessarily so for the individual, who must endure significant suffering as a result of this emancipation. Nevertheless, Kant considers the abandonment of the state of nature a moral duty. Unlike Herder, he does not see history as a regression but rather as a path of advancement.

In contrast, Friedrich Nietzsche, in *On the Genealogy of Morals* (2006), argues that the process of civilization is a form of disguise that suppresses the intrinsic values of human nature. As cultural production increases, the innate essence of the human being becomes ever more obscured and stifled. The process of emancipation from nature is therefore regarded by the German philosopher as an act of control over the human being—a true repression aimed at normalization—whereby civilization becomes the history of human decline. For Nietzsche, emancipation from nature is a lie and a form of degeneration, because the human being is not so much a rational and social animal as an instinctual one. A full acceptance of this instinctual nature is necessary if we are to avoid dehumanization.

Jean-Jacques Rousseau, in his *Discourse on Inequality* (1988), had already contrasted a state of nature—where the human being was self-sufficient and in harmony with the environment—with a civil state, or emancipation from nature, based on oppression, superfluous needs, and falsehood. Thus, Rousseau and Nietzsche agree in identifying emancipation from nature as a harmful development and the natural state as the only true model of freedom. Of course, there are many differences between the two thinkers—not least the fact that, while for Rousseau emancipation from nature resulted in inequality among human beings, for Nietzsche it meant that the human being had been caged and domesticated.

4. *The concept of alienation from nature*

The concept of human alienation from nature is based on assumptions that, as we have seen, rest on foundational pillars of Western philosophical thought. The Aristotelian conception of the human being as a rational animal, over time, has led to the overshadowing of our shared condition of animality, placing increasing emphasis on the attribute of rationality to the point that Aristotle's definition became an oxymoron: being rational makes man a non-animal. The feeling of alienation experienced by humans in relation to nature has, however, taken on a particular significance in recent years, especially as ecological issues have forcefully entered the cultural debate. It is a form of estrangement in which the human being does not feel involved in the unfolding destruction, but is merely an anxious spectator. This is not only a process of denial—a refusal to become fully aware—but a loss of belonging, indeed a sense of alienation, as if ecological problems were simply ruining the party.

Some authors emphasize that we should speak of a constitutive separation between humans and nature (Hailwood 2015)—a perspective that holds nature to be fundamentally foreign to us, so that alienation is constitutive rather than a byproduct of civilizational development. A similar view is proposed by Peter Reed who, in his essay *Man Apart: An Alternative to the Self-Realization Approach* (1989), argues that nature is inevitably a mysterious otherness to humans. As such, our attitude toward nature should be one of reverent contemplation carried out, as much as possible, from a distance. Simon Hailwood, in his essay "Alienation and Nature in Environmental Philosophy" (2015), stresses that this alienation is an integral part of being human and should not be seen as something to regret or overcome.

According to other authors, estrangement from nature is not inherent but the result of a gradual historical process. Advocates of this view often promote a kind of human rewilding or a return to primitivism. They call for greater awareness of the damage caused by this separation and propose a lifestyle rooted in simplicity and alignment with the human being's natural tendencies. One prominent voice in this tradition is philosopher John Zerzan, who, in his essay *Future Primitive* (1994), argues for the clear superiority of the hunter-gatherer lifestyle. He rejects the idea that post-Neolithic cultural systems and technologies are neutral, viewing them instead as tools designed to keep human beings in a state of enslavement. A similar perspective can be found in Paul Shepard's *Nature and Madness* (1982), which also draws heavily on Rousseau's work. Another important work in this field is Stanley Diamond's *In Search of the Primitive* (2017), which begins with a self-critical reflection on anthropological methods. Diamond particularly challenges the uncritical idealization of modern civilization, calling for a deeper examination of its underlying assumptions.

Among the authors who have challenged the dichotomy between alienation as an original condition and as a product of civilization—and who have especially highlighted the contradictions between these two views—is Steven Vogel. In his essay *Against Nature* (1996), Vogel argues that: (i) if alienation were an original condition of the human being, then it would not make sense to speak of it as alienation; (ii) if, on the other hand, it were the result of a historical process, it is unclear how such a separation could have occurred, given the human being's original unity and consubstantiality with nature. According to Vogel, the concept of nature poses serious interpretive challenges that continue to affect contemporary environmental philosophy. Addressing these challenges, he argues, requires taking seriously the idea that nature is a socially constructed concept. His critique sheds light on the tendency to hypostatize nature—a point echoed, though from a different angle, by Timothy Morton in *Dark Ecology* (2016). At the same time, it's important to recognize that alienation should not be understood as a literal separation, but rather as a loss of relationship—a felt distance from all forms of biological otherness. This is not a real break, since the human being remains inseparably connected to the rest of nature. In Arne Naess' *Ecology, Community and Lifestyle* (1989), we find this very idea of interconnectedness: a relational field that links all natural entities in their uniqueness and diversity. This interconnectedness includes human beings. Alienation, in this sense, is not a real separation but the loss of a felt affinity within the biosphere.

5. *The humanist view*

The humanist tradition represents a fundamental moment in the history of Western thought, effectively marking the cultural beginning of the modern age (Garin 1976). The announcement of a complete liberation of human beings from the natural boundaries of their rank emerges clearly in *Oratio de hominis dignitate* by Giovanni Pico della Mirandola:

> We have given you, Oh Adam, no visage proper to yourself, nor any endowment properly your own, in order that whatever place, whatever form, whatever gifts you may, with premeditation, select, these same you may have and possess through your own judgement and decision. (Pico 1996, 7)

This passage clearly reveals the mandate given to human beings to be self-determining both ontologically and in the selection of their goals. The oration continues:

> The nature of all other creatures is defined and restricted within laws which We have laid down; you, by contrast, impeded by no such restrictions, may, by your own free will, to whose custody We have assigned you, trace for yourself the lineaments of your own nature (Ibid.)

Here, a clear separation is established between humans and other living beings—a distinction no longer based on specialization or even a hierarchical position over other beings, as it was in Aristotle, but rather on an ontological difference. And again: "I have placed you at the very center of the world, so that from there you may with greater ease observe everything that exists in the universe" (Ibid., modified translation). Pico's conception of human beings refers first and foremost to their ontological condition, but it also inaugurates a new era for humanity—no longer theocentric, but anthropocentric. He concludes:

> We have made you a creature neither of heaven nor of earth, neither mortal nor immortal, in order that you may, as the free and proud shaper of your own being, fashion yourself in the form you may prefer. (ibid.)

There is no doubt that this is a manifesto for the future. Pico della Mirandola compares the human being to a chameleon, capable of assuming different forms according to their intentions and actions. This invites us to consider not only the sense of complete disjunction from nature that permeates the writings of Renaissance humanists, but also the vision of human totipotency that would characterize the entire modern age, albeit in different forms. First, the scientific revolution of the seventeenth century laid the foundation for a mechanistic view of nature, then the Enlightenment of the eighteenth century celebrated human reason above all else; and finally the industrial revolution of the nineteenth century transformed societies from agrarian and artisanal systems into industrial systems powered by machines and fossil fuels.

The modern age appeared to follow Pico's principles to the letter, becoming in every respect the age of anthropocentrism. What defines this era is not merely a detachment from the perimeter of nature—a kind of coming of age in which humanity is fully aware of its abilities and asserts its autonomy from the natural world—but also an unbounded faith in human potential and progress. The few critical voices that emerged remained weak and failed to dent the prevailing climate of optimism, which endured well into the early twentieth century.

To understand this trajectory, it is necessary to examine its beginnings, which shaped the entire epoch. For example, it is essential to acknowledge the principles of alienation from nature embedded in humanism, which served as the ideological driving force of the modern age. The humanist manifesto, in fact, posits that the achievements of humanity—whether in the realms of art, science, technology, or culture in the broadest sense—should be attributed to: (i) an upward progression or departure from nature, rather than inspiration drawn from it or convergence with it; (ii) human ingenuity alone, overlooking the relational, mimetic, and transformative capacities through which individuals are constantly shaped by their environment.

The intentions of humanists extended beyond merely marking a difference between humans and every other animal; they argued for a sharp ontological break between humans and the entire community of living beings. This vision not only justified specific lines of conduct but also nullified any moral consideration, even just in terms of responsibility or good governance, toward the biosphere. Ethical anthropocentrism, in fact, stems from ontological anthropocentrism; thus, criticizing the former without addressing the latter would be futile. From a broader perspective, encompassing aesthetics and epistemology, *thinking ecologically* primarily involves challenging anthropocentrism (Iofrida 2019; Iovino 2022; Amoroso 2023). This isn't a straightforward task, because this way of thinking is deeply ingrained in the human approach to knowledge—which is why we speak of an *anthropocentric perspective*, a form of species-level egocentrism (Marchesini 2018). Moreover, anthropocentrism must also be understood as an ideology: its roots stretch across multiple cultural traditions and converge, both programmatically and conceptually, in humanism.

Indeed, the Promethean view of humans as liberated from nature, masters of their own destiny, capable of rising above and bending the biosphere to their will, reached its symbolic peak at the dawn of the modern age. This perspective is visible not only in Giovanni Pico della Mirandola but also in Giannozzo Manetti, Marsilio Ficino, and, most famously, in Leonardo da Vinci's *Vitruvian Man*. As said, the legacy of the humanist fathers, embraced by Renaissance scholars, shaped an age of intellectual awakening and human emancipation, paving the way for the Scientific Revolution, the Enlightenment, and idealism, and remaining influential into the early twentieth century. The technoscientific and industrial revolutions further solidified this unwavering faith in humanity and its limitless progress. Humanism, then, was not only a framework for interpreting human nature and its attributes—it was also a programmatic blueprint that spanned the entire modern age. I argue that a conscious reexamination of this cultural manifesto—rooted in the elevation and emancipation of humanity—is crucial to understanding the ecological crisis of the twenty-first century.

6. *The ontological distinction between humans and the biosphere*

As previously mentioned, humanity's gradual estrangement from the biosphere—part of a deliberate effort to emancipate itself from nature—gained significant momentum during the modern age (Latour 1993). While other contributing factors must not be overlooked, it thus seems reasonable to trace this definitive detachment back to the modern world itself (Shepard 1982). This calls for a critical reassessment of the foundational principles that define this era. When I argue that the ecological crisis should be understood through the lens of how humans perceive themselves, I am referring specifically to the mythopoeic narrative handed down by humanism—a vision grounded in the heroic ideal of a self-determining and autonomous humanity (Marchesini 2024). Cultivating ecological thought requires confronting this self-image, because the way we understand ourselves directly shapes how we relate to the biosphere.

One of the key moments in the construction of this worldview was the rise of Cartesian thought, which laid the groundwork for the Western notion of emancipation from nature. René Descartes (1596–1650) introduced a radical divide between humans and the biosphere by reducing all life phenomena to the same status as non-living, abiotic entities. In this view, the behavior of animals became no more complex or meaningful than the flow of water over a waterfall (E. Thomas 2020). With Descartes, a rationalist conception of the world and knowledge took hold—one grounded in mathematics and oriented toward precision, predictability, and certainty. By redefining nature as *res extensa*—something extended and measurable—Cartesian philosophy cemented a fundamental divide between humans and the rest of nature. The human being, defined by *res cogitans*, was not only seen as qualitatively different but also as the only entity capable of affirming its own existence: for while everything else may be doubted, the very act of doubting cannot.

Consequently, according to Descartes, man is essentially an exile in a world entirely foreign to him—cut off from any meaningful connection to the biosphere (Skirry 2005). As Thomas Hobbes argued, Descartes had fallen into a solipsistic trap, a kind of vicious circle. The Cartesian revolution intensified the disjunctive elements already present in Pico della Mirandola's humanism, launching a long and gradual process of detachment that would continue until the dawn of the nineteenth century (Garin 1976). If humanism already contained the seeds of humanity's estrangement from nature, it was Descartes who radicalized this separation through the opposition of *res cogitans* and *res extensa*. Human beings were now interpreted as a hybrid of these two entities—a view that resonates with Pico's vision of man as an intermediate being between the earthly and the spiritual, and with Leonardo's Vitruvian Man, placed at the center of a square (symbolizing Earth) and a circle (symbolizing Heaven). Within this framework, the body came to be likened to a clock, and illness understood as the malfunctioning of a mechanical system. Cartesian dualism thus reinforced a form of analogical anthropocentrism with reductionist overtones: if a mechanical model could describe a biological phenomenon, it was assumed that it could also explain it. In this way, a methodological tool became the ontological foundation of life itself.

Nature thus came to be seen as a vast mechanism governed by rigid, deterministic laws—leaving no room for individuality or agency. Life itself was interpreted through the lens of automation. This mechanistic view extended to non-human animals, which were conceived as machines: their capacity for suffering was denied, since they were not considered to possess a subjectivity capable of experiencing pain. In this framework, animal intentionality and presence in the here and now were dismissed as mere illusions—like the rest of the *res extensa*, animals are mere puppets moved by invisible strings. As we will see, between the late nineteenth and early twentieth centuries, explanations of animal behavior continued to follow this Cartesian imprint. And clearly, if an animal is nothing but a machine as there is "no one" inside, there is no need to worry about how it is treated. For Descartes, animals were automatons—immeasurably more complex than anything humans could design, since they were divine creations—but lacking reason and consciousness, they could not feel pain. Even when they appeared to suffer, he argued, they were merely reacting mechanically to external stimuli, like a clock's hands moving when a spring is pressed.

This view would later be harshly criticized by philosophers such as Voltaire and Auguste Comte, who warned against its potential to justify and encourage cruelty toward animals. Yet, at the same time, the eighteenth century saw a growing fascination in France with automata. The most renowned inventor in this field was Jacques de Vaucanson, who not only developed the first automatic loom in 1745, but also created the famous digesting duck, designed to replicate the process of digestion. Humans thus began to see themselves as modern-day demiurges—capable not only of dominating nature by reducing it to mechanical principles, but also of reproducing the very expressions of life through technology. Beginning with Jacques de Vaucanson, who appeared to fulfill the ideas of Hero of Alexandria (first century CE) as outlined in *On Making Automata*, the human ambition to build mechanical entities that could mimic the functions of living beings took shape. This ambition placed humanity in a position of clear superiority over all other life forms, as if, through technology, humans had assumed a divine status.

Mechanicism deeply influenced the devaluation of non-human life, feeding a worldview in which the control and exploitation of the biosphere became humanity's prerogative. In this view, the human being was seen as the only entity possessing intrinsic value—a premise that established a definitive divorce between humanity and the rest of the biosphere. After all, the seeds of this worldview are already found in Genesis 1:26:

> Then God said, "Let us make man in our image, after our likeness. And let them have dominion over the fish of the sea and over the birds of the heav-

ens and over the livestock and over all the earth and over every creeping thing that creeps on the earth." (ESV)

In other words, to understand the persistent failure to recognize a shared destiny between humans and the biosphere, we must examine the paradigm of human supremacism that has defined human culture, particularly since the fifteenth century (Padwe 2013). From its earliest formulations, this model shaped a specific interpretation of human identity and destiny—one in which we were meant to cast off our final ties to nature and leave the "island" of the biosphere behind, rising above the world to reach the only dimension truly congenial to us.

7. *The idea of the human as a being without nature*

As we have seen, both in the foundational myth of the two Titans and later in Pico della Mirandola's *Oration*, the human being is portrayed as distinct from other living creatures due to a lack of natural rank. This idea resurfaces in Martin Heidegger's philosophy, where he distinguishes between the "world-poor" animal, confined within a fixed dimension of life, and the "world-forming" human, whose uniqueness lies in the absence of boundaries and the drive to shape his own existence, as discussed in *The Fundamental Concepts of Metaphysics* (1995). This understanding of the human being gave rise in the twentieth century to a transdisciplinary field known as philosophical anthropology. Its goal is to capture the human experience in its entirety by integrating humanistic traditions with scientific insights, offering a comprehensive image of humanity (Coreth 1973).

Among the authors who pioneered philosophical anthropology, the philosopher Max Scheler stands out. In his essay *L'homme et l'histoire* (1955), he emphasizes the intention to establish a science focused on the essence and eidetic structure of human beings. For Scheler, the human being is an exceptional entity, profoundly different from all other living beings due to the presence of a spiritual principle that contrasts with the vital principle shared with other animals. To understand the cultural climate that gave rise to this twentieth-century neo-humanism, we must not forget the narcissistic wound left by Darwin's evolutionary ideas, as well as the sense of an identity loss following the development of technology, mass society, and the emergence of psychological theories. Philosophical anthropology, therefore, emerged as a form of barrier against the risk of a reductionist view of the human being. Among the most important authors of this new current of thought is Arnold Gehlen, who, in his work *Man: His Nature and Place in the World* (1988), highlights the essential distance between humans and animals, considering the former, though a product of nature, as the result of a special project.

Gehlen, drawing on certain morphological features of the human being—though with a bias far removed from objective biological evidence—concludes that humans are deficient and incomplete, lacking organs and functions specialized for any particular environment. However, as we have seen, this claim does not hold up under a descriptive analysis of human morphology. Our species shows numerous significant specializations, such as the structure of the foot, the characteristics of the spinal column, the shape of the pelvis, the absence of a baculum (penile bone), the shape of the skull, our unique sweating patterns, the loss of visible estrus, the lowered larynx, the placement of the foramen magnum, the differing lengths of forelimbs and hindlimbs, the structure of the hand, and the development of the neocortex—especially the prefrontal area—just to name a few.

Nonetheless, Gehlen's perspective serves to frame the human being as maladapted to any specific environment, and therefore compelled to construct a "second nature"—the cultural and technological world—as compensation. In this way, humans remain open to the world rather than confined within a fixed ecological niche. This idea clearly draws on mythopoetic themes, particularly the myth of Prometheus, who compensates for Epimetheus' oversight by giving humanity fire and *techne*. At the same time, it closely echoes themes from Pico della Mirandola's *Oration*. To ground his view in naturalistic terms, Gehlen continues his examination of humanity through comparisons with animals. However, his arguments are often speculative, lacking scientific backing, and ultimately tautological—intended less to provide objective insight than to affirm the supposed superiority of our species and the necessity of culture as a means of adaptation.

Gehlen argues that while animals are subject to a limited number of drives and sensory stimuli—responding to them in a deterministic, fixed, and automatic manner—human beings, by contrast, are flooded with a torrent of impulses and sensory inputs. This abundance, on the one hand, allows for broad freedom of action, but on the other, if not kept in check, can lead to dangerous and maladaptive behaviors. According to the German philosopher and anthropologist, unlike animals—which respond with immediate, blind reactions—humans do not merely react. Instead, they act through conscious, deliberate, and planned behaviors, enabled by what he calls the principle of *exoneration*: a vague concept referring to the capacity to distance oneself from the flow of instinctual and sensory pressures. This mechanism of exoneration allows humans to avoid being dominated by these reactive pressures and to channel their energy toward more complex, higher-level pursuits. To fully realize their existential dimension, humans require institutions—particularly educational processes—capable of guiding and shaping their psychological traits, modifying and stabilizing them according to the surrounding cultural environment.

For Gehlen, human incompleteness is thus paired with the need for culture, which acts as a crutch to support natural deficiencies. This gives rise to various forms of compensation and support: technology, as a set of tools designed to assist human beings in their conquest of the world; institutions, understood as political and social structures that organize and regulate social life; and moral systems, which establish rules of conduct based on abstract principles of good and evil. These cultural mechanisms serve not only to enhance and extend human capabilities but, above all, to compensate for their deficiencies. Gehlen's neo-humanism, along with that of other thinkers in this philosophical current, rests—as mentioned—on a mistaken understanding of both human beings and other animals. Nevertheless, it draws from a cultural tradition that bears a specific responsibility: it sustains a disjunctive view of humanity and justifies the indiscriminate exploitation of nature. In doing so, it interrupts and effectively reverses the glimpse of continuity and ontological reconciliation between humans and other life forms that had begun to emerge with the Darwinian revolution.

8. *The autarkic and autopoietic vision of humanity*

The neo-humanist image of the human being as defined primarily by its cultural dimension—set against the animal world, seen as entirely confined within its phylogenetic inheritance—can seem justified if we consider certain factors:

(i) The strong sense of dependence our species feels toward technical and cultural supports, and the perceived vulnerability when deprived of them;

(ii) The idea that the technical-cultural dimension has become a kind of second reality, enveloping the human and separating it from the natural world;
(iii) The perception of a unified, animal condition opposed to the human, fundamentally instinctive and confined within its ecological niche;
(iv) The belief that all cultural and technological achievements are the product of human ingenuity alone, reflecting a solipsistic expression of our species.

To understand the human-biosphere divide that lies at the heart of today's crisis—both ecological and anthropological—we must first question these assumptions.

Let's begin with the first point (i). It is undeniable that our species now experiences a sense of deprivation at the thought of being without its technological and cultural apparatus. The mistake lies in assuming that this apparatus evolved to compensate for some original deficiency in the human condition. Technopoiesis—the development of technological devices—did not arise from lack, but from the drive to expand the human horizon. It does not fill a void; it creates a new need. In this sense, deficiency is not the cause of technopoiesis, but rather its consequence. Gehlen's explanation therefore falls into one of the most common explanatory biases: mistaking the effect of a process for its cause. Indeed, if technopoiesis were merely compensatory, we would expect to see it slow down over time—whereas what we observe is the opposite: a rapid acceleration in technological growth.

The second aspect (ii) is even more significant and leads me to reject Sloterdijk's idea that the technosphere has a disjunctive effect on the human being. In the essay *Not Saved—Essays after Heidegger*, Sloterdijk states that human beings do not descend from the ape (which is, moreover, an absurdly generic phylogenetic claim) but rather "from the stone [...] provided we are in agreement with the view that it was the use of stones that opened up the horizon of prototechnics" (2016, 114). Once again, in this author, the hybridizing meaning of the tool—that is, its power to enhance our connection with the world—is mystified by claiming that technology allows us to "deactivate bodily contact and replace it with stone contact" (ibid.) when, on the contrary, it extends the feedback we receive from the world, shaping our body plastically.

Setting aside the unfounded notion that nature is something external to the human being, and that technological devices bypass selective pressures—when in fact they merely modify them—the most glaring error remains the idea that technology distances us and shields us, like a kind of glove, from contact with the world. On the contrary, technology intensifies our relationships with and dependencies on the external world, precisely because it dramatically expands our operational horizon. This is evident when we compare hyper-technological Western societies with more traditional ones: the demand for external inputs is incomparably higher in the former. Technology is not a glove, for countless reasons—I'll mention three: because it is not simply added onto our body but becomes somatized; because it does not preserve the body's purity but contaminates it; and because it does not separate but unites.

As for the third aspect (iii), I won't dwell on it here, since it will be addressed in a dedicated chapter on animal agency. Still, it is clear that the interpretation of animals as entirely driven by instinct suffers from numerous biases, including: the perspectival bias, which leads us to construct fictitious categories within a complex reality—such as the concept of "barbarians"—and to exaggerate the distance between "us" and "them"; and the cherry-picking bias, which leads us to prejudicially select only those examples that confirm our theory.

The most significant issue, however, concerns the point (iv): the idea of human self-sufficiency. This reflects an essentialist view of human nature—one that treats human expressive capacities as innate emanations rather than as emergent properties of our species' fundamentally relational character. Rather than seeing human creativity as something entirely internal and opposed to nature—portrayed, in this view, as a hostile stepmother—we should embrace a relational perspective; this approach recognizes nonhuman entities as epiphanies of possible existential dimensions. Our connection to the biosphere should not be confined to biology or the environment, nor reduced to issues of resources and survival. It must instead be redefined in ontopoietic terms (Tymieniecka 1990).

Ecological thinking requires us to abandon the arrogance of human autopoiesis and to acknowledge the debts we have accumulated with the biosphere throughout our cultural evolution. This awareness must not be confined to a historical perspective—it is a pressing, present reality. Our relationship with otherness is the very oxygen of human ontopoiesis. From an ecological standpoint, we are constituted by the relationships we establish with what lies outside ourselves. The human condition is thus deeply intertwined with non-human others, demanding an attitude of openness and recognition. What seems to be missing in much of today's discourse is precisely the realization that we are inseparable from the biosphere. What we arrogantly claim as our own—as if we were ontologically self-sufficient—is in fact the hybrid product of our interactions with the world. Ecological thinking, then, is not simply about preserving the biosphere as a resource; it is about acknowledging its active role in shaping our existential reality. What we need is a true paradigm shift—one that drops any claim to solipsism and autonomy. We must re-evaluate the entire process of human-becoming, or anthropopoiesis (Remotti 1996), not as an autarchic and self-determined endeavor, but as a relational and hybrid process. Only by relinquishing the myth of human self-foundation and purity can we recognize the eco-ontological ties that bind us to the biosphere.

Beginning with the way humans think—particularly since the humanist revolution and its project of emancipating humanity from nature—we can start to rethink the assumptions of philosophical ecology (Horstmann 1984). The tendency to view human cultural traits as mere emanations of individual genius, rather than as emergent from our capacity to relate to the non-human and to acquire new models, reveals the humanist paradigm's inability to think ecologically.

2.4 The Great Anthropological Depression

The ongoing ecological crisis is bringing to light the contradictions inherent in this anthropology of divergence, which has led human beings to see themselves as a special project, entirely separate from the biosphere. It is therefore appropriate to also speak of an ongoing anthropological crisis, one that is spreading across multiple spheres—among them, a sharp decline in humanity's confidence in itself, manifesting in attitudes of discouragement and resignation, and at times, in blind fatalism or sterile self-condemnation. What emerges in today's cultural landscape, however, cannot be simply defined as a form of pessimism, such as we might find in the writings of twentieth-century authors haunted by the threat of a thermonuclear war that could wipe out humanity as a whole. Rather, it appears more as a state of depression or of awareness that finitude is no longer looming, but already underway—an unstoppable trajectory to which we can only surrender. For instance, in their book *The Ends of the World* (2016), Déborah Danowski and Eduardo Viveiros de Castro ask whether a world-to-come still exists, implicitly assuming that such a possibility may already be irreparably compromised.

The major difference between the late-twentieth-century dystopias—by now a well-established movie genre, as seen in Ridley Scott's *Blade Runner* or George Miller and George Ogilvie's *Mad Max*—and the imaginary of the twenty-first century lies in the prevailing sense of resignation. Instead of dystopia, we are presented with the total destruction of the planet, as in Lars von Trier's *Melancholia*, Abel Ferrara's *4:44 Last Day on Earth*, or Adam McKay's *Don't Look Up*. The anthropological depression that characterizes the twenty-first century—ushered in by the wound inflicted at the heart of the West with the attacks on New York's Twin Towers—goes hand in hand with a dreamlike suspension of time, as though we are facing the omen of an already unfolding catastrophe—the annihilating comet or the planet-destroyer—and a sudden reversal, or even end, of the course of history. Whereas, for Heidegger, the death of the individual human being encapsulated the meaning of existence, the end of humanity itself strips the present of meaning, erasing not so much the continuity of the species, but the consoling narrative of humanity's adventure.

In this sense, Danowski and de Castro speak of a *dysphoric blossoming* in contemporary discourse, evident in an artistic production—such as the aforementioned films and Cormac McCarthy's *The Road* (2006)—which stands in stark contrast to the humanist euphoria that characterized the last five centuries of Western history. We are facing a genuine anthropological crisis: the idea of a future that is no longer available, and therefore no longer translatable into anthropopoietic terms. This is compounded by a condition in which ecological degradation becomes unthinkable, because, as Timothy Morton suggests in his essay *The Ecological Thought* (2012), ecological changes involve time spans and spatial-ranges that we struggle to envision. The human being of the twenty-first century is no longer the cognitive measure of the world—partly due to the limited processing capacity and interpretative neutrality that computer systems have been highlighting for at least four decades—but also appears no longer capable of observing the world from the outside. This is due to our total immersion in it and the massive distribution of effects, both in global reach and in their durations, which exceed an anthropocentric timescale. According to Morton, catastrophic events are *hyperobjects* (2013)—entities that go beyond our experiential capacities.

In this regard, I wish to recall a book by Enzo Tiezzi from 1983 titled *Tempi storici, Tempi biologici*, in which the author emphasized humanity's inability to comprehend ecological and evolutionary changes due to our tendency to reason using the metrics of historicity. Bruno Latour, in his essay *Facing Gaia: Six Lectures on the Political Theology of Nature* (2017), underscores this sense of disconnection that characterizes the contemporary age, causing a kind of paralysis in the face of events, rooted in the awareness of a lack of usable time and space. Isabelle Stengers, in *In Catastrophic Times: Resisting the Coming Barbarism* (2015), states that the psychological and ecological spaces have collapsed into one another, erasing the sense of human control over the environment and turning instead into a condition of panic, given the distance between scientific understanding of the crisis and political powerlessness.

Indeed, we are faced with a condition of *freezing* or radical nihilism, exemplified by the idea that the human being is a terminal illness for the biosphere—as argued in Bruno Sebastiani's essay *Il cancro del pianeta* (2017)—a kind of metastatic neoplasm that must be eradicated, as advocated by the Voluntary Human Extinction Movement (VHEMT), a non-governmental organization founded in 1991 by Les U. Knight. According to the extinctionist line of thought, human beings should stop reproducing so that the species gradually goes extinct. Some authors suggest that the extinctionist proposal reflects precisely this condition of final surrender in the face of the ecological crisis. More moderate positions define the VHEMT project not so much as aiming at extinction per se, but rather at reducing birth rates, in the belief that no more human beings should be brought into the world under current conditions (Carlson 2005). Among the scholars close to this movement—though not necessarily sharing identical views—one may mention Paul R. Ehrlich, author of *The Population Bomb* (1968), Alan Weisman, author of *The World Without Us* (2007), and Serge Latouche, author of *Farewell to Growth* (2009).

There is, moreover, a connection between the idea of humanity's end and a somewhat more moderate vision of *wilderness*, that is, the withdrawal of humans from natural spaces in order to reduce their impact to zero. In both cases, the human being is not seen as an integral part of nature but as an alien entity that must step back. This re-emergence of the human–nature dualism, which frames the human perspective from the outside and affirms a fundamental discontinuity—even if it reverses the traditional hierarchy—can still be considered a form of anthropocentrism, albeit in a negative sense. Even here, the anthropological crisis appears as a form of disconnection: a vision of a world without human beings, which is often mirrored by the opposite idea—an age of humans without the world.

There is no doubt that, faced with such a stark vision of disjunction between the human being and the biosphere, one can no longer ignore the ongoing anthropological crisis. I have already mentioned Weisman's essay, but in a similar vein are Cal Flyn's *Island of Abandonment: Nature Rebounding in Post-Human Landscapes* (2021) and Telmo Pievani's *La terra dopo di noi* (2019). What characterizes these works is their focus on a future Earth where the retreat of the human species has given rise to a new world—one not only capable of recycling the fossils of human civilization but also of fostering the evolution of new species, as imagined in Dougal Dixon's *After Man: A Zoology of the Future* (1981). Naturally, a world without us lacks observers and, from an anthropological standpoint, is defined by the absence of perspective. In a similar way, Ray Brassier in his essay *Nihil Unbound: Enlightenment and Extinction* (2007) envisions a distant future, beyond the time of anthropological correlation—an age in which everything is extinguished: human beings, life, the Earth, and even the Universe—a time that is, in any case, certain in its work of annihilation. At the very moment one becomes aware of this inevitable end, one reaches the conclusion that everything is already dead, stripping all physical and metaphysical positions of their foundation. For Brassier, a new Enlightenment must begin precisely with the awareness of this end, which calls into question any search for meaning.

An approach that starts from different premises yet still arrives at the idea of the accidental and, above all, limited significance of human experience, focuses on everything that preceded it. The twentieth century saw a surge of interest in prehistoric eras—so strong it became almost an obsession—as illustrated by Boria Sax's essay *Dinomania* (2018) and the success of Steven Spielberg's film *Jurassic Park* (1993). On the one hand, this fascination encouraged the imagination of a world without human beings; on the other, it fostered a connection between the great mass extinctions of previous eras and the current ecological crisis. Following this interest in a pre-human reality is Tom Higham's book *The World Before Us* (2021), which discusses early hominids and sister species such as Denisovans and Neanderthals, who, long before our species appeared, had already developed lifestyles marked by cultural forms. Here too we encounter a disorienting image of Earth, characterized by its own experiential dimension even in our absence.

In a similar vein, in his essay *After Finitude* (2009), Quentin Meillassoux employs the device of taking us back to a time before the emergence of the human species—such as the formation of the solar system or the origin of life on our planet. These events, which he calls "arche-fossils," left traces of reality that, though predating humanity, are still accessible to knowledge. This, he argues, undermines correlationism—the idea that we can only access the correlation between thought and being, never either term in isolation. At the same time, what loses consistency is the very meaning of life, which comes to appear as a fortuitous contingency—rendering even the end of the world meaningless.

As mentioned, the other side of the anthropological disjunction is the idea of a humanity without Earth—either because it has been relocated to another planet (as countless science fiction films depict), or through the transhumanist vision advanced by some authors, who consider the biological condition of the human as merely a phase in humanity's history—a transitional state that will lead to our emergence in new, non-organic forms. This school of thought maintains that the real priority is not to find a solution to the ecological crisis, but rather to secure enough time to bring about the Singularity—that quantum leap which, in every respect (including material ones), will allow us to finally emancipate ourselves from nature. The transhumanist vision promoted by thinkers like Ray Kurzweil (1999), Nick Bostrom (2016), and Vernor Vinge (1992) is, in effect, a form of anthropocentrism without the human: a material transcendence, a techno-soteriology that inherits and amplifies the humanist legacy to its highest degree.

Leaving aside the feasibility of post-organic transhumanization—the practical possibility of mind uploading, and the likelihood that such interventions might evolve from Dickian imagery to realizable processes within a useful timeframe—these visions still make one thing clear: they underscore a profound anthropological crisis. This is a crisis that reveals humanity's struggle to find a lifeboat in a world that seems to be falling apart—borrowing the stylistic register of Philip K. Dick. The core difficulty that emerges from all these reflections remains the absence of an integrative anthropology, one capable, once and for all, of resolving the false opposition between the human and nature.

As previously mentioned, the main difficulty that still undermines the debate on the ecological crisis lies in how we interpret the human condition—often viewed as detached from the fate of the biosphere. Even when relational ontology is discussed, the focus typically remains confined to interpersonal dynamics within the human realm (Simondon 2020), rarely venturing to include eco-ontology or eco-humanism. Any ontopoietic dialogue, when acknowledged, tends to remain confined within the human perimeter—restricted to internal exchanges such as those between individuals or cultures (Fabietti 1995). Anything beyond this boundary is usually considered a resource for human use, never something capable of shaping human predicates themselves. It is man who tames the wolf; there is no place for the reciprocal possibility whereby the wolf also changes the anthropological dimension of certain populations. This reflects a form of ontological anthropocentrism grounded in human autarchy and ontological impermeability to the world. Consequently, the biosphere is viewed either as a home or as a resource—an asset to be protected, but fundamentally unrelated to the human condition itself.

Positing a disjunction between humanity and nature means denying the ecological dimension of our existence, rejecting the reciprocal permeabilities that have always allowed us to be inspired by the encounters with non-human others. And yet, non-humans have always shaped our existential perspective, beyond Claude Lévi-Strauss' important, yet limited, notion of the animal as good to think (1963). Claiming that human expression is a pure emanation of our internal qualities blurs the line between human predicates and the species *Homo sapiens*, isolating the anthroposphere from everything else. Essentialism relies on this separation, where the human being is viewed as uncontaminated, ascending from the world into a different, self-contained dimension (Sloterdijk 2001). This stance inevitably leads to a devaluation of the world.

By highlighting the multifactorial causes behind humanity's estrangement from nature, I do not intend to diminish the pivotal role of capitalist society in the current planetary crisis. Rather, I aim to situate this socio-economic model within a broader value framework. Capitalism, in fact, is itself a product of anthropocentrism, the archetype of all hierarchical and exclusionary systems. In turn, capitalism amplifies human supremacism, perpetuating a recursive dynamic. Thus, the ecological disaster cannot be traced back to a single causal factor; it must be understood as a complex and profound disruption in our relationship with the biosphere as a whole (Delort and Walter 2001). In short, in my view, a key element contributing to our neglect of nature is the anthropological model shaping our self-definition. We have grown accustomed to thinking of the human being as fundamentally separate from the natural order, and this mindset prevents us from envisioning a shared destiny with the biosphere. *A fortiori*, this longstanding attitude makes it even harder to recognize a simple but vital truth: that we are, and have always been, an integral part of nature itself.

The environmental crisis can no longer be ignored or underestimated; it is a severe ailment affecting the entire biosphere, which is not merely our home but the organism to which we belong. However, I believe the environmental crisis reflects a closely related anthropological crisis (Bonato and Kirchmayr 2023). Indeed, it represents the culmination of misguided assumptions that have led us astray, preventing a proper understanding of the gravity of the situation. The slumber we struggle to awaken from reveals our inability to recognize its urgency. Stubbornly, we prefer to believe that our goals are within reach and that our current trajectory is correct, despite overwhelming evidence to the contrary. It is imperative that we change our assumptions, but not to blindly follow a different path or pursue the same goals through alternative means. It certainly would not help to trade the spiritual elevation of early humanism for the technological verticalization of transhumanism: simply altering our course is insufficient if the underlying principles remain unchanged. An anthropological revolution demands a profound reexamination, challenging the Cartesian foundations of what we define as the human dimension.

In other words, it is essential to transition from a solipsistic and divergent view of humanity, where human qualities are seen as expressions of individual ingenuity, to a perspective that places relationships with the world at the forefront (Marchesini 2017). To paraphrase the title of a famous work by Kant, these considerations on human ontopoiesis can be viewed as *prolegomena to any future ecological thinking*.

References

Amoroso, Prisca. 2023. *Ecologia*. Roma: Castelvecchi.

Andersen, Gregers. 2020. *Climate Fiction and Cultural Analysis*. London: Routledge.

Archer, David. 2010. *The Global Carbon Cycle*. Princeton: Princeton University Press.

Armstrong, Edward. 1947. *Bird Display and Behaviour*. London: Lindsay Drummond.

Art, Jason. 2022. *Bias Cognitivi*. Turin: Amazon Print.

Balbi, Gabriele. 2024. *The Digital Revolution*. Oxford: Oxford University Press.

Beck, Ulrich. 1997. Subpolitics: Ecology and the Disintegration of Institutional Power. *Organization & Environment* 10 (1): 52–65. https://doi.org/10.1177/0921810697101008.

Benasayag, Miguel, and Gérard Schmit. 2003. *Les passions tristes*. Paris: Éditions La Découverte.

Benjaminsen, Tor A., and Hanne Svarstad. 2021. *Political Ecology: A Critical Engagement with Global Environmental Issues*. London: Routledge.

Bevilacqua, Piero. 2011. *Il grande saccheggio. L'età del capitalismo distruttivo*. Bari-Roma: Laterza.

Bonato, Beatrice, and Raoul Kirchmayr, eds. 2023. *La filosofia e la crisi ecologica. Atti della winter school 2021–2022*. Sesto San Giovanni: Mimesis.

Bostrom, Nick. 2016. *Superintelligence: Paths, Dangers, Strategies*. Oxford: Oxford University Press.

Brassier, Ray. 2007. *Nihil Unbound: Enlightenment and Extinction*. Basingstoke: Palgrave Macmillan.

Carlson, Tom. 2005. More Moderate Positions Define the VHEMT Project Not So Much as Aiming at Extinction Per Se, But Rather at Reducing Birth Rates, in the Belief That No More Human Beings Should Be Brought into the World under Current Conditions. *Voluntary Human Extinction Movement*.

Celli, Giorgio. 1989. *Le farfalle di Giano*. Milano: Feltrinelli.

Chauvin, Rémy. 1968. *Animal Societies from the Bee to the Gorilla*. New York: Hill and Wang.

Chauvin, Rémy. 1973. *Le comportement social chez les animaux*. Paris: Presses Universitaires de France.

Chazan, Michael. 2017. Toward a Long Prehistory of Fire. *Current Anthropology* 58 (S16): S351–S359. https://doi.org/10.1086/691988.

Coreth, Emerich. 1973. *Was ist der Mensch?: Grundzüge einer philosophischen Anthropologie*. Innsbruck/Wien/München: Tyrolia.

Danowski, Deborah, and Eduardo Viveiros de Castro. 2016. *The Ends of the World*. Cambridge: Polity Press.

Dark, Stephen M. 2019. *Contemplating Climate Change*. London: Routledge.

Darwin, Charles. 1871. *The Descent of Man, and Selection in Relation to Sex*. London: John Murray.

Darwin, Charles. 1872. *The Expression of the Emotions in Man and Animals*. London: John Murray.

Delort, Robert, and François Walter. 2001. *Histoire de l'environnement européen*. Paris: Presses.

Descartes, René. 2008. *A Discourse on the Method: Of Correctly Conducting One's Reason and Seeking Truth in the Sciences*. Oxford: Oxford University Press.

Descola, Philippe. 2005. *Par-delà nature et culture*. Paris: Gallimard.

Diamond, Jared. 1998. *Guns, Germs, and Steel: The Fates of Human Societies*. New York: W. W. Norton & Company.

Diamond, Jared. 2011. *Collapse: How Societies Choose to Fail or Succeed*. London: Penguin Books.
Diamond, Stanley. 2017. *In Search of the Primitive: A Critique of Civilization*. New York: Routledge.
Dixon, Dougal. 1981. *After Man: A Zoology of the Future*. New York: St. Martin's Press.
Ehrlich, Paul R. 1968. *The Population Bomb*. New York: Ballantine Books.
Eibl-Eibesfeldt, Irenäus. 2007. *Human Ethology*. London: Routledge.
Eliade, Mircea. 2004. *Shamanism: Archaic Techniques of Ecstasy*. Princeton: Princeton University Press.
Fabietti, Ugo. 1995. *L'identità etnica. Storia e critica di un concetto equivoco*. Roma: Carocci.
Farnham, Connor. 2020. Misguided Misanthropy: Why the "Humans Are the Virus" Mindset Is Damaging to the Environmental Movement. https://blogs.bard.edu/cepblog/?p=11973
Flyn, Cal. 2021. *Island of Abandonment: Nature Rebounding in Post-Human Landscapes*. Glasgow: William Collins.
Gallese, Vittorio, and Maria Alessandra Umiltà. 2006. Cognitive Continuity in Primate Social Cognition. *Biological Theory* 1:25–30.
Ganapathy, Deepti. 2021. *Media and Climate Change*. London: Routledge.
Garin, Eugenio. 1976. *Italian Humanism*. Westport: Praeger.
Gehlen, Arnold. 1988. *Man: His Nature and Place in the World*. New York: Columbia University Press.
Giraldo, Omar Felipe, and Ingrid Fernanda Toro. 2020. Afectividad ambiental: Sensibilidad, empatía, estéticas del habitar. Chetumal/Quintana Roo/Messico: El Colegio de la Frontera Sur e Universidad Veracruzana.
Godelnik, Raz. 2021. *Rethinking Corporate Sustainability in the Era of Climate*. New York: Palgrave Macmillan.
Godfrey-Smith, Peter. 2020. *Metazoa. Animal Minds and the Birth of Consciousness*. Glasgow: William Collins.
Goudsblom, Johan. 1992. *Fire and Civilization*. London: Allen Lane.
Hailwood, Simon. 2015. *Alienation and Nature in Environmental Philosophy*. Cambridge, MA: Cambridge University Press.
Han, Byung-Chul. 2009. *Duft der Zeit. Ein philosophischer Essay zur Kunst des Verweilens*. Bielefeld: Transcript Verlag.
Harari, Yuval N. 2011. *From Animal into Gods*. Israel: Kinneret Zmora-Bitan Dvir.
Hecht, Susanna B., and Alexander Cockburn. 2011. *The Fate of the Forest: Developers, Destroyers, and Defenders of the Amazon*. Chicago: University of Chicago Press.
Heidegger, Martin. 1995. *The Fundamental Concepts of Metaphysics: World, Finitude, Solitude*. Bloomington: Indiana University Press.

Herder, Johann Gottfried. 2024. *Ideas for the Philosophy of History of Mankind.* New York: Princeton University Press.

Higham, Tom. 2021. *The World Before Us: The New Science Behind Our Human Origins.* New York: Viking.

Horkheimer, Max, and Theodor W. Adorno. 2002. *Dialectic of Enlightenment: Philosophical Fragments.* Stanford: Stanford University Press.

Horstmann, Rolf-Peter. 1984. *Ontologie und Relationen: Hegel, Bradley, Russell, und die Kontroverse über interne und externe Beziehungen.* Königstein: Athenäum Hain Verlag.

Ingold, Tim. 2000. *The Perception of the Environment: Essays on Livelihood, Dwelling and Skill.* London: Routledge.

Iofrida, Manlio. 2019. *Per un paradigma del corpo: una rifondazione filosofica dell'ecologia.* Macerata: Quodlibet.

Iovino, Serenella. 2022. *Paesaggio civile.* Milano: Il Saggiatore.

Jamieson, Dale. 2014. *Reason in a Dark Time: Why the Struggle Against Climate Change Failed and What It Means for Our Future.* Oxford: Oxford University Press.

Jonas, Hans. 1984. *The Imperative of Responsibility: In Search of an Ethics for the Technological Age.* Chicago: University of Chicago Press.

Jones, Clive G., John H. Lawton, and Moshe Shachak. 1994. Organisms as Ecosystem Engineers. *Oikos* 69 (3): 373–386. https://doi.org/10.2307/3545850.

Kant, Immanuel. 2006. *Toward Perpetual Peace and Other Writings on Politics, Peace, and History.* New Haven: Yale University Press.

Kant, Immanuel. 2007a. *The Critique of Judgement.* Oxford: Oxford University Press.

Kant, Immanuel. 2007b. Conjectural Beginning of Human History. In *Anthropology, History, and Education*, ed. Günter Zöller and Robert B. Louden, 160–175. Cambridge: Cambridge University Press.

Kempf, Hervé. 2008. *How the Rich Are Destroying the Earth.* Vermont: Chelsea Green Publishing.

Kurzweil, Ray. 1999. *The Age of Spiritual Machines: When Computers Exceed Human Intelligence.* New York: Viking Press.

Latouche, Serge. 2009. *Farewell to Growth.* Cambridge: Polity Press.

Latour, Bruno. 1993. *We Have Never Been Modern.* Cambridge, MA: Harvard University Press.

Latour, Bruno. 2004. *Politics of Nature: How to Bring the Sciences into Democracy.* Cambridge, MA: Harvard University Press.

Latour, Bruno. 2017. *Facing Gaia: Lectures on the New Climatic Regime.* Cambridge, UK: Polity.

Latour, Bruno. 2019. *Essere di questa terra.* Torino: Rosenberg & Sellier.

Leopardi, Giacomo. 2013. *Zibaldone.* New York: Farrar, Straus and Giroux.

Lesch, Harald, Cecilia Scorza, and Katharina Theis-Bröhl. 2023. *Understanding Climate Change*. Heidelberg: Springer Nature.
Lévi Strauss, Claude. 1963. *Totemism*. Boston: Beacon Press.
Lopez, Barry. 1978. *Of Wolves and Men*. New York: Charles Scribner's Sons.
Loreau, Michel. 2023. *Nature That Makes Us Human*. Oxford, UK: Oxford University Press.
Lorenz, Konrad. 1966. *On Aggression*. New York: Harcourt, Brace & World.
Lorenz, Konrad. 1974. *Civilized Man's Eight Deadly Sins*. New York: Harcourt Brace Jovanovich.
Lorenz, Konrad. 1987. *The Waning of Humaneness*. Boston: Little, Brown and Company.
Lovelock, James E. 1979. *Gaia: A New Look at Life on Earth*. Oxford: Oxford University Press.
Lovelock, James. 2006. *The Revenge of Gaia: Earth's Climate Crisis & the Fate of Humanity*. New York: Basic Books.
Lucretius Carus, Titus. 2020. *On the Nature of Things*. Independently Published.
Manetti, Giannozzo. 1975. *De dignitate et excellentia hominis*. Padova: Editrice Antenore.
Marchesini, Roberto. 2002. *Post-Human. Verso nuovi modelli esistenziali*. Torino: Bollati Boringhieri.
Marchesini, Roberto. 2017. *Over the Human: Post-Humanism and the Concept of Animal Epiphany*. Cham: Springer International Publishing.
Marchesini, Roberto. 2018. *Beyond Anthropocentrism. Thoughts for a Post-Human Philosophy*. Sesto San Giovanni: Mimesis International.
Marchesini, Roberto. 2023. *Etologia del desiderio. Riscoprire la propria animalità*. Torino: Lindau.
Marchesini, Roberto. 2024. *Posthumanist Manifesto*. New York: Lexington Books.
McCarthy, Cormac. 2006. *The Road*. New York: Alfred A. Knopf.
Meijer, Lourens J. J., et al. 2021. More than 1000 Rivers Account for 80% of Global Riverine Plastic Emissions into the Ocean. *Science Advances* 7 (18): eaaz5803. https://doi.org/10.1126/sciadv.aaz5803.
Meillassoux, Quentin. 2009. *After Finitude: An Essay on the Necessity of Contingency*. Londra: Continuum.
Moffett, Mark W. 2019. *The Human Swarm: How Our Societies Arise, Thrive, and Fall*. London: Head of Zeus.
Moore, Jason W. 2015. *Capitalism in the Web of Life: Ecology and the Accumulation of Capital*. London: Verso.
Morgan, Richard P. C. 2005. *Soil Erosion and Conservation*. 3rd ed. Malden: Blackwell.
Morton, Timothy. 2012. *Ecological Thought*. Cambridge, MA: Harvard University Press.
Morton, Timothy. 2013. *Hyperobjects. Philosophy and Ecology after the End of the World*. Minneapolis: University of Minnesota Press.

Morton, Timothy. 2016. *Dark Ecology. For a Logic of Future Coexistence*. New York: Columbia University Press.
Næss, Arne. 1989. *Ecology, Community and Lifestyle: Outline of an Ecosophy*. Cambridge: Cambridge University Press.
Nietzsche, Friedrich. 2006. *On the Genealogy of Morality*. New York: Cambridge University Press.
Nietzsche, Friedrich. 2014. *Beyond Good and Evil*. Stanford: Stanford University Press.
Odling-Smee, John. 1988. Niche Construction: The Neglected Process in Evolution. In *The Role of Behavior in Evolution*, ed. Henry C. Plotkin, 98–134. Cambridge, MA: MIT Press.
Padwe, Jonathan. 2013. Anthropocentrism. In *Oxford Bibliographies in Ecology*, ed. David Gibson. New York: Oxford University Press.
Parker, Christopher H., et al. 2016. The Pyrophilic Primate Hypothesis. *Evolutionary Anthropology: Issues, News, and Reviews* 25 (2): 54–63. https://doi.org/10.1002/evan.21475.
Pascal, Blaise. 2023. *Pensées*. Washington, DC: The Catholic University of America Press.
Pico della Mirandola, Giovanni. 1996. *Oration on the Dignity of Man*. Washington, DC: Gateway Editions.
Pievani, Telmo. 2019. *La terra dopo di noi*. Roma: Contrasto.
Plessner, Helmuth. 2019. *Levels of Organic Life and the Human*. New York: Fordham University Press.
Quigley, Carroll. 1979. *The Evolution of Civilizations: An Introduction to Historical Analysis*. New York: Liberty Fund.
Raveendranathan, Devadasan. 2018. *Developments Lead to Pollution and Depletion of Natural Resources*. Chennai: Notion Press.
Reed, Peter. 1989. Man Apart. An Alternative to the Self-Realization Approach. *Environmental Ethics* 11:56.
Remotti, Francesco. 1996. *Contro l'identità*. Bari: Laterza.
Rosa, Hartmut. 2010. *Alienation and Acceleration*. Aarhus: Aarhus Universitetsforlag.
Rosen, Devan. 2022. *The Social Media Debate*. London: Routledge.
Rousseau, Jean-Jacques. 1988. *A Discourse on Inequality*. Harmondsworth: Penguin.
Rousseau, Jean-Jacques. 1991. *Emile, or On Education*. London: Penguin Books.
Rozzoni, Stefano, and Plinio Limata, eds. 2022. *The Economy of Francesco. Un glossario per riparare il linguaggio dell'economia*. Roma: Città Nuova.
Rushkoff, Douglas. 2013. *Present Shock: When Everything Happens Now*. New York: Current.
Safran Foer, Jonathan. 2010. *Eating Animals*. New York: Back Bay Books.

Sahlins, Marshall D. 1972. *Stone Age Economics*. Chicago: Aldine-Atherton.
Sax, Boria. 2018. *Dinomania. Why We Love, Fear and Are Utterly Enchanted by Dinosaurs*. London: Reaktion Books.
Scheler, Max. 1955. *L'homme et l'histoire*. Paris: Aubier Éditions Montaigne.
Scheler, Max. 1961. *Man's Place in Nature*. New York: Noonday Press.
Scheler, Max. 2009. *The Human Place in the Cosmos*. Evanston: Northwestern University Press.
Schilthuizen, Menno. 2018. *Darwin Comes to Town: How the Urban Jungle Drives Evolution*. New York: Picador.
Sebastiani, Bruno. 2017. *Il cancro del pianeta*. Roma: Armando Editore.
Sennett, Richard. 2011. *The Corrosion of Character: The Personal Consequences of Work in the New Capitalism*. New York: W. W. Norton & Company.
Servigne, Pablo, Raphaël Stevens, and Gauthier Chapelle. 2021. *Another End of the World Is Possible: Living the Collapse (and Not Merely Surviving It)*. Cambridge: Polity Press.
Shepard, Paul. 1973. *The Tender Carnivore and the Sacred Game*. New York: Scribner.
Shepard, Paul. 1982. *Nature and Madness*. San Francisco: Sierra Club Books.
Simondon, Gilbert. 2020. *Individuation in Light of Notions of from and Information*. Minneapolis: University of Minnesota Press.
Skirry, Justin. 2005. *Descartes and the Metaphysics of Human Nature*. London: Bloomsbury Publishing.
Sloterdijk, Peter. 2001. *Nicht gerettet. Versuche nach Heidegger*. Frankfurt am Main: Suhrkamp Verlag.
Sloterdijk, Peter. 2013. *You Must Change Your Life: On Anthropotechnics*. Cambridge: Polity Press.
Sloterdijk, Peter. 2016. *Not Saved: Essays After Heidegger*. Malden: Polity Press.
Stengers, Isabelle. 2015. *In Catastrophic Times: Resisting the Coming Barbarism*. Lüneburg: Meson Press.
Thomas, Keith. 1983. *Man and the Natural World*. London: Penguin books.
Thomas, Evan. 2020. Descartes on the Animal Within, and the Animals Without. *Canadian Journal of Philosophy* 50 (8): 999–1014.
Tiezzi, Enzo. 1983. *Tempi storici, tempi biologici*. Milano: Garzanti.
Tymieniecka, Anna-Teresa. 1990. *Logos and Life: The Passions of the Soul and the Elements in the Onto-Poiesis of Culture. Book 3*. Dordrecht: Kluwer.
Vinge, Vernor. 1992. *A Fire Upon the Deep*. New York: Tor Books.
Vogel, Steven. 1996. *Against Nature: The Concept of Nature in Critical Theory*. Albany: State University of New York Press.
Weisman, Alan. 2007. *The World Without Us*. New York: St. Martin's Press.
Wilson, Edward O. 1978. *On Human Nature*. Cambridge, MA: Harvard University Press.

Wilson, Edward O. 2017. *The Origins of Creativity*. New York: Liveright Publishing Corporation.

World Meteorological Organization. 2025. *State of the Global Climate 2024*. Geneva: WMO.

Zerzan, John. 1994. *Future Primitive and Other Essays*. New York: Autonomedia.

CHAPTER 3

The Individualistic Obsession

3.1 Premise

The disorientation of postmodernity regarding a meaningful foundation for the human condition—marked by a drift into self-referentiality and epistemological relativism (Lyotard 1984)—reflects a deeper lack of reference points, a kind of drift toward nothingness that deconstructs without constructing anything. The postmodern climate of skepticism, irony, and ideological disillusionment—challenging the grand narratives of modernity—has paved the way for a subjectivism of values. Everything has become tied to the immediate and the contingent—a view that influenced not only philosophy and art but eventually became a widespread cultural trend (Habermas 1987). This loss of shared ideals gradually pushed the individual to retreat inward, into a private center—the chiasm of perspectival relativity, no longer a personal comfort zone but rather a locus of overwhelming ontological responsibility.

This condition inevitably gave rise to what Spinoza called "sad passions" (Benasayag and Schmit 2003). As psychiatrists working with children and adolescents, Benasayag and Schmit have shown how young people are often the first victims of this value vacuum and lack of future prospects. This is evidenced by rising demands for psychological support and the increasing prevalence of mental health disorders among youth. In *Freedom's Right: The Social Foundations of Democratic Life* (2014), German philosopher Axel Honneth underscores that individual freedom

R. Marchesini, *Philosophical Ecology*,
https://doi.org/10.1007/978-3-032-04987-2_3

can only be fully realized through social connectedness and institutional support. We can therefore say that the individual begins to unravel when the web of relationships that binds them to others starts to fray. Social philosophy, with its long tradition from Rousseau to the Frankfurt School, has long sought to understand the trajectories of societies and forms of government through a critical lens. Today, however, it is also tasked with diagnosing society's psychological pathologies. These manifest as anxiety, depression, compulsive behaviors, and dysfunctional stress regulation—so much so that, according to some authors, the ecological crisis should no longer be understood only in terms of its external consequences but also in terms of its psychological impact on individuals (Levine and Leven 2014). In *Collapsing Consciously* (2013), Carolyn Baker urges us to prepare emotionally, cognitively, and spiritually for a future marked by profound psychological strain—a call echoed by Clive Hamilton in *Requiem for a Species* (2010) and by Inka Weissbecker in *Climate Change and Human Well-Being* (2011). Today, for example, we're beginning to talk about *eco-nostalgia*, a disorder that cuts across human populations: from Australian farmers who can no longer predict weather patterns, to European residents who no longer see fireflies lighting up summer nights or hear frogs croaking in the dark.

As I mentioned in the previous chapter, the anthropological crisis of the twenty-first-century strikingly mirrors the ongoing ecological crisis. It is plausible to suggest that this resemblance stems from a common root: a shared sense of disjunction—a kind of disintegration of the human experience that spreads from a feeling of alienation from the biosphere into the fabric of society itself; this is evident in works such as *In the Swarm* by Byung-Chul Han (2017a) and *Liquid Life* by Zygmunt Bauman (2005). The effects also permeate the individual. Despite what the word's etymology implies, the individual is now subject to a severe fragmentation of the self (Fisher 2017; Rushkoff 2013)—a disintegration that signals the breakdown of the relational "cement" once meant to give predicative form to their existence. In *Alienation* (2016), Rahel Jaeggi describes this as a deterioration of human life, particularly in the sphere of affectivity: a narrowing of our capacity for experiential openness in relationships with others, which leads to a state of persistent distress that ultimately culminates in depression. Alain Ehrenberg, in *The Weariness of the Self* (2009), goes even further, identifying depression as a psychological pandemic—a rising condition directly linked to the social transformations of contemporary society, particularly the growing emphasis on self-realization as a personal responsibility.

Contemporary individualism clashes with the forces of globalization, and it does not always result in the forms of escape imagined by Ulrich Beck in *Power and Global Age* (2005). On the contrary, this excessive burden of self-responsibility ends up weighing even more heavily on the individual than the guilt-driven moral codes of earlier times. The demands of the new capitalism have had severe consequences on the individual psyche, with drastic effects on personal character and identity. For instance, the requirement for permanent flexibility in the workplace, as Richard Sennett highlights in *The Corrosion of Character* (1998), undermines long-term planning and erodes the coherence needed to build a sustainable life perspective. It's clear that the individual, increasingly disconnected and burdened with an unbearable existential weight, is drifting. The only apparent way out—though far from salvific—seems to be a retreat inward.

Andrea Millefiorini, in *L'individuo fragile* (2015), examines this process of individualization, which is rooted in a progressive detachment from all forms of heteronomy, now perceived as limitation and control. Over recent decades, we've seen the shift from a mass society to a society of individuals, seemingly more autonomous and self-made, but in truth more fragile and alone. This solipsistic drift has a multifactorial etiology, as we will explore further in this chapter, and an equally complex pathogenesis. Indeed, it would be simplistic to trace this disintegration of the self to a single cause. As noted earlier regarding the coercive demand for flexibility, social disconnection organically leads to personal fragmentation. This existential disintegration replicates itself like a fractal: from humanity's separation from the biosphere, to the divide between culture and nonhuman others; from technology as a medium to retreat into virtual reality; from individual emancipation from society to the splintering of the self.

This unraveling of all bonds and organismic connectivity produces paradoxical effects: the subject, in response, intensifies their search for existential meaning within, by collapsing time into the present moment and amplifying hedonistic and experiential pursuits. The erosion of both inner and outer cohesion is a clear symptom of a deeper eco-ontological disorder. As Gilles Lipovetsky writes in *Le bonheur paradoxal* (2006), we are now facing two major drifts. On one hand, the disjunctive tendency of modern ontology has reached an extreme, where the individual is expected to find meaning solely within themselves, accelerating a solipsistic closure toward any form of otherness. On the other hand, the consequences of this "sentimental miseducation"—a progressive loss of relational thresholds and connections, leaving people adrift in a void of meaning—are, so to speak, beginning to come to a head.

In his essay *La società dell'estinzione* (2022), scholar Matteo Pietropaoli presents a coherent thesis that retraces the evolution of the individual's emancipation from the collective, a process that ultimately leads to the individual's liberation from themselves—i.e., extinction. While it is undeniable that the causes of this disintegration are multifactorial and manifest in a variety of symptoms, these manifestations are ultimately just expressions of a deeper unraveling of the individual's connective tissue and a degradation of relationships that culminates in an implosion of the self. Much like in an ecosystem, where a lack of internal variety and reciprocal connections creates a vulnerable habitat, the exaltation of individualism—yet another symptom of anthropocentrism—is driving a profound existential crisis. Western society, propelled by capitalist economics and a culture increasingly focused on elevating the self-referential individual—two converging factors that elevate the consumer subject as the prototypical identity—has embarked on a problematic path. This trajectory threatens both the holistic development of the individual and the stability of their social and ecological affiliations. Indeed, I believe it is crucial to reaffirm the absolute necessity of being-in-relationship in all its forms, within a dimension of sharing and conviviality. Connection is the very foundation of existence itself.

3.2 Individuation and Existence

To understand the eco-ontological condition of the individual, we must consider two fundamental aspects: (1) the process of individuation, that is, the emergence of an individual with a self—an ontological boundary through which the individual recognizes themselves in relation to the other; (2) the individual's path of existence as a subject, which refers to the way the individual, as an entity with partial orientation in the world (both affective and cognitive), builds selective interactions with others. The first point to note in the eco-ontological view of the individual is that the two concepts—(1) selfhood, as an outcome of individuation realized through diachronic relationships with the world, and (2) subjectivity, as the expression of existential electivity or the individual's Being-there in the world—are not interchangeable. The second key point is that both individuation and existence are not realized in isolation; they occur through the principle of ontopoiesis, which: (i) in individuation, tells us that the self emerges through relationships with otherness, which can be oppositional, inclusive, or positional, leading to a self-recognition of the individual along their journey through the world; (ii) in existence, highlights the metastable nature of subjectivity—that is, its continuous becoming through the otherness the subject encounters along their Being-there, with this otherness offering specific expressive fields that influence the subject's internal configuration. This demonstrates that the individual's nature is heteronomous, never autonomous.

The process of individuation begins from a pre-individual condition that cannot be considered emergent *ex-nihilo*. Just as the zygote already contains the information—albeit not exhaustive—necessary for the formation of the embryo, the phylogenetic heritage equips the mind with previous memories, unacquired knowledge, and unexperienced recollections (Lorenz 1977; Popper 1972; Campbell 1974). Children are not thrown into a world that is foreign to them, because everything they encounter through their senses—including their mother's face—exactly matches what their minds expected. It's no coincidence that the infant responds with a smile. As Plato says in *Meno*, knowledge is anamnesis—it is the result of reminiscence, the recalling of what was already known. This seemingly counterintuitive idea, already present in the Orphic-Pythagorean view, reappears in Kant's concept of the a priori. As Lorenz suggests in his essay *Behind the Mirror* (1977), this is not about the soul and ideas, but about true representations of the world that speak to us about the diachronic continuum of life, because the cognitive a priori is nothing more than the phylogenetic a posteriori. We can therefore imagine this pre-individual as a quasi-subsumptive entity of the phylogenetic trajectory that preceded it. In fact, it carries fragments of distant epochs—such as its having once been a fish (Shubin 2008)—which appear as ghosts of the past coming to life when its senses meet the world. In a way, it is as though the pre-individual is revived and sparked by the excitement of rediscovering the world just as it remembers it. I have called this feeling *Having-already-been-there* as opposed to Heidgeer's *Dasein* or *Being-there* (Marchesini 2024, 105), to emphasize the eco-ontological nature of the individual that carries within itself previous relationships.

The entire embryonic and fetal development of a human being unfolds through a constant relationship with the mother, of whom the pre-individual is an integral part. The pre-individual is not merely contained within the mother's body—it literally partakes in her physiological metabolism and in the relationships she establishes with the external world. From this perspective, the fetus is not only the result of a relationship, but its entire development takes place through and within relationship. This integrated dimension does not end at birth but continues throughout the neonatal phase—the first six months of life. This period is known as exogestation and is divided into two three-month stages: (i) Adaptation, during which the infant adjusts to external conditions, including the exogenous biorhythm, tactile interaction with the mother, and feeding (regulative heteronomy); (ii) Equilibrium, where a symbiotic relationship takes shape, fostering trust and mutual recognition (attachment process). The relationship with the caregiver then plays a central role in the early years of life by shaping the individual's internal working models, that is, their first mental representations of the Self and of relationships with others. According to attachment theory, the bond with the mother—or with another primary caregiver—serves as an emotional archetype for the individual, influencing: (i) the cohesion of the Self in terms of resilience and stability; (ii) the type of relationships the individual will form with others; (iii) the sense of security in relating to the outside world, especially the expectations they'll develop; (iv) autobiographical memory, and thus the sense of coherence and continuity of the Self; and (v) openness to external experiences, and thus the capacity to engage in new ones (Hinde 1982; Bowlby 1969; Ainsworth et al. 1978). All this indicates that individuation is a relational process.

Beyond the process of attachment, it is important to note that the first six years of life are fundamental for the rhizomatic construction of the individual—using the metaphor presented by Gilles Deleuze and Félix Guattari in their essay *A Thousand Plateaus* (1987). This mode of Being-there extends into the surrounding world like the roots of a tree into the soil. During these formative years, the child forges genuine existential symbioses, such that everything they come into contact with becomes an integral part of the self. We should not imagine this process of socialization as a simple acquaintance with the surrounding reality, but rather as an inclusion of otherness, a translation of what is other-than-self into one's own somatic-identitarian configuration, converting it into other-within-self. Just as nitrogen-fixing bacteria, mycorrhizae, and symbiotic invertebrates are no longer external to the tree but become part of it, transforming it into a holobiont (Bordenstein and Theis 2015), identity itself must be conceived as a process of incorporation occurring on all levels—from the immune system to the endocrine and neurobiological systems (PNEI hypothesis). In this period, the relationship with the world and with non-human otherness is crucial to shaping the individual's lived experience and their emotional bond with the world. I am referring, for example, to the scent of flowers, the rustling of plants, the song of birds, the colors of nature, the cycle of the seasons, the smell of rain, the feeling of the wind, the rumble of a thunderstorm, the murmur of a stream. The relationship with nature is fundamental in defining the individual's experience of, and rootedness in, the world.

Individuation continues during adolescence, especially in terms of the formation of social identity. This process begins during early attachment, develops further in the pre-pubertal period through interactions with peers, and matures through deeper relationships: friendships, romantic connections, group affiliations, and the adoption of a cultural identity. The role of human otherness in shaping the individual—both ontologically and ethically—was explored by two foundational thinkers who paved the way for the concept of *transindividuality*, later developed by Gilbert Simondon in 1964 through his doctoral thesis, and by Étienne Balibar in his readings of Spinoza. The first of these thinkers is Martin Buber, who, in *I and Thou* (1937), emphasizes that life is above all intersubjectivity: at the root of every act of individuation lies relationship. For Buber, the human being is essentially dialogue—a constant exchange with others, grounded in reciprocity and responsibility. To engage in dialogue with another is to enter their reality and take it on as one's own. The "I" and the "Thou" do not exist separately, because the human being cannot detach from this plane of encounter. One cannot live an authentic life without relating to another. The second thinker is Emmanuel Lévinas, who, in *Entre Nous* (2000), explores the fundamentally ethical nature of intersubjective relationships, which bind each person to their fellow humans. This relationship is not chosen but happens by encountering the other—which engenders a deep call that demands love, sincerity, empathy, wisdom, and responsibility. For Lévinas, the full realization of the self does not lie in turning inward, but in *being-for-the-other*: a kind of captivity in which the face of the other loses its merely phenomenal dimension and becomes an *epiphany*—a visitation, an involvement, a conferral of meaning, a revelation rather than just an appearance. It is an event not meant to be understood but to be responded to.

In the essay *Over the Human* (2017a), I argued that individuation should not be limited to relationships with other human beings. It involves all the forms of otherness the subject encounters, including non-humans—such as animals and plants. It is therefore misleading to imagine an ontogenetic Rubicon that confines relational space exclusively to the interhuman realm. In reality, humans have always engaged in formative relationships with the rest of the biosphere, not just with other people. This human exclusivism is a legacy of ontological anthropocentrism. This mindset may tolerate interpersonal individuation or intercultural dialogue, but ideologically—and despite all evidence—denies the possibility of meaningful relationships with non-human beings. As discussed in the previous chapter, human culture has long absorbed influences from the animal world. Yet anthropocentrism still tends to frame these relationships as acts of consumption or use, following the principle of instrumentalization expressed in the idea of the animal as "good-to" something. Today, however, we know that a child who grows up with a dog or a cat is shaped by that relationship in meaningful ways. The process of individuation rests on the human susceptibility to external influence—an openness that recognizes no boundaries and extends to all entities the subject comes into contact with. This calls for a deeper reflection on the social vocation of our species and its innate tendency toward cooptation.

The social dimension of human beings has been explored across various biological disciplines, including Darwin's evolutionism: *The Descent of Man* (1871) highlights the importance of prosocial behaviors in the phylogenetic history of humans. For its part, ethology, emphasizes the complexity of human social instincts and the need to take them seriously (Lorenz 1968; Eibl-Eibesfeldt 2007). Edward O. Wilson's sociobiology, as presented in *On Human Nature* (1978), examines certain human tendencies and cultural expressions as the result of interactions between genes and the environment, rooted in our species' evolutionary past. Indeed, human ethology suggests the coexistence of two opposing tendencies within our species. First, there is an affiliative tendency, which fosters a sense of belonging to a group and leads individuals to share its values and beliefs. This can generate internal conformity and homogenization, but also promotes participation and cooperation. Then there is a conflictual tendency, which drives individuals to see others as threats to their happiness, goals, or even survival—leading to antagonism, competition, or open conflict. Such a view challenges both the Plautine notion of *homo homini lupus* from *Asinaria* (211 BCE) and Hobbes' *bellum omnium contra omnes* in *De Cive* (1642). Rather than being inherently solitary or aggressive, humans should be seen as a social species that always organizes itself within a group. Even conflictual behavior tends to manifest collectively, not individually.

In *The Human Swarm* (2018), Mark W. Moffett emphasizes that human societies do not form in a generic or indistinct way. Instead, they are based on a shared sense of belonging—distinct groups of individuals embedded in a complex web of relationships. For this mutual recognizability to exist, a supporting cognitive system is essential. Anthropologist Robin Dunbar proposed that in our species, the size of the neocortex co-evolved with the size of the social group (1993, 681–735). The so-called Dunbar number defines a theoretical cognitive limit to the number of people with whom one can maintain stable social relationships. By comparing the neocortex size of primates with the nature of their social bonds, Dunbar estimated that for humans, this number is around 150 individuals. Research into how social groups regulate themselves and interact with others was further developed in the UK by Henri Tajfel and John C. Turner beginning in the 1970s, culminating in the Social Identity Theory (SIT). This theory seeks to explain the organizational structures and functions within social groups—such as cohesion and internal conformity—as well as intergroup dynamics, which often involve contrast and competition (Hogg 2016). Tajfel's early experiments in the 1950s explored how people self-segregate into so-called minimal groups. He would divide participants arbitrarily—such as by their preference for the paintings of two different artists—and observed how individuals quickly began to perceive their own group as superior and opposed to the other (Tajfel 1974). These studies reveal a dual relational mechanism at work: one of belonging (in-group dynamics) and one of opposition (out-group dynamics). It turns out that even trivial distinctions can give rise to a spontaneous psychological divide between "us" and "them."

According to SIT, the formation of social identity occurs through three main processes: (i) Categorization—creating sharply defined, homogeneous groups that minimize internal differences and exaggerate them between groups; (ii) Identification—developing identity through both self-perception and internal alignment with the group; (iii) Social comparison—perceiving one's own group as superior to others, often via evaluative bias (Brown 2000). From these considerations, it becomes clear that the process of individuation is, in every respect, a product of relationship, which can manifest in three main ways:

(i) Opposition—the tendency to build one's identity by casting the other as antithesis or dark mirror, resulting in a process of distancing that eliminates continuity and shared traits, and turns the other into a contrasting reference point for self-recognition;
(ii) Inclusion—when otherness is incorporated by the individual through assimilation of qualities or by complementarity, as in co-evolutionary processes, where a trait develops in response to the characteristics of the other—such as the shape of a flower adapting to its pollinator;
(iii) Positionality—where the predicates of an entity are defined by the type of relationship it establishes with otherness (e.g., being a parent, child, citizen) or by the expressive field that the other makes available at a given time.

These three modalities define not only the process of individuation but also the existential trajectory of the individual, who maintains a metastable structure—open to partial reconfiguration and therefore adaptable to external conditions. However, the term *adaptation* captures only part of this existential ontopoiesis, mistakenly suggesting passivity in the manifestation of the Self through the Other. In truth, individuation does not result in an autonomous product but in a relational one—marked by an inherent need for the Other. The eco-ontological nature of the individual pertains not only to individuation but also to the phenomenology of the individual, who can only appear *in relation*—that is, as a being-for-relationship.

As paradoxical as it may seem, the isolation of a person within a dimension of existential autonomy—one that demands self-sufficiency and self-determination—represents the greatest threat to the full expression of their individuality and their ability to give meaning to their life. It renders the individual far more fragile, less vital, significantly more unhappy, inevitably drawn toward nihilism, and much less resilient to processes of social conformity. It is precisely through connection that the individual is strengthened: continual interaction with others enables both the assumption of an ontological framework and the opening of osmotic thresholds that enrich the self with new content. As philosopher Rahel Jaeggi (2016) suggests, the construction of the self fundamentally requires otherness in order to access the meaning of one's individual imprint. This idea is also present in Husserl, particularly in his *Cartesian Meditations* (1960), where he emphasizes that the formation of individuality is not the result of a closed monad but rather a relational process in which the individual coexists with others. Empathy is the key to this openness—a window opened wide for the Other. In *Zur Phänomenologie der Intersubjektivität* (1973), Husserl further argues that the solipsistic dimension is entirely inadequate for grasping the reality of the individual. Instead, we must take a different path and turn our attention to the others we encounter in the external world, because individual reality can emerge only through intersubjectivity.

In his essay *Individuation in Light of Notions of Form and Information* (2020), Gilbert Simondon explores the relational nature of the individual—a theme also developed by other thinkers such as John Bowlby in attachment theory (Holmes 1993), and Lev Vygotsky in his theory of the zone of proximal development (Yasnitsky and Van der Veer 2016). For Simondon, it is essential to give an ontological dimension to relationships, moving beyond the traditional view that treats them merely in logical terms. This shift is crucial, as it significantly alters the very principle of individuation. In the French philosopher's own words: "relation does not spring forth between two terms that would already be individuals; relation is an aspect of the *internal resonance of a system of individuation*" (Simondon 2020, 8). Thus, according to Simondon, attention should be redirected away from the individual as a fixed ontological entity—whether conceived in substantialist or hylomorphic terms—and toward the process of individuation itself: that is, toward what gives rise to the individual. This perspective also draws on scientific insights, beginning with the theories of Ernst Haeckel as outlined in his 1874 work *Anthropogenie*. Individuation, for Simondon, not only occurs through relationships but also gives rise to a metastable system—one that is locally stable yet partially open to change—which is nourished by and expressed through its external relations. This view places Simondon's philosophy in close dialogue with complexity theory. Just like a complex system, the metastable individual is highly sensitive to external perturbations. And crucially, it is through these perturbations that emergent properties arise—features that cannot be explained by looking inward, in essentialist terms, but only through the network of relationships the individual engages in.

The inherently relational nature of the individual is reaffirmed by biologist Richard Lewontin in *The Triple Helix: Gene, Organism and Environment* (2000), where he argues that individualism rests on a false assumption—namely, that the human mind is an autonomous and independent entity. For his part, Étienne Balibar, in his essay *Spinoza, the*

Transindividual (2020), delves into Spinoza's ontology and develops the concept of transindividuality to describe the individual not as an isolated atom, but as an entity that exists only through its relations with others. He applies the principle of individuation to Spinoza's philosophy and his view of mind–body as part of a web of ontological relations. For Spinoza, the individual is never self-sufficient but exists precisely as a being-in-relation with others. Identity is thus formed and transformed across all dimensions of relation—social, political, and emotional. We can say, then, that connection is what brings the individual into being and protects them from existential erosion. As Spinoza suggests, the power of the individual—in all its expressions—is the product of relation and interweaving. From this perspective, ecology cannot be reduced to the study of the environment or the description of ecosystems alone. *Eco-logy* must be understood first and foremost as a way of thinking that applies to ontology itself.

3.3 The Decline of the Convivial Dimension

When we examine contemporary individualism, a clear trend emerges: a gradual shift from heteronomy—that is, dependence on forces external to the individual—toward existential autonomy (Pietropaoli 2023). This is not, of course, a biological or ontogenetic autonomy, given our nature as mammals, but rather a process through which individuals claim emancipation from both convivial participation—that feeling of belonging to something greater than oneself—and the resulting sense of responsibility toward society. This path of disconnection began in the nineteenth century and has since accelerated exponentially. As Philipp Mainländer observed in 1876, every step away from connection becomes a slippery slope, leading to an ever-greater presumption of autonomy (2017).

In *Culture Shift in Advanced Industrial Societies* (1989), Ronald Inglehart highlights the contrast between the materialist generations born before World War II and the post-materialist generations that emerged after it—generations shaped by drastically different values and preferences. This transformation of postwar Western societies coincided with a shift from rural culture—based on the extended family and strong ties to nature—to urban culture, defined by the nuclear family and increasing detachment from agrarian and communal life. The result has been a profound anthropological metamorphosis, one that has increasingly confined the individual to their private world. The most striking difference between the two generations compared by Inglehart is the shift from a way of life rooted in participation in communal living—a cycle of convivial stages such as anticipation, seeking, preparation, production, and enjoyment—that once shaped everyday experiences like farm work, artisanal labor, daily chores, celebrations, and the agapic ritual, to a lifestyle centered on consumption.

Urban culture, in fact, broke this closed-cycle dynamic, leading to a fragmentation of community and the erosion of its connective fabric. This results in a kind of individual alienation, with the family itself becoming an increasingly less cohesive unit due to the loss of shared convivial experience. Over time, what emerged is a "society of individuals," where people seek constant relief from social uprootedness through surrogates. Meanwhile, economic actors—through sophisticated marketing strategies—encourage compulsive consumption (Hoffmann and Akbar 2023). Tellingly, the word *consumption* carries synonyms such as "exhaustion," "dissipation," and "waste,"—terms that inherently point to a depletion of value. There is a meaningful difference between "enjoyment," which includes the entire backstory that made it possible—the convivial trajectory from production to community access—and "consumption," which highlights the act of burning through a resource, exhausting a good, squandering or wasting it, using without regard for the future. Postwar generations were encouraged to consume, to become above all consumers, because—as Jean Ziegler argues in *Les nouveaux maîtres du monde et ceux qui leur résistent* (2002)—consumer society rests on a fundamental principle: its members are first and foremost clients, driven to buy, use, and discard more and more goods without reflecting on what came before or what will come after.

Consumption relies on the transient life of goods because they are produced with the expectation of: (i) obsolescence—for example, computers or smartphones that can no longer support new updates, or whose spare parts are discontinued; and (ii) an intrinsic shutdown (design life; programmed end-of-life; decommissioning)—a kind of engineered apoptosis that might be called planned decay, engineered suicide, or functional death. Another factor is the ever-accelerating pace of generational fashions since the postwar period—a kind of new tribalism that creates deep divides between generations, where something quickly comes to appear outdated, obsolete to the point of being ridiculous.

Consumerism not only detaches the individual from the production process but also pushes for a continuous renewal of one's existential wardrobe. Thus, there is not only the imperative to buy but also the equally strong pressure to throw away, without considering where the waste from this lifestyle ends up. The rise of plastics was a true revolution in transforming consumption patterns. On one hand, it accelerated the throwaway culture due to the low cost of the material; on the other, it filled the world with non-biodegradable pollutants (Tozzi 2025). The history of plastic begins in the late nineteenth century with chemists like Georges Audemars, Alexander Parkes, and John Wesley, who developed rayon and celluloid. But it was especially from the 1950s onward—with polyethylene by German chemist Karl Ziegler and polypropylene by Italian chemist Giulio Natta, both Nobel laureates in 1963—that plastics spread into all aspects of daily life (Weissermel and Arpe 2003). Today, about 390 million tons of plastic are produced annually, much of which has accumulated in the oceans, causing severe damage to aquatic wildlife and harming ecologically and naturally significant beaches. For example, the Maldives have microplastic concentrations in their waters of about 1000 particles per kilogram of sand; the Cocos Islands have reached staggering levels, with 414 million plastic items weighing 283 tons; and the Pacific atolls have seen the National Oceanic and Atmospheric Administration collect 57 tons of waste.

This lifestyle, welcomed in the early 1950s as a positive renewal for people's lives, has in fact brought not only extremely serious ecological problems—which we now know also impact human health (vom Saal and Hughes 2005) even at the stage of embryonic development (von Goetz et al. 2010)—but it has also polluted the very mindset of individuals, conditioning them to not reuse goods and therefore to adopt a lifestyle focused solely on immediate satisfaction. It was, in other words, a kind of consumerist intoxication that made everything easily accessible, devaluing the things themselves and convincing the individual that autonomy was the goal of their existence. The culture of disposability creates a condition of immediacy, an over-scaling of the present, as Douglas Rushkoff emphasizes in his essay *Present Shock* (2013), producing not only a space-time compression that erases the narrative dimension of existence, but also a real shock that prevents the individual from accessing their presence in a full and critical sense. This condition recalls the prophetic words of Theodor Adorno in his 1951 essay *Minima Moralia*, where the philosopher stressed how consumerism is the greatest obstacle to forming a full existence illuminated by critical consciousness, because this type of economy tends to make us lonely—interrupting social conviviality—and steering us toward the superfluous, thus obscuring real needs. For Adorno, consumption becomes a source of all kinds of gratification, providing individuals not only with comforts and means of enjoyment but also with a sense of meaning and values. In a similar vein, Herbert Marcuse, in his book *One-Dimensional Man* (1991), links freedom to convivial participation, arguing that the creation of false needs, inducing an obsessive demand for consumption up to wastefulness, not only causes total stupefaction but produces a deceptive freedom.

The concept of *Homo consumens*, based on the principle of consuming more and more to compensate for feelings of emptiness, passivity, loneliness, and anxiety, was introduced by Erich Fromm in *The Sane Society* (1990) and *To Have or To Be?* (2013). However, we could invert this perspective by considering loneliness not as the cause of consumerist frenzy, but rather as its consequence. Zygmunt Bauman, in his essay *Consuming Life* (2007), starts precisely from this premise: waste is no longer seen as a sign that something is wrong—as in Fromm's compensatory view—but has instead become a symbol of wealth and hope. In this sense, Pier Paolo Pasolini's *Scritti Corsari*, written between 1973 and 1975, denounce the darker sides of Italy's postwar economic boom and its social transformations. Pasolini uses the disappearance of fireflies as a metaphor for the end of a civilization—namely, the rural one—that preserved a sense of the sacred in its relationship with nature. For the Italian writer, a society based on the superfluous inevitably transforms even individual lives into something superfluous and devoid of value. Pasolini views the peasant world with a kind of nostalgia, seeing it as an expression of authenticity. The decline of rural culture profoundly changed people's lives and emptied the countryside of birdsong—just as Rachel Carson noted in her essay *Silent Spring* (1964). It was an anthropological shift that affected every aspect of the individual, beginning with the most essential: the emotional education of the person—that is, the kind of existential bond the individual forms with the world.

A significant difference between these two worlds concerns, for example, the approach to food. It shifts from being an agapic ritual—in the sense of a fraternal gathering and banquet—rooted in a convivial, local, and seasonal dimension, to the opposite condition: mere solipsistic consumption, delocalized and detached from any temporal context. It is no coincidence that pre-war generations, even within large urban centers, still retained a rural mindset that regarded food as: (i) something with not only high material value, but also symbolic significance; (ii) a moment that ritually brought the group together, often marked by communal celebrations;

(iii) something that reflected the territory—through local produce—and the seasons, with the year divided into distinct phases based on the food served at the table; (iv) something deliberately prepared rather than simply consumed, thus infused with lived experience and a deep connection to the world.

We can say, then, that food was once a relational element, whereas today it often marks a moment of individual solitude. The deep relationship with nature in the rural world—although not as immersive as in hunter-gatherer communities—while it may not have entailed particular attentiveness to non-human animals, at least affirmed respect for them through the attribution of a ritual, rather than purely economic, value. Moreover, this intense connection to the organic world, experienced from early childhood through interactions with animals and plants, fostered a stronger rootedness in nature—a sense of belonging and relationship with the world.

To summarize, we can identify several specific phenomena in this process of individual disjunction:

(i) Infantilism—a tendency to seek lightness, to remain in an adolescent state, to prioritize playfulness and carefree living, and to reject all forms of responsibility.
(ii) The supremacy of the here and now—the present as the only temporal dimension; becoming is no longer about planning or purpose, but about fleeting opportunities and the expansion of the moment, accompanied by a desire to live as long as possible in order to experience everything.
(iii) Hedonism—rooted in egocentrism and narcissism, where the only meaning of life is to enjoy every possible opportunity to the fullest; a relentless pursuit of pleasure.
(iv) Existential disenchantment—disillusionment with ideals, truth, or collective participation; the demystification of all narratives, leaving only the self as the sole focus of meaning and action.
(v) The drive for self-actualization—leading to competitive distress, burnout, obsession with perfect physical form, and an exaggerated emphasis on performance.
(vi) Consumerist bulimia—a compulsive accumulation of objects, devices, and gadgets that turns the individual into a black hole, swallowing and nullifying everything; desire is interpreted as a lack to be endlessly filled.

(vii) Emancipatory detachment—the loss of all forms of relational bonds, with others reduced to tools for personal ends, and no sense of emotional responsibility toward them.
(viii) Social isolation—retreating into one's own world, transforming life into a solitary dream and withdrawing from relationships with others, in a form of existential onanism.
(ix) Chronic dissatisfaction—a constant state of unease, marked by swings in arousal, depressive episodes, boredom, and overexcitement—often numbed by widespread use of psychiatric medication, alcohol, or various drugs.
(x) Phobic-anxious states—a deep fear of the future that leads the individual to build walls around themselves, to cling to the comfort zone, and to avoid any form of risk, preferring passive existence.

A solipsistic drift has thus taken hold, in the name of a false vision of freedom and self-determination. This has led to a neoplastic conception of the human being in relation to the planet that leaves the individual detached from any meaningful connection with the surrounding world. In line with this worldview, just as a cancer cell is consumed by its own frenzy of replication, so too does the individual live to consume, in a complete absence of broader meaning. It is therefore essential to understand this paradox, and above all, to analyze the features and causes of this drift. I believe the current ecological emergency cannot be overcome unless human beings first confront and resolve their own inner crises.

If one were to identify a metaphor for the ontology of our time, it might resemble a billboard: a glossy image exalting individual life, glorifying the aspiration to be "a winner." The person is depicted as—and urged to become—a model of strength and seductiveness, surrounded by success and admiration, set against an extraordinary backdrop. Existential burnout lies precisely in this imperative of self-realization, in the constant public exposure of the self that erodes all private space, in the demand to always win and—above all—to handle every challenge alone. This image is not merely an invitation to optimism, an osmotic transmission of positivity meant to spread like wildfire. It morphs into a categorical imperative to act immediately, and a harsh judgment on one's life. Its message overflows with calls for transformation, urging individuals to turn their existence into a kind of bodybuilding regimen, to reject all limits, and to adopt a lifestyle akin to a never-ending obstacle course—as in Peter Sloterdijk's

You Must Change Your Life (2013). We are faced with a relentless exhortation to take charge of our lives and fill them with experiences, because time is short and everything is moving fast. There is no room to fall behind—an inversion of Horace's carpe diem, for here the present cannot be gently "seized" but only snatched on the fly.

The injunction to turn life into a work of art—echoing the pages of the Italian poet Gabriele D'Annunzio—or spectacle (Debord 1995) appears to be the core imperative of the individualistic society. Yet, despite its apparent celebration of vitality, it more closely resembles a *danse macabre*, where *horror vacui*—the fear of emptiness—remains the constant dance partner. The shadow of meaninglessness becomes the driving force behind this frenzy (Tenenti 1989). In truth, it's a rather hollow masterpiece, one that seems to dissolve even as it is being created. What remains of the self is a bitter, paradoxically forgetful subject, filled with regrets. Self-affirmation hides its opposite: the loss of subjectivity, masked by conformity to an increasingly standardized, mainstream way of thinking.

The reason is simple: the uniqueness of the individual is the result of a complex system of relationships the person establishes with the world from early childhood—especially through the setbacks and frustrations that encounters with otherness impose. The more effort the individual puts into adapting, seeking connections, negotiating, engaging in dialogue, and—crucially—being challenged by unexpected perspectives, corrections, and shifts in viewpoint—in short, the more they experience decentering—the more resilient they become against the seductive techniques of the mass media. By contrast, an individual raised in an environment where their every whim is indulged, shielded from difficulty, and kept in a bubble of constant reassurance without ever stepping outside their comfort zone, will not be prepared to face such seductions—and will be completely overwhelmed by them.

Individualism rests on several aspects that must be taken seriously if we hope to restore an eco-ontological sense of existence. Without this, any appeal for attention to the ecological dynamics of the biosphere will remain entirely futile. The first point I want to highlight is the pursuit of happiness based on a specific understanding of desire—as a lack that must be compensated through appropriation. This perspective leads to the so-called consumerist syndrome, which continually raises the bar of expectations and increases what Gilles Lipovetsky (2006) and Zygmunt Bauman (2000) might call the "consumption of the world." It is important to shift the concept of desire away from the sense of *penia*—that is, lack—and toward the Dionysian idea of exuberance: desire as motivation to act. I explored this in my essay *Etologia del desiderio* (2023a). If desire is understood as a verbal predicate, a drive to engage in particular activities—an intrinsic motivation—then no object can ever truly satisfy it. Objects will never bring lasting fulfillment, satiety, or resistance to new temptations.

The second aspect to consider is the overstimulation individuals are subjected to today. This leads to fluctuations in arousal, emotional swings, and disruptions in attention and concentration. These effects not only cause emotional dysregulation but also trigger distress, which in turn fuels compulsive and compensatory behaviors. This dynamic was explored by psychologist Alexander Lowen in his essay *Narcissism: Denial of the True Self* (1983). Overstimulation is also tied to a false sense of limitless possibility, which leads to what Byung-Chul Han, in *Psychopolitics: Neoliberalism and New Technologies of Power* (2017b), describes as a psychopolitical power—not the power of direct control, but the creation of perceived possibilities. This, in turn, leads to what Alain Ehrenberg, in *The Weariness of the Self* (2009), calls existential burnout: a state that ultimately results in the depletion of the individual.

This transformation of the individual into an *animal laborans* is mirrored by a surrogate need to consume—along with the excessive use of psychiatric drugs, such as dopamine reuptake inhibitors (like methylphenidate, known as Ritalin), serotonin reuptake inhibitors (like fluoxetine, known as Prozac), anxiolytics like benzodiazepines or buspirone, amphetamines for performance enhancement, and opioids such as oxycodone and fentanyl. It is as if we are facing a kind of psychotic pandemic, in which people no longer take these drugs solely to treat specific pathologies, but to actively alter their mood—a form of psychological doping. In this condition of existential precariousness, the individual is only seemingly autonomous. In reality, they are increasingly drugged, ever more dependent on exogenous substances to manage their emotional state.

Society, while raising individual expectations and promising a horizon full of possibilities, reveals itself instead as a society of constant risk. This risk is ever more global and thus harder to contain, the result of increasingly chaotic dynamics and therefore less predictable—leaving the individual more exposed, as they now lack the social safety nets that once offered protection. This is also a consequence of the breakdown of the social contract—of the disintegration of the connection between the mechanisms of production and those of consumption, and, at the same time, of the fracture between politics and citizenship, as Christopher Lasch pointed out in *The Revolt of the Elites and the Betrayal of Democracy* (1995). In this context, individuals suffer the consequences of community fragmentation caused by delocalization and globalization. Fear becomes a kind of background radiation, leaving people feeling increasingly powerless and unable to influence public affairs. This erosion of a shared "we" deprives individuals of any real political impact, as Byung-Chul Han notes in his *In the Swarm* (2017a). This lack of influence, in turn, creates a compensatory need to be seen and recognized, often leading to forms of social exhibitionism—further fueling individual investment in visibility and the pursuit of fame.

On top of that, the educational environment shaping individuals in recent decades has increasingly moved in the direction of drastically reducing opportunities for real-life experiences. Some speak of a true extinction of encounters with the world during formative years. As a result, individuals are becoming less capable of critical thinking and more vulnerable to external influences. In this sense, it is fair to speak of an emotional miseducation: this inward retreat—where the subject is wrapped in a bubble of comfort, affection, and protection—hinders the balanced and conscious development of their emotional landscape. In the shift from rural to urban culture, we witness a marked loss of emotional connection to the natural world. Despite widespread intellectual agreement—though far less universal than one might expect—on the importance of protecting nature, there is a notable lack of emotional involvement. Yet this kind of affective connection is actually the primary source of value attribution and a key driver of proactive behavior. The creation of a techno-imaginary sphere that envelops the individual while severing them from the world—celebrated by some philosophers as the destiny of the human being—is, in truth, one of the greatest obstacles to fully expressing our human predicates. If, as Marsilio Ficino's *Homo copula mundi* suggests, being human means engaging in a process of union with the world, then this drift toward disconnection represents a threat to the very fulfillment of our humanity.

Today, we are witnessing two increasingly invasive phenomena in people's daily lives. First, indoor life—the tendency to live within confined, technologically modified spaces, cutting off direct contact with natural environments, physical and atmospheric phenomena, and natural referents. In response, many educators have called for outdoor education (Farnè et al. 2018). Second, the so-called *onlife*—a concept emphasized by Luciano Floridi in *The Onlife Manifesto* (2014), referring to the tendency to live constantly connected to the internet, immersed in the digital infosphere. It is clear that these two phenomena capture adults' time and attention, distancing them from genuine social interaction—based on physical presence and multisensory experience—as well as from their relationship with nature. What is even more concerning, however, is the impact this totalizing experience is having on young people, creating an emotional disconnect from the natural world and contributing to serious issues in emotional development, social behavior, and cognitive growth (Greenfield 2015). It is often assumed that the human world of imagination surpasses the natural world—but this is a serious mistake: the former depends on the latter and, paradoxically, possesses even greater diversity.

3.4 The Influence of Technology on the Individual

Various factors have progressively led to the rise of a solipsistic attitude in individuals—a condition I have defined as the loss of the convivial dimension of life. Some of these are social or economic in nature, but we must not overlook the profound impact that twentieth-century technological development has had on people's lifestyles. Certainly, the evolution of urban living cannot be understood—nor would it have been possible—without the advent of household appliances, which have fundamentally transformed the home. On one hand, they have freed human beings from burdensome tasks; on the other, they have turned the home into a place of comfort and relaxation. The domestic installation of a range of everyday service devices has made the nuclear family increasingly independent, but also increasingly less convivial.

Furthermore, the "short century" saw mass communication technologies play an ever-growing role in shaping people's tastes, opinions, habits, and lifestyles: in 1922, the BBC was founded in the UK as the world's first radio station; in 1936, NBC in the United States began television broadcasts, simultaneously with the BBC; finally, and perhaps most importantly, in 1991, the World Wide Web was developed, forever transforming how people work, relax, and access information. From the arrival of radio to today's social media, this process has progressively captured people's attention—bringing countless benefits, especially in terms of information, but also fostering conformism, alienation, and passivity. The digital revolution, and especially the emergence of social media platforms—starting with Facebook in 2004—and on-demand entertainment platforms like

Spotify in 2008 for music and Netflix in 1997 for films and TV shows, has dramatically eroded the convivial dimension of life. The individual has been increasingly transformed into a compulsive spectator, locked into a space of solipsistic voyeurism. In 2006, the first smartphone with a touchscreen was launched, and the following year Apple released the first iPhone. This marked the beginning of a rapid expansion of similar products by competing companies and the creation of what has arguably become the most powerful tool of social isolation in the twenty-first century.

The evolution of mass media technologies over the past hundred years and their influence on changes in individual experience and social configuration clearly show that technology is by no means neutral with respect to the directions human existence will take. The entertainment industry has leveraged these shifts to strengthen its power as a social catalyst. This was a multi-phase process, which took on a massive scale around the 1950s, when the widespread adoption of household appliances not only transformed people's lifestyles but also the way they interacted with the "other-than-self." The fortress-like home, a product of urbanization, on one hand was enriched by machines, but on the other increasingly aimed at sterilizing the living environment—not only through an obsession with cleanliness and disinfection but also by producing a kind of intolerance toward anything organic. This brought about marked changes in aesthetics, risk perception, individual experience, and also in the way of thinking and interacting with otherness. Indeed, while in rural culture the other was represented by an organic referent—a plant or an animal—with which one had to relate by respecting biological cycles, latency periods, and species-specific processes, in the relationship with machines everything takes place through automatisms, but above all through the immediate performance guaranteed by the on-off button.

The creation of a technological system around the individual, based on increasingly less shareable services, has established a tendency toward personal consumption that leads to growing social fragmentation. Over the last twenty-five years, this has culminated in the widespread condition of living alone. Existential atomism, fueled by the rise of digital technologies in recent decades, has caused social withdrawal. This solitude is often mistaken for self-determination and the gaining of more freedom from the social obligations that once formed the connective fabric of communities. Several other processes have accompanied this lifestyle shift, resulting in a loosening of civil and emotional bonds between people and creating a state of relational liquidity.

Material hardship, which was deeply traumatic in the early twentieth century, became the catalyst for a moral redemption of the individual through consumerism. This shift was supported by a strong industrial recovery across all sectors, including agriculture, through the historic Green Revolution between the 1940s and 1960s. Manufacturing growth furthered this trend in two stages: the first, in the latter half of the twentieth century, can be called the analog phase, where people were surrounded by many tools such as household appliances and recreational devices; the second, in recent decades, is the digital phase, where analog devices were replaced by immersive digital experiences that absorb all devices and forms of use into a new reality. In the analog phase, the connection with natural entities was replaced by artificial substitutes, causing an initial separation from reality. With the arrival of the digital era, the allure of the inorganic (Perniola 2017) transforms into an attraction toward the immaterial—essentially, the annihilation of direct experience.

These transformations have primarily had significant effects on people's experiences in two main areas: socially, in how people organize their daily lives, reducing opportunities for socializing and civic participation; and experientially, in their relationship with the real world—especially nature—by increasing projection tendencies and decreasing the insights and challenges that come from interacting with otherness. This disconnect, which affects people from a very young age (it is estimated that young people spend over eight hours a day online), impacts not only the individual's experience but also their emotional and cognitive development.

To understand the risks brought about by this progressive loss of connection—between humans and the biosphere, and between individuals and social relationships—we need to ask whether we are indeed facing a form of techno-solipsism. This question is justified by numerous authors who warn of a growing tendency for people to retreat into their technological shells, as part of the multifactorial breakdown of social cohesion I have been discussing. Therefore, it is essential to reflect on the causes and medium-term effects of this trend. If we accept a relational view of human ontopoiesis, it becomes clear that any event causing disconnection and withdrawal from interactive and social relationships should be considered a threat to human self-creation. Once again, the problem of a possible techno-solipsistic drift concerns two aspects: experience, related to the human condition and its ability to sustain meaning when the osmotic connection with the world and biosphere is interrupted; and the social aspect, related to the individual's condition and their capacity to develop amid decreasing public social interactions. These are two distinct but related problems caused by the rise of an anthropocentric culture that distances people from direct experience, shifting them from connection with nature to immersion in virtual realities and the dominance of social models that emphasize individualism, closely linking consumerism and technology. We must move beyond the prejudice that individualism's rise is solely the result of socio-economic evolution in Western societies—technology is also playing a significant role.

We've seen that we are facing two interconnected issues: one experiential and the other social (B. Fogg 2005). The experiential issue raises the question of whether humanity's withdrawal into its own technological and cultural sphere—becoming impermeable to the biosphere—could result in the erosion of the very qualities we define as inherently human (Marchesini 2017). This concern can be expressed through a series of questions: Can a human being, enclosed within their technological world, still express the same creative energy that defines humanity? Or will they become like a Prometheus who holds the fire, yet lacks the references needed to fuel imagination and cultural expression? Can there be a form of Prometheanism that excludes otherness—that is, the gift of Epimetheus, bestowed upon non-human others, whose presence has inspired human ontopoietic imagination? In other words, can humanity, from an ontological standpoint, exist independently of the biosphere? If we adopt a humanistic view of anthropopoiesis—where human becoming is seen as a self-referential process—then perhaps we need not worry about this rupture (Sloterdijk 2016). But does such a self-contained vision truly reflect reality?

The social issue, instead, concerns whether we are witnessing a general decline in human relationships, exacerbated by emerging technologies and leading to a state of what might be called *liquid affectivity*. This culture of disconnection, fueled by the individual's hedonistic tendencies, finds reinforcement in digital culture. As mentioned, the result is a kind of techno-solipsism, driven by immersive technologies. In this scenario, not only can the human being disconnect entirely from the biosphere, but they could also become independent from all emotional references—locked in a monadic closure, entirely self-contained (Gilbert and Teglborg 2023). While this process is not yet complete, its harmful effects are already becoming visible. Relationships seem to have lost their strength, as if the glue binding us to friends, partners, parents, and colleagues no longer holds. The result is a fragile society—emotionally impoverished and lacking the solid foundations that once sustained human connection and conviviality.

The final question—one that touches both the experiential and social aspects—is whether the driving force behind this phenomenon lies primarily in technological evolution itself, or in deeper cultural roots that shape how we interpret technology (Hughes 2004). Clearly, new technologies play a significant role by offering an alternative to the real world. They present an easy escape that reinforces a trend already set in motion by the consumerist society: the drift away from communal, real-world engagement (Haddon 2000). But we must ask ourselves: are we truly condemned to an unstoppable techno-solipsism, a path of growing separation from nature and deeper entrenchment in egocentric isolation? Or is this trend shaped by the kind of anthropocentric culture that influences not only how we use technology, but how we value and conceptualize it?

The question is not simply how we use technology—whether well or poorly, productively or destructively, whether to further exploit the biosphere or to mitigate harm and adopt more sustainable practices. Rather, it's about how we *understand* technology. In this sense, we can distinguish between two contrasting paradigms. The Promethean view sees technology as a way to compensate for a biological deficiency in humans (Gehlen 1988), or as a tool that inserts a separating layer between humans and the world (Sloterdijk 2016). The Epimethean view, on the other hand, highlights how technopoietic development increases human dependence on nature—if only because of the rising need for energy and raw materials it generates. This dependence exposes humanity *more* to the world, positioning technology not as a barrier but as an amplified interface with it. According to Günther Anders (2009), we might even speak of human subjugation to machines, resulting in a growing imbalance between humans and technology that borders on submission.

From its earliest Promethean interpretations, technology has been seen as a tool of redemption for humanity—a means to assert dominance over the biosphere and a controllable instrument in our hands. Within this framework, our technological arsenal becomes a kind of exoskeleton, enabling the human being to function as a self-sufficient monad, cleansed of all worldly contamination. Enclosed in this hall of mirrors, humanity can finally contemplate its own essence, fulfilling the Vitruvian dream of purging every non-human residue. This *ontological* anthropocentrism—rarely discussed in philosophy due to a focus on *ethical* anthropocentrism, which is merely its consequence—seeks a purified human ideal, calling on technology to assist in this alchemical process. There is a deep connection between ontological anthropocentrism and how we conceive of *techne*. It reflects an autarchic vision of humanity that translates into a solipsistic ritual, made possible through technological instruments. In this view, the technological device becomes: a glove that shields us from the world, supposedly protecting the purity of human essence, and a tool that amplifies our power over the world. However, as I argued in *Technophysiology* (2023b), every technological device is actually embodied: it reshapes the biological organization of the body—altering synaptic wiring, hormonal systems, sensory structures, the musculoskeletal framework, and even the immune system. It is therefore entirely misleading to think of technology as something external to us.

Interpreting technology through an Epimethean lens means considering a range of variables that are often overlooked or pushed to the background in humanistic culture. First, we must ask how a given technology impacts the individual—physiologically, developmentally, educationally, and psychologically—especially by acting on certain predispositions of human nature. Second, we ought to examine the effects of technology on social structures, particularly in terms of in-group and out-group dynamics. It has become increasingly evident—starting with mass media technologies—that human societies are profoundly shaped by the technological devices they use. Third, we need to understand the kind of interface a given technology establishes between human beings and the world—both in terms of how it affects our perception of other living beings and our actual exposure to them. Finally, now more than ever, we must ask about a technology's ecological footprint—weighing its potential benefits (like reducing certain types of consumption or pollution) against its drawbacks, especially in terms of its demand for raw materials and energy. An Epimethean interpretation of technology—viewing it as something that increases our dependency on nature and deepens our entanglement with it—challenges the Promethean narrative, built on the opposite assumption: that technology represents humanity's revenge on nature, a path to separation and purification from it, and a total mastery not only over nature but also over technology itself—seen as an external artifact the Vitruvian man could dispose of at will without losing his sense of self.

I would like to make it clear, however, that questioning Prometheanism does not mean falling into some form of anti-humanism, but rather entails a connective re-evaluation of the human-machine-world relationship. The point is not to dissolve the human into the indistinctness of nature, nor to treat the human dimension as an illusion—a face drawn in sand (Foucault 2001). Instead, the rediscovery of human rootedness in the world can only happen by reaffirming human qualities through a stronger connection with the biosphere. The Promethean mirage, by severing this bond and isolating the individual within themselves, disrupts the vital flow of relationships on which human expression has always relied—relationships often mediated and enhanced by technology. The risk is a kind of human desertification, a withering and regression of the human condition due to a lack of vital sap, a falling out of step with the process of becoming.

It is clear that culture and technology have shaped a defining niche and a particular kind of interface from which we cannot detach. Yet there is a profound difference between conceiving of this kind of "sphere" as a disjunctive space rather than a connective one. I believe that it is often the mental images we form of reality that lead us astray—like the idea of the enclosing bubble (Sloterdijk 2011). Certainly, fetal membranes resemble a sphere that encloses the embryo, just as personal space, intimacy, the home, the circle of friends and acquaintances, the mother tongue and culture, and even an ecosystem—our *Umwelt*, the immersive and operative world of a species—can appear spherical. So too do the egg, the cell, and the organism. But the disjunctive view of this space is misleading: every dimension of the self in relation to otherness is not the product of opposition, but of connection. It is this very principle of connection that gives rise to the sphere—not a need to lock oneself inside it. We have a home because we live in a society: it is the principle of social connection that makes the home possible. Human beings do not inhabit a central point around which they draw concentric circles of separation—they always dwell on the threshold of the connections they are in the process of forming.

The risk of human beings progressively retreating into a self-contained technosphere is far from exaggerated—it is, in fact, becoming alarmingly evident. We are witnessing a pathological trend toward ego-centered isolation that spans multiple dimensions of life:

1. The emotional sphere—Individuals are becoming less sociable and increasingly exposed to insecurity. Relationships grow weaker and more fleeting, destabilizing people's emotional grounding.

2. The educational sphere—Learning environments are increasingly shaped by isolation, with individuals retreating into their own worlds. This has led to a sharp decline in empathy and interpersonal negotiation skills.
3. The experiential sphere—Direct engagement with the world is diminishing. People now often prefer retreating into artificial realities over real-world interaction.
4. The cultural sphere—There is a growing dominance of cultural models recycled by humans or derived from technological systems, while natural content is increasingly less popular.
5. The moral sphere—Both societies and individuals are leaning into self-interest and consumerism, aligning with the logic of the capitalist economy.
6. The aesthetic sphere—Aesthetic preferences are increasingly biased toward what is man-made or modified by humans compared to natural forms of otherness, not under human control.

This trend toward isolation can be linked to certain intrinsic human traits and vulnerabilities: laziness and a preference for what is convenient, selfishness and the tendency to prioritize personal needs, fear and distrust of the unfamiliar. These tendencies intensify when individuals are confined within their own "bubble," have limited relational experiences, and rarely engage with difference. If we trace human behavior over the successive waves of technological connectivity in the past century, a clear trajectory emerges: one of gradual withdrawal and self-enclosedness.

This social disconnection—and more broadly, disconnection from the world—has unfolded behind the illusion of tech-mediated relationships. Social media is a prime example: it gives individuals the unfounded impression that they are at the center of a rich, complex web of interpersonal connections. In reality, we are witnessing a subtle and insidious process of substitution, where the feeling of being connected to many people masks a growing state of isolation. Digital technologies have undoubtedly accelerated this trend, offering ever more refined tools and justifications for building a kind of techno-solipsism. With each upgrade, their countless gadgets train individuals in the practice of personal hedonism. At the same time, we are seeing a steady erosion of our ability to engage with others, to cope with setbacks, and to manage frustration. Take, for example, what is happening with artificial intelligence. Increasingly, AI is being used to guide users through common digital experiences—whether entertainment or information seeking. It may seem incredibly convenient to rely on an entity that makes choices for us, adapting to our tastes and preferences: curating a playlist, selecting films we might enjoy, or filtering information that aligns with our political views. We are handing over to technology the role of comforting us, caring for us, indulging us in a lazy, pseudo-amniotic environment—smooth, frictionless, and without contradictions. But behind this tech-mediated comfort lies a trap: one of passivity and social isolation.

As mentioned, social media platforms lead us to believe we have rich and complex social lives, while the infosphere offers the promise of unlimited access to every source of information. This creates the sensation that the world now comes to us—effortlessly—making our own movement, our stepping outside the cocoon, seem unnecessary. And that cocoon, over time, begins to harden. Every day brings new entertainment platforms offering pleasant distractions that gradually expand until they turn us into compulsive consumers of images, videos, and games, transforming our lives into a passive dissipation of time. And yet, despite their pleasurable veneer, these experiences are only surface-level; as substitutes for real life, they ultimately deepen our restlessness and trap us further inside our personal bubbles. We are thus increasingly seduced by the allure of a nymphal life—one that is voyeuristic, passive, and in some ways dreamlike or psychedelic. It's a kind of existence defined not by participation but by sensory shortcuts, and sometimes even by physiological doping that allows us to feel without truly engaging.

The problem is that we risk believing that satisfaction lies in comfort—an easy life where everything is within reach, where we can binge on sensations obtained in a diluted form, without the need for direct, participatory experience. But this leads to a bleak state of never feeling satisfied, because real fulfillment doesn't come from the outcome alone—it's born from the effort and commitment we invested to achieve it. Life is more like a rollercoaster: joy follows effort, and satisfaction grows from sacrifice. What merely indulges us may appear to offer joy and fulfillment, but in reality, it often deepens discomfort and dissatisfaction. Moreover, this tendency to isolate ourselves feeds on itself: the more we get used to being indulged, the more we seek out confirming environments; the more unsatisfied we feel, the more we consume; the fewer solid relationships we have, the more we retreat into narcissism; the less contact we have with nature, the more we fear it—and the more we fall back into the human-made world.

It's important to emphasize that the way individuals relate to technology is heavily shaped by how technology itself is interpreted within a given culture. This goes beyond mere usage—it concerns the human *stance* toward technology, which today is largely framed by a Promethean model. Indeed, the ecological crisis is often blamed on technological development, leading to technophobic or even neo-Luddite attitudes. But this again reflects a tendency to shift responsibility away from humans, when what truly needs to be reexamined is the role we assign to technology—an attitude rooted in our Promethean interpretation of it. If we were to view technopoiesis as a form of niche construction—that is, one of nature's ways of modifying environments (Odling-Smee 1988)—we would begin to understand the interactive nature of technological systems. To reflect on our stance means to ask how technology integrates with human nature and interacts with the cultural framework of a specific historical and social context. In this light, one can argue that the rise of digital technology encountered a human being embedded in a capitalist society grounded, since the post-World War II decades (well before the advent of the internet, the web, or social and entertainment platforms), in consumerism. Digital technology didn't create individualism; it simply brought to the surface an individualism that was not just latent but already deeply rooted.

We can say, then, that the new digital techno-mediations have had a significant impact in two main ways: (i) on certain intrinsic human vulnerabilities, such as the need for self-affirmation and the pursuit of comfort, offering the individual a life that appears rich and fulfilling; (ii) on an anthropocentric and individualistic attitude, rooted in the humanist cultural tradition but increasingly reinforced by the capitalist socio-economic model. Techno-solipsism is therefore, on one hand, the result of a preexisting condition of postmodernity, shaping how humans have steered these technologies toward consumerist applications; and on the other, it recursively amplifies that very condition. The feeling of living inside a neural simulation—like the characters in Lana and Lilly Wachowski's *The Matrix* (1999)—resonates with many, beyond its special effects, as a vague yet persistent intuition in a spectacle-driven society where everyone is reduced to the role of spectator (Debord 1995). Spending many hours a day immersed in virtual reality seems to trigger a progressive chrysalis effect in the individual, who, like a larva, ceases to move and begins to spin a nymphal cocoon around themselves.

On one hand, digital technologies are triggering a kind of techno-addiction, where people now spend most of their time glued to their phones, losing a continuous connection with the world around them. On the other, they are fueling a growing solipsistic drift that isolates individuals within their own dissatisfactions. The technosphere is becoming a sort of cocoon, secreted by the individual from their own body—growing thicker and harder with time—while the person remains trapped in a state of isolation. Techno-mediation thus turns into an exudative niche: it contains and isolates the individual, caters to them, offering small gratifications, but at the same time causes a gradual withdrawal into the self. It's as if the experience of the world had turned into a walk through a hall of funhouse mirrors—certainly entertaining, but ultimately nothing more than reflections of the observer's own image. We are facing a projective experience—meaning a false one, in some ways a tailor-made simulation (Zhao et al. 2008; Wood and Scott 2016). This stands in stark contrast to encounters with natural otherness, which create a hard, disorienting experience capable of unsettling the subject—a resistant, untamed, and disturbing experience, but precisely for that reason, one that reveals an "elsewhere" and ignites a vital tension within the individual.

Every technology should be evaluated not objectively or merely in terms of its use, but according to the impact it has on specific human vulnerabilities—vulnerabilities that can lead to dangerous drifts. In this light, we must not overlook two key risks that make today's culture of technology—not technology itself—a contributing factor to a solipsistic drift: (i) the ego-driven tendency of the human being, which leads them to prefer a domesticated and anthropomorphized reality, centered entirely around the self; (ii) the Pygmalion-like impulse of the Promethean mindset, which drives humans to favor the products of their own ingenuity over those that arise from natural otherness. Withdrawing from a wild, unpredictable reality—one that challenges us and compels projective thinking—offers a kind of comfort to the individual who prefers to be indulged and to seek affirmation. Yet this leads to a gradual narrowing of coexistence, where the individual becomes unable to tolerate even minimal disagreement, differences in values or opinions, or the demands of dialectical engagement and negotiation. At the root of this is an inability to cope with even the slightest frustration—an issue that carries alarming social consequences.

The break from direct and shared engagement with non-human otherness marks only the beginning of a regressive process—one that drives individuals to seek sameness, to chase consensus, and to demand that every experience be affirming and comforting, never unsettling. This growing tendency for people to withdraw into their own self-contained worlds has disastrous consequences for the individual and threatens precisely those qualities that define us as human—relational qualities. The solipsistic drift distances people from real, tangible relationships, eroding their social, empathetic, collaborative, and communal capacities. Increasingly, the human being seems unable to respond to the face of the Other that calls to them, to borrow the words of Emmanuel Lévinas (1999). The individual's interface thresholds appear hardened, rendering them impermeable to their surroundings and entirely at the mercy of a virtual world which is fully capable of absorbing their entire waking attention. We are witnessing, quite plainly, the danger of a profound disconnection—one where individuals are no longer capable of paying attention not only to the natural environment but also to the social and emotional relationships that once formed the very foundation of their existence.

Acknowledging human fragility in the face of technological possibilities does not mean passively accepting decline as inevitable. Rather, it means attempting to transform that weakness into a strength. While we must avoid the temptation of fatalism, we also cannot ignore the specific vulnerabilities of the human being. That was the mistake of modernity: it not only denied the co-presence of otherness but also erased the tragic dimension at the heart of the human condition. Maintaining an attitude of "trust with caution" therefore means keeping a high level of awareness about human predispositions—neither denying them nor excusing their excesses—and avoiding that anthropocentric arrogance which unconsciously fuels decline. Technology, being itself a product of the human capacity for relationship rather than a mere expression of ingenuity, can help mitigate human fragilities and turn them into strengths—because even the awareness of one's fragility is a virtue. But it can also magnify those weaknesses into dehumanizing vices. Indeed, our worst behaviours are characterized by several features: they build upon existing vulnerabilities or predispositions, such as fear and distrust of otherness, amplifying them; they depend on specific contextual conditions—social and environmental—that either enable or steer their development, which is why they often become recursive; they feed on the absence of dialogue, relationship, and encounters—in a non-relational view of existence.

Contrary to the views of those who dream of a return to wilderness and adopt a survivalist stance (Glover and Carr 2023), technology can be an important tool not only for addressing the current ecological crisis but also for shifting people away from solipsism and restoring an eco-ontological perspective. This will only be possible, however, if we reclaim the gift of Epimetheus and reorient our approach to technopoiesis—toward connection, not separation.

References

Ainsworth, Mary D. Salter, Mary C. Blehar, Everett Waters, and Sally Wall. 1978. *Patterns of Attachment: A Psychological Study of the Strange Situation*. Hillsdale: Lawrence Erlbaum Associates.

Anders, Günther. 2009. *Die Antiquiertheit des Menschen*. Vol. I. München: C.H. Beck.

Baker, Carolyn. 2013. *Collapsing Consciously: Transformative Truths for Turbulent Times*. Berkeley: North Atlantic Books.

Balibar, Étienne. 2020. *Spinoza, the Transindividual*. Edinburgh: Edinburgh University Press.

Bauman, Zygmunt. 2000. *Liquid Modernity*. Cambridge: Polity Press.

Bauman, Zygmunt. 2005. *Liquid Life*. Malden: Polity Press.

Bauman, Zygmunt. 2007. *Consuming Life*. Cambridge: Polity Press.

Beck, Ulrich. 2005. *Power in the Global Age: A New Global Political Economy*. Cambridge: Polity Press.

Benasayag, Miguel, and Gérard Schmit. 2003. *Les passions tristes*. Paris: Éditions La Découverte.

Bordenstein, Seth R., and Kevin R. Theis. 2015. Host Biology in Light of the Microbiome: Ten Principles of Holobionts and Hologenomes. *PLoS Biology* 13 (8): e1002226. https://doi.org/10.1371/journal.pbio.1002226.

Bowlby, John. 1969. *Attachment and Loss. Volume I: Attachment*. New York: Basic Books.

Brown, Rupert. 2000. Social Identity Theory: Past Achievements, Current Problems and Future Challenges. *European Journal of Social Psychology* 30 (6): 745–778.

Buber, Martin. 1937. *I and Thou*. Edinburgh: T. & T. Clark.

Campbell, Donald T. 1974. Evolutionary Epistemology. In *The Philosophy of Karl Popper*, ed. Paul Arthur Schilpp, 413–463. La Salle: Open Court.

Carson, Rachel. 1964. *Silent Spring*. New York: Fawcett.

Darwin, Charles. 1871. *The Descent of Man, and Selection in Relation to Sex*. London: John Murray.

Debord, Guy. 1995. *The Society of the Spectacle*. New York: Zone Books.

Deleuze, Gilles, and Félix Guattari. 1987. *A Thousand Plateaus: Capitalism and Schizophrenia*. Minneapolis: University of Minnesota Press.

Dunbar, Robin I. M. 1993. Coevolution of Neocortical Size, Group Size and Language in Humans. *Behavioral and Brain Sciences* XVI:681–735. https://doi.org/10.1017/S0140525X00032325.

Ehrenberg, Alain. 2009. *The Weariness of the Self: Diagnosing the History of Depression in the Contemporary Age*. Montreal: McGill-Queen's University Press.
Eibl-Eibesfeldt, Irenäus. 2007. *Human Ethology*. New York: Routledge.
Farnè, Roberto, Alessandro Bortolotti, and Marcella Terrusi. 2018. *Outdoor Education: Prospettive*. Rome: Carocci Editore S.p.A.
Fisher, Janina. 2017. *Healing the Fragmented Selves of Trauma Survivors*. New York: Routledge.
Floridi, Luciano. 2014. *The Onlife Manifesto*. Cham: Springer International Publishing.
Fogg, B. J. 2005. *Persuasive Technology: Using Computers to Change What We Think and Do*. San Francisco: Morgan Kaufmann.
Foucault, Michel. 2001. *The Order of Things: An Archaeology of the Human Sciences*. London: Routledge.
Fromm, Erich. 1990. *The Sane Society*. New York: Holt Paperbacks.
Fromm, Erich. 2013. *To Have or To Be?* London: Bloomsbury Academic.
Gehlen, Arnold. 1988. *Man: His Nature and Place in the World*. New York: Columbia University Press.
Gilbert, Patrick, and Ann-Charlotte Teglborg. 2023. Human-Technology Symbiosis: Digital Transformation and Employee Empowerment Renewing Each Other. *LSE Business Review*, March 7.
Glover, Mike, and Jack Carr. 2023. *Prepared: A Manual for Surviving Worst-Case Scenarios*. New York: Penguin Random House.
Greenfield, Susan. 2015. *Mind Change: How Digital Technologies Are Leaving Their Mark on Our Brains*. London: Rider.
Habermas, Jürgen. 1987. *The Philosophical Discourse of Modernity: Twelve Lectures*. Cambridge, MA: MIT Press.
Haddon, Leslie. 2000. Social Exclusion and Information and Communication Technologies: Lesson from Studies of Single Parents and the Young Elderly. *New Media & Society* 2 (4): 387–406.
Hamilton, Clive. 2010. *Requiem for a Species: Why We Resist the Truth about Climate Change*. London: Earthscan.
Han, Byung-Chul. 2017a. *In the Swarm: Digital Prospects*. Cambridge, MA: MIT Press.
Han, Byung-Chul. 2017b. *Psychopolitics: Neoliberalism and New Technologies of Power*. London: Verso.
Hinde, Robert A. 1982. *Ethology: Its Nature and Relations with Other Sciences*. Oxford: Oxford University Press.
Hoffmann, Stefan, and Payam Akbar. 2023. *Consumer Behavior: Understanding Consumers – Designing Marketing Activities*. Wiesbaden: Springer Fachmedien Wiesbaden.

Hogg, Michael A. 2016. Social Identity Theory. In *Understanding Peace and Conflict Through Social Identity Theory*, ed. Shelley McKeown, Reeshma Haji, and Neil Ferguson, 3–17. Cham: Springer International Publishing. https://doi.org/10.1007/978-3-319-29869-6_1.

Holmes, Jeremy. 1993. Attachment Theory: A Biological Basis for Psychotherapy? *The British Journal of Psychiatry* 163 (4): 430–438. https://doi.org/10.1192/bjp.163.4.430.

Honneth, Axel. 2014. *Freedom's Right: The Social Foundations of Democratic Life*. New York: Columbia University Press.

Hughes, James. 2004. *Citizen Cyborg*. New York: Basic Books.

Husserl, Edmund. 1960. *Cartesian Meditations: An Introduction to Phenomenology*. The Hague: Martinus Nijhoff.

Husserl, Edmund. 1973. *Zur Phänomenologie der Intersubjektivität: Texte aus dem Nachlass. Erster Teil: 1905–1920*. The Hague: Martinus Nijhoff/Springer Netherlands.

Inglehart, Ronald. 1989. *Culture Shift in Advanced Industrial Society*. Princeton: Princeton University Press.

Jaeggi, Rahel. 2016. *Alienation*. New York: Columbia University Press.

Lasch, Christopher. 1995. *The Revolt of the Elites and the Betrayal of Democracy*. New York: W. W. Norton & Company.

Levinas, Emmanuel. 1999. *Alterity and Transcendence*. New York: Columbia University Press.

Lévinas, Emmanuel. 2000. *Entre Nous*. New York: Columbia University Press.

Levine, Daniel S., and Samuel J. Leven. 2014. *Motivation, Emotion, and Goal Direction in Neural Networks*. New York/London: Psychology Press.

Lewontin, Richard C. 2000. *The Triple Helix: Gene, Organism, and Environment*. Cambridge, MA: Harvard University Press.

Lipovetsky, Gilles. 2006. *Paradoxical Happiness: Essays on the Hyperconsumption Society*. Paris: Gallimard.

Lorenz, Konrad. 1968. *On Aggression*. London: Methuen.

Lorenz, Konrad. 1977. *Behind the Mirror: A Search for a Natural History of Human Knowledge*. Londra: Methuen.

Lowen, Alexander. 1983. *Narcissism. Denial of the True Self*. New York: Simon & Schuster.

Lyotard, Jean-François. 1984. *The Postmodern Condition: A Report on Knowledge*. Minneapolis: University of Minnesota Press.

Mainländer, Philipp. 2017. *Die Philosophie der Erlösung*. Hamburg: Hanse.

Marchesini, Roberto. 2017. *Over the Human: Post-humanism and the Concept of Animal Epiphany*. Cham: Springer International Publishing.

Marchesini, Roberto. 2023a. *Etologia del desiderio. Riscoprire la propria animalità*. Torino: Lindau.

Marchesini, Roberto. 2023b. *Technophysiology, or How Technology Modifies the Self.* Cambridge: Cambridge Scholars Publishing.

Marchesini, Roberto. 2024. *Posthumanist Manifesto.* New York: Lexington Books.

Marcuse, Herbert. 1991. *One-Dimensional Man: Studies in the Ideology of Advanced Industrial Society.* Boston: Beacon Press.

Millefiorini, Andrea. 2015. *L'individuo fragile: Genesi e compimento del processo di individualizzazione in Occidente.* Santarcangelo di Romagna: Maggioli Editore.

Moffett, Mark W. 2018. *The Human Swarm: How Our Societies Arise, Thrive, and Fall.* Bisignano: Apollo.

Odling-Smee, John. 1988. Niche Construction: The Neglected Process in Evolution. In *The Role of Behavior in Evolution*, ed. Henry C. Plotkin, 98–134. Cambridge, MA: MIT Press.

Perniola, Mario. 2017. *The Sex Appeal of the Inorganic: Philosophies of Desire in the Modern World.* London: Bloomsbury Publishing.

Pietropaoli, Matteo. 2022. *La società dell'estinzione. Liberazione dell'essere umano da se stesso.* Santarcangelo di Romagna: Orthotes.

Pietropaoli, Matteo. 2023. *Individualism and the Rise of Egosystems: The Extinction Society.* Cham: Springer Nature Switzerland.

Popper, Karl R. 1972. *Objective Knowledge: An Evolutionary Approach.* Oxford: Oxford University Press.

Rushkoff, Douglas. 2013. *Present Shock: When Everything Happens Now.* New York: Current.

Vom Saal, Frederick S., and Claude Hughes. 2005. An Extensive New Literature Concerning Low-Dose Effects of Bisphenol. A Shows the Need for a New Risk Assessment. *Environmental Health Perspectives* 113:926–933.

Sennett, Richard. 1998. *The Corrosion of Character: The Personal Consequences of Work in the New Capitalism.* New York: W.W. Norton & Company.

Shubin, Neil. 2008. *Your Inner Fish: A Journey into the 3.5-Billion-Year History of the Human Body.* New York: Pantheon Books.

Simondon, Gilbert. 2020. *Individuation in Light of Notions of from and Information.* Minneapolis: University of Minnesota Press.

Sloterdijk, Peter. 2011. *Bubbles: Spheres I: Microspherology.* Los Angeles: Semiotext(e).

Sloterdijk, Peter. 2013. *You Must Change Your Life: On Anthropotechnics.* Cambridge: Polity Press.

Sloterdijk, Peter. 2016. *Not Saved: Essays After Heidegger.* Malden: Polity Press.

Tajfel, Henri. 1974. Social Identity and Intergroup Behaviour. *Social Science Information* 13 (2): 65–93. https://doi.org/10.1177/053901847401300204.

Tenenti, Alberto. 1989. *Il senso della morte e l'amore della vita nel Rinascimento: Francia e Italia.* Torino: Einaudi.

Tozzi, Mario. 2025. *Prove tecniche di estinzione. Istruzioni per salvare il salvabile.* Milano: Touring Club Italiano.

Von Goetz, Natalie, et al. 2010. Bisphenol A: How the Most Relevant Exposure Sources Contribute to Total Consumer Exposure. *Risk Analysis* 30:473–478.

Weissbecker, Inka. 2011. *Climate Change and Human Well-Being: Global Challenges and Opportunities.* New York: Springer.

Weissermel, Kurt, and Hans-Jürgen Arpe. 2003. *Industrial Organic Chemistry.* Weinheim: Wiley-VCH.

Wilson, Edward O. 1978. *On Human Nature.* Cambridge, MA: Harvard University Press.

Wood, Heater C., and Holly Scott. 2016. Social Media Use in Adolescence Is Associated with Poor Sleep Quality, Anxiety, Depression and Low Self-Esteem. *Journal of Adolescence* 51:41–49.

Yasnitsky, Anton, and René Van der Veer. 2016. Vygotsky's Published Works: A(n Almost) Definitive Bibliography. In *Revisionist Revolution in Vygotsky Studies*, ed. Anton Yasnitsky and René van der Veer, 243–260. London/New York: Routledge.

Zhao, Shanyang, et al. 2008. Identity Construction on Facebook: Digital Empowerment in Anchored Relationships. *Computers in Human Behavior* 24 (5): 1816–1836.

Ziegler, Jean. 2002. *Les Nouveaux maîtres du monde et ceux qui leur résistent.* Paris: Fayard.

CHAPTER 4

The Anthropocenic Transition

4.1 Premise

To think ecologically means to consider entities not as isolated elements, but as parts of systemic dynamics that give them a certain predicative expression—the qualities an entity manifests at a given moment. We know, for instance, that the same genome can produce different phenotypes precisely because it is the epigenetic factors, whether endocellular or exocellular, that influence one protein translation over another (Nijhout 1999). A similar argument can be made for individuals: in many social species—chimpanzees, wolves, lions—individuals develop different personalities depending on the group to which they belong (Hinde 1982). When we observe an organism, we tend to believe its nature stems from a well-defined essence, but in reality, we should speak of a *holobiont*—a consortium of symbionts that together constitute the organism (Rosenberg et al. 2007). Even a species is not a rigidly defined entity, but displays different morphological, physiological, and behavioral traits depending on its environment (West-Eberhard 2003). An ecosystem, in turn, exhibits certain characteristics that make it recognizable based on internal interactive dynamics among its various species—symbiotic relationships and trophic chains, for example, such as mycorrhizae that enable nutrient exchange between plants (Philip et al. 2010). At the same time, ecosystems must be considered metastable entities, because even when they reach their *climax*—the final stage of evolution—they still maintain a degree of

R. Marchesini, *Philosophical Ecology*,
https://doi.org/10.1007/978-3-032-04987-2_4

fluctuation, due to network changes among their components and ongoing interactions with their external environment (Smith and Smith 2000).

Thinking of the biosphere ecologically means understanding the predicative nature of relationships across the various levels outlined above. It requires rejecting the notion that biological entities possess untouchable intrinsic qualities, and instead recognizing that the manifestations of an entity are expressions of its intrinsic potentialities. Speaking of a virtuality inherent in an entity, as mentioned, does not mean saying that it lacks its own consistency or is entirely at the mercy of external factors, nor that it can take on any form whatsoever. Rather, it means the entity possesses a range of possibilities (*inherence*) that become actualized in specific forms depending on the external relationships it engages in. For this reason, for instance, a locust (*Schistocerca migratoria*) can take on either gregarious or solitary morphology and behavior depending on the population density during the nymph stage (Tawfik et al. 1999). Likewise, a turtle (*Trachemys scripta*) can develop as male or female depending on the incubation temperature of its eggs (Crain and Guillette 1998). The essentialist view of the biosphere is based on a notion of species as fixed, self-contained entities, untouched by their relationships with the external world. This view leads to the belief that the biosphere is merely a common home that hosts various species—species that are already fully defined in and of themselves. In this perspective, a species' interaction with the biosphere concerns only resources (what to eat to survive) and location (where to live). Thinking ecologically, by contrast, means acknowledging that one cannot extract an entity from the web of relationships it maintains and expect its qualities to remain unchanged (Rescher 1996).

In an eco-ontological perspective, every expression of an entity is always relational. This means we do not speak of essential qualities that grant stable predicates, but rather of a range of predicative actualizability. An entity, therefore, does not have a single way of manifesting itself, but a finite set of possibilities (inherence)—a spectrum of possible forms determined by the entity's internal organization. This repertoire of potential expressions is then realized in a specific form—how the entity manifests—depending on the relationships it establishes at a given time. We can thus say that in its interaction with the external world, the entity is situated within a particular systemic field, which brings forth—or enables—only certain predicative possibilities (Ladyman and Ross 2007). So we should not think of the entity as something detached, but neither should we imagine it as immersed in some generic whole, a flat holistic network of relations. As a relational entity, it participates in multiple systemic fields—for example: an individual within a population, that population within the dynamics of a species, the species within an ecosystem, the ecosystem within a biome, a biome within a geographic region, and ultimately, the biosphere as a whole (Odum and Barrett 2005). Maintaining a systemic approach does not mean adopting a generalized holism in which every element relates equally to everything else. Rather, it means recognizing the existence of systemic bubbles that contain the entity and give rise to hierarchical levels of relation—and thus, of predication (Koestler 1990).

Ultimately, we can say that an entity does not express a stable and clearly defined intrinsic nature—something untouched by external relations—as posited in essentialism (Armstrong 2010). Rather, it possesses inherence: an internal organization, a particular structure that does not produce a fixed set of predicates, but instead offers a range of potential expressions that are actualized depending on the specific external relations the entity engages in. On the other hand, external relations should be seen neither as accidental, as suggested by David Armstrong (1989), nor as strictly necessary, but rather as carrying their own gradients of probability (Lewis 1986). An entity is situated within specific fields of external relations, not in a vague *mare magnum* of relational possibilities. In nature, a species forms a fairly precise relationship with the external world, which is what we refer to as its ecological niche. This also applies to migratory and generalist species. Every organism is thus structured to complement a set of co-constitutive relationships based on its belonging to a specific environment lifestyle (Cain et al. 2011).

A niche has its own internal organization and a range of external interactions, structured through bubbles of proximity and inclusion. Within each of these relational bubbles, entities are situated in closer connection to one another, forming what can be described as an organizational system. This proximity imposes certain expressive constraints and shapes a specific gradient of relationality, effectively narrowing the range of ways in which an entity can manifest. The organization of these systems gives rise to emergent qualities—predicates that do not belong to the entity in isolation but emerge only through its participation in the relational structure. These emergent features would disappear were the entity extracted from its systemic context. Furthermore, the expressive field made possible by each bubble helps define the entity's probable trajectories of development, both in terms of individual growth (ontogeny) and evolutionary change (phylogeny). These systemic bubbles also maintain their own internal homeodynamics—dynamic equilibria sustained through internal processes of restoration and compensation that preserve the system's integrity. At the same time, they remain open to external influences, and thus subject to fluctuations in predicative expression depending on their interactions with other systems, whether through proximity or inclusion.

Thinking in systemic terms means looking at the biosphere—and in particular the presence of the human being within it—with a very different lens from that shaped by the usual metaphor of dwelling. As haunting as the image of a house on fire may be, we still tend to believe that human qualities are expressions of internal predicates, rather than the result of our relationships with everything non-human—which not only surrounds us but also plays a profound role in defining what we are. The ecological crisis we are experiencing does not concern only what we might call functional aspects—such as deforestation, CO_2 emissions, biodiversity loss, soil erosion, pollution, ocean acidification, or resource depletion—but also the very way we think: the paradigmatic framework that shapes our reflections on the relationship between humanity and the biosphere.

It's important to stress that the current massive destruction of the biosphere, along with the breakdown of the most vital homeodynamic cycles that have characterized the latest phase of life on Earth, does not signal the beginning of a new geological era—some mythical Anthropocene—but rather the twilight of the Cenozoic era. This is the era that saw the rise not only of humans, but also of mammals, birds, and numerous insect orders such as beetles and hymenopterans. If we wish, we can refer to this period as an Anthropocenic *transition*, with two points made clear: (i) this is not the dawn of a new era, but a shift that may mark the end of the previous one—just as it happened during the Permian-Triassic or the Cretaceous-Paleocene transitions; and (ii) this transition will not bring about the triumph of humankind, but rather its decline.

Ultimately, we can say that the ecological catastrophe we are witnessing today—manifesting in various forms, with climate change representing only the tip of the iceberg—is an epochal crisis that affects the entire biosphere, and therefore humanity itself. But this emergency cannot be reduced to causes arising solely in the past three centuries of Western history, as Jason W. Moore (2016) suggests, although it is undeniable that during this period, the anthropocentric disease of Western culture—intensified by the rise of capitalism—reached its peak in terms of destructiveness. In truth, its roots lie in a specific conception of the human that emerged in the modern age, because it is from that moment onward that we begin to detect the seeds of an attitude marked by distance, neglect, and what we might call a devaluation of nature. In this sense, we must indeed reckon with social and economic phenomena that significantly accelerated this process—the Industrial Revolution, globalization, the rise of capitalist society, and finally the technological developments of the last hundred years. Nevertheless, it is essential to understand how the inability to think in a systemic and eco-ontological way has contributed to the unfolding of this crisis.

4.2 Our Planet Today: A Product of the Biosphere

To think of Earth as a form of paradise that can host life due to its specific conditions, which supposedly arose prior to the biosphere, is a gross mistake rooted in mythological narratives. It was life itself, starting from the planet's extreme conditions, that initiated the evolutionary process that gradually shaped the biosphere we now inhabit (Lovelock 1979; Lovelock and Margulis 1974; Margulis 1998; Clarke and Dutreuil 2022). Presently, our planet exhibits a specific homeodynamic configuration capable of supporting diverse organisms and the physiological exchanges they engage in. This state of relative equilibrium relies on intricate processes—ocean currents (Siedler et al. 2013; Marshall and Plumb 2007), the gradient of thermal energy accumulation (Steinmann 2022), the albedo phenomenon associated with glaciers (Qu and Hall 2005; Ouyang et al. 2022), water acidification levels (Gattuso and Hansson 2011), and the cycles of compounds like carbon and nitrogen. Any disruption to these processes can lead to chain effects, the severity of which is hard to quantify.

Without a doubt, altering just a few fundamental parameters could lead the planet to assume different configurations that might not necessarily support life, or at least not the life forms we are familiar with (Weisman 2007). The precarious balance of retained and radiated energy highlights the fragility of the climatic range suitable for biological processes, almost as if it were delicately upheld by a historical scaffolding of events culminating in a climax state (Cain et al. 2011). Analyzing Earth's conditions and comparing them to values found on other planets in the solar system reveals that this outcome is the result of a convergence catalyzed by the evolution of the biosphere—not the other way around (Cain et al. 2011). Naturally, our planet boasts favorable preconditions conducive to the development of cyclic processes and the stabilizing of gradients. Consider, for instance, the moon's impact on Earth's rotational motion or the favorable climatic regime—averaging a range between −15 °C and 110 °C—that permits the presence of liquid water, a crucial element for biochemical reactions. Indeed, a consistent heat source is essential for life, enabling thermodynamic processes, the appropriate mix of chemical elements (especially carbon, oxygen, and hydrogen), and the succession of mantle renewal processes, such as volcanic eruptions, facilitating a geological turnover of various essential minerals. Undoubtedly, Earth meets these prerequisites and this is no small feat.

However, it is equally evident that the current state of our planet is also the outcome of biological events that transpired over the past four billion years, since the advent of the first living forms. In other words, life has not merely occupied a pre-existing niche that allowed for its evolution (Lewontin 1983; Haraway 2016); rather, through its evolution, it actively contributed to shaping the conditions we observe today. What we call Earth, which is not merely a celestial body revolving around a star in one of the numerous constellations within our galaxy, is, in fact, *the expression of the biosphere itself*. The biosphere represents the transformative influence that life has exerted over billions of years. So, today we witness vast biomes as a result of the niche construction and microclimatic self-regulation carried out by organisms themselves. Examining the undergrowth of a European temperate forest, for instance, one can observe a significant amount of organic matter serving as a living environment for fungi, lichens, ferns, and a diverse pedofauna consisting of burrowing and detrivore invertebrates. Indeed, one fundamental characteristic of living organisms is their ability to modify the environment through niche construction, so that the joint action of multiple species engenders a veritable ecological remodeling (Odling-Smee 1988).

Ecology demonstrates the impossibility of evaluating each biological entity solely based on essential qualities, separating the organism from its relationships within the community of living beings (Odum and Barrett 2005). The qualities exhibited by each life form are in fact contingent on its interactions with the environment, making it impossible to comprehend species-specific characteristics and the metapredicates of life through a simple internal assessment. Additionally, the phenotypic translation of phylogenetic heritage, whether genetic or epigenetic, is subject to modulation by the environment (Carroll 2005). Thus, we can only understand ecological thinking by applying an ontopoietic operator with a relational focus. Moreover, life cannot be comprehended from an isochronic or synchronic perspective, disregarding references to past events. This is because life, with its intricate internal organization, relies not only on the negentropic action facilitated by radiant energy (Schrödinger 1940; Prigogine and Nicolis 1977) but also on the information accumulated by living entities in numerous adaptive systems. In essence, a living being is not merely a system of thermodynamic dissipation but also involves diachronic or informational dissipation, meaning the utilization of accumulated information (Marchesini 2022).

Thinking in ecological terms, therefore, is counterintuitive and, more importantly, sharply contrasts with the traditional anthropocentric view whereby, adopting an *ex-post* narrative similar to Rudyard Kipling's *Just So Stories*, the biosphere is believed to be *designed* for human beings. Instead, Darwinist thinkers, despite engaging in vigorous debates and taking antithetical positions—consider the opposition between Richard Dawkins (1976) and Stephen J. Gould (1990)—have consistently furnished us with a plethora of interpretative tools capable of unveiling the inconsistencies of anthropocentric explanations. Some argue that science has drained poetry from nature, relegating it to a dimension of disenchantment (Weber 2004), but I vehemently disagree. Few things are more captivating than the gradual progression of geological eras on Earth and the ebb and flow of diverse generations of living beings.

To understand the fundamental principles governing the physiology of organisms and the mechanisms of self-organization, two aspects must be considered: (i) life's ability to maintain its internal order through the dissipation of incoming radiant energy from the Sun; (ii) life's capacity to accumulate information and transmit it to future generations through natural selection and inheritance. Although the principle of differential replication among living things, upon which Darwinian theory is founded, might be assumed to have an individualistic effect, it actually determines the mutual dependence of living entities (Margulis 1998; Fleming 2024). Each ecosystem consists of interdependent entities, much like a house of cards; multiple ecosystems constitute a biome, which, in turn, is a segment of the biosphere characterized by specific climatic and biological features such as tundra, taiga, steppe, savanna, rainforest, etc. Different biomes are connected, and life has been able to construct its complexity precisely through extensive interconnection. Thus, the biosphere as a whole has developed into increasingly improbable forms (Dawkins 1996).

Analyzing the history of life on our planet reveals that everything is interrelated in both spatial and temporal terms. Energy dissipation isn't the sole phenomenon characterizing life processes; the endergonic reaction of chlorophyll photosynthesis enables the accumulation of solar energy within carbon chains (Stewart and Globig 2012). This results in exergonic availability that isn't always exploited, as seen in the Carboniferous fossil deposits. Moreover, the vast amount of organic compounds on our planet allows for the development of interrelationship systems between different species, as exemplified by the trophic chain. In essence, life doesn't merely exploit favorable preconditions; it produces them (Pigliucci and Müller 2010). This is why the evolution of life on Earth has progressively increased the complexity of living things, leading to the development of organisms with increasingly articulated morphologies, ecosystems forming networks of inextricable biological relationships, and ultimately giving rise to living hyperobjects, such as a rainforest (Morton 2013). Each entity, while thermodynamically unstable, is capable of developing mutual dependencies, leading to the formation of autopoietic structures.

Approximately three billion years ago, the development of cyanobacteria, capable of chlorophyll photosynthesis, led to the diffusion of oxygen into the atmosphere. This event facilitated the evolution of other life forms dependent on this gas. The proliferation of prokaryotes with various dissipative specializations in the period preceding the Ediacaran set the stage for the emergence of eukaryotes, likely through endosymbiosis (Margulis 1998). The subsequent Cambrian revolution witnessed the development of the fundamental arrangements of the different phyla present today (Gould 1990). During the Paleozoic era, life on Earth expanded across the globe, establishing systems of mutual dependence among different organisms and giving rise to the first ecosystems. Throughout Earth's history, life has faced periodic disruptions caused by high-impact events, including mass extinctions that seriously threatened its continuity (Benton 2003). However, life appears to generate resilience mechanisms capable of mitigating these shocks and recovering by retrieving scattered information from residual areas of the biosphere. Consequently, new types of animals and plants have emerged, replacing those of previous generations while still building upon the ecological legacies of the past.

After Darwin—and especially from the second half of the twentieth century onward—the idea of a selfish paradigm in the evolution of living beings took hold, becoming more broadly a leitmotif of life on Earth. This perspective can be seen as a biological reworking of Hobbesian thought

(Hamilton 1964; Dawkins 1976). Viewing life as an eternal struggle for existence, and natural selection as the sole force shaping living beings—assigning it the role of primary agent in defining the characteristics of different species—was the dominant view during the rise of a particularly rigid form of neo-Darwinism. This view was characterized by the idea that selection and adaptation were so comprehensive in their explanatory power that life on Earth could be reduced to a kind of algorithm (Dennett 1995), despite opposition from some authors (Gould and Lewontin 1979). Also, it posited that the evolution of living beings proceeded through small, gradual mutations accumulating along a particular phylogenetic line (Nilsson and Pelger 1994; Dawkins 1976)—an idea also challenged, especially in paleontology, due to the evident, sudden discontinuities among taxa, notably addressed by the theory of punctuated equilibria (Gould and Vrba 1982). On the other hand, both hypotheses—the dominance of selection and the notion of gradualism—converged in their portrayal of living organisms as essentially passive in their relationship with the environment. What was missing was a historical view of life on Earth, one that includes narratives, agency, complexity, and thus singularity—an approach we do find in Stephen J. Gould's *Wonderful Life* (1990). The supposed passivity of the living would later be challenged by the aforementioned theory of niche construction, and especially by the critique of the vision of constant struggle among organisms—an ultra-Darwinist conception that had already been opposed by Russian philosopher Pëtr Kropotkin in his essay *Mutual Aid, A Factor of Evolution* (2011).

The idea that twentieth-century biology is linked to an ideological project and a specific cultural vision—both rooted in determinism and reductionism, and oriented toward the control of biological mechanisms, not only to serve colossal economic interests (such as the comprehensive mapping of the human genetic heritage) but also to complete the project of separating the individual from the environment—is denounced by Richard Lewontin in his essay *Biology as Ideology* (1991). For the American geneticist, this model of disjunction aligns with a broader project of individualism and the detachment of humans from their environment. Undoubtedly, interactions among living organisms in nature are far from being based on mutual love and harmonious coexistence. Just as the principle of a war of all against all is incorrect, so too is the idyllic image of a world defined by mutual support, systemic balance, and the unique, purposeful role of every single organism misleading. In nature, we observe several examples that disprove this harmonious image, including sexual competition that is far from chivalrous or ritualized, contrary to what Konrad Lorenz once suggested, but often involves killing or castrating a rival; also, reproductive behaviors in many species include gang rape, infanticide to bring the female back into heat (as in lions), or the killing of males, as seen in hymenopterans and mantises. There are also instances of predation that does not favor the old or sick—on the contrary, newborns are often the first targets; and there are parasitic processes so disturbing they can make one's skin crawl, such as fungi infecting ants or solitary wasps laying eggs in caterpillar larvae. It is therefore safe to conclude that the image of nature as harmonious and provident—above all as a loving mother to her creatures—is a legacy of romantic visions or Edenic myths.

That said, it is equally clear that a disjunctive and individualistic conception is just as mistaken. What is needed is a shift toward an ecological perspective—one based on the connections between organisms, a principle that goes beyond both the idea of mutual aid and that of perpetual struggle. Ecology is the science that studies the interactions between organisms and their environment—understood also as the community of other organisms. These interactions also determine the geographic distribution of organisms and the abundance of different species (Cain et al. 2011). Public opinion often holds incorrect assumptions about ecological principles, such as: (i) the belief in a natural balance, which leads to the idea of stable ecosystems and biomes, when in reality they undergo strong fluctuations as well as transformative processes; (ii) the belief that natural systems tend to return to their original state after a disturbance, whereas many disruptions lead to significant changes that cause systems to evolve in new directions; (iii) the belief that each species holds a fixed and stable role within a system, when in fact it is evident that species respond differently to disturbances and often evolve independently. The fundamental aspect of ecological thinking is interconnection, which can be translated into two general principles. First, the mutual dependence of systems in nature, meaning that a change in one part of an ecological system can alter other parts of the same system or even other systems. Second, the constitutive significance of relationships, meaning that life would be impossible without the interactions among species, which depend on each other for a range of factors—not least of which is energy. For this reason, we must speak of a *web of life*, not isolated entities inhabiting the biosphere.

Alongside the processes of competition, predation, and parasitism—which highlight the conflictual side of existence, an undeniable aspect shaped by environmental selective pressures—we must also remember that the web of life is equally sustained by symbiosis, mutualism, cooperation, and alliances. These collaborative relationships occur not only within the same species but also between different ones. Such cooperative dynamics are just as fundamental as conflict, revealing that life's web is built above all on the ability of organisms to seize opportunities—sometimes at the expense of others, sometimes through the support of others. *In this light, the guiding principle is not simply struggle or mutual aid, but interconnection.* A prime example is symbiosis, which can occur within an organism—endosymbiosis, either extracellular or even within cells, forming what is called a holobiont—or between different species. Take, for instance, the microorganisms living in the gut of a mammal, known as the microbiome. These microbes not only support digestion and nutrient absorption but can also influence the host's mood. Many of them exist on a spectrum between symbiosis and parasitism, depending on their interactions with other microbes, the host's immune conditions, and their own population size. Even co-evolution—when two or more species evolve adaptively in response to each other—blurs into arms-race dynamics, where adaptation in one species provokes counter-adaptation in another.

Ecological relationships, which underpin the dynamics of the biosphere, are often so intricate and nonlinear that humans struggle not only to predict them, but sometimes even to perceive them. A paradigmatic example of ecological interconnection—both a strength and a vulnerability of life on Earth—is the ongoing decline in amphibian populations (Wake and Vredenburg 2008). As of now, around 41% of the approximately 8474 known amphibian species are in decline—a figure likely underestimated by at least 40%, given that since 1985 the number of recognized species has increased by over 60% (Mazzotti 2024). This decline is driven by multiple factors, including the spread of fungal diseases, pollution, and habitat destruction. But at its core, the vulnerability of amphibians can be traced to their dual lifestyle—both aquatic and terrestrial—which makes them particularly sensitive to changes in both environments. Their skin further compounds this sensitivity: it is highly permeable, vulnerable to pollutants and radiation, and reactive to climate fluctuations, making them effective environmental indicators.

Amphibians also offer a window into the complexity of ecological networks. Studies of their population crashes in the United States since the 1980s (Sessions and Ruth 1990; Johnson et al. 1999; Kiesecker 2002) revealed that what initially appeared to be a simple parasitic problem—caused by *Ribeiroia ondatrae*, a flatworm that induces limb malformations—was in fact part of a much more intricate web of interactions. Three key elements emerged. First, pesticides in water weakened the immune systems of tadpoles, making them more susceptible to parasites. Second, runoff from excessive fertilizer use triggered algal blooms, which in turn attracted *Helisoma tenuis*, a snail that acts as an intermediate host for the parasite and releases infective larvae into the water. Finally, the deformed limbs of infected amphibians made them easier prey for herons, which not only increased predation pressure but also reintroduced the parasites into the water through their droppings. Moreover, fungal infections once confined to limited regions have now spread globally, aided by human activity—particularly the global trade of goods. The result is a current extinction rate estimated to be 45,000 times higher than the natural background rate (McCallum 2007). As Elizabeth Kolbert recounts in *The Sixth Extinction* (2014), these pathogens have been disseminated in part by the global transport of African clawed frogs (*Xenopus laevis*) used in pregnancy tests, and American bullfrogs (*Lithobates catesbeianus*) farmed for food—unwitting agents in the unraveling of amphibian biodiversity.

Ecological phenomena, thus, speak to us in the language of relationships and interconnections, where even a seemingly isolated event can ripple outward in unexpected ways. A single action may set off a chain reaction whose effects emerge only after several intermediate steps—for example, a fertilizer might ultimately cause malformations in amphibians through a cascade of ecological changes. Such events often give rise to complex outcomes shaped by multiple, interacting factors, rather than a single, linear cause. In many cases, feedback loops come into play, amplifying the original phenomenon rather than containing it. The consequences can extend far beyond the initial context, and their full scope often becomes clear only in hindsight. Sometimes, these processes initiate what scientists call singularities—moments of instability in which a system, once nudged in a certain direction, can no longer return to its original state. In short, we are dealing with complex systems that behave according to the dynamics of complexity itself: chaotic turbulence, emergent behavior, field fluctuations, and threshold effects. Yet our ability to interpret these processes is often limited by deeply rooted cognitive biases. Humans tend to favor simple, immediate cause-and-effect explanations and struggle to connect distant or delayed outcomes. We also find it difficult to navigate the non-linear, unpredictable nature of complex systems. Compounding the issue is a broader cultural divide—between a reductionism that overlooks systemic complexity, and a form of holism distorted by New Age notions of harmony, balance, and providence.

4.3 An Unknown and Threatened Biodiversity

The term *biodiversity* refers not only to the variety of species inhabiting our planet but also to the differences among them. It encompasses both the number of known species—classified into taxonomic groups—and the variability within and between those groups. In recent decades, the scientific conversation around biodiversity has evolved, gaining greater depth and awareness compared to early conservationist approaches. A key realization has emerged: the vast majority of species on Earth are still unknown. Most living organisms are bacteria, and among eukaryotes, the dominant forms include fungi, nematodes, and arthropods. This has shifted the view of biodiversity from a simple inventory of disconnected species to a dense network of relationships—interconnections so intricate that attempting to isolate them often proves impossible. The biosphere, in this light, challenges us to rethink the way we interpret natural phenomena. Traditional modes of thought—those that rely on clear-cut distinctions, rigid categories, and a search for fixed essences—fall short in grasping the fluidity and complexity of life. These include: (i) dualism, which views beings as isolated; (ii) categorization, which ignores blurred boundaries; (iii) essentialism, which claims to derive qualities by internal reconnaissance; (iv) determinism, which relies on the linearity of processes; (v) synchronicity, which nullifies the time factor; and (vi) perspectivism, which makes differences less visible the further we move away from an entity. These tendencies inevitably distort our understanding of the biosphere, which leads to serious repercussions.

Today, biodiversity is understood on three interconnected levels: genetic diversity, which encompasses the variety of genes within living organisms; species diversity, referring to the range of different species; and ecosystem diversity, which includes the variety of natural environments. Within this framework, particular attention is given to two types of environments with special significance for biodiversity. One includes hotspots, areas recognized for their exceptionally high levels of documented biodiversity, and the other includes darkspots, regions believed to harbor a large number of yet-undiscovered species. Thanks to the work of researchers like Lynn Margulis (1967), our understanding of life has shifted dramatically. As mentioned, instead of seeing nature as a constant battleground of individual struggles and fierce competition, we now recognize—alongside conflict—the central role of cooperation. One key example is endosymbiosis, the process through which eukaryotic cells originated about 2.1 billion years ago, when an archaeal cell hosted a bacterium that eventually became the mitochondrion—paving the way for animal life—and later a second bacterium that evolved into the chloroplast, giving rise to plant cells. Life on Earth itself likely began around four billion years ago through the union of organic matter, as hypothesized by biochemist Aleksander Oparin and biologist John Haldane. Modern evolutionary processes are no longer viewed simply as direct competitions—such as the "arms race" between two species—or through sexual selection alone. Instead, evolution is understood as the outcome of complex networks of interactions involving many species, often dozens or hundreds, each potentially influencing the evolutionary paths of others. In some cases, particularly for highly specialized species, the indirect influence of distant partners can be even stronger than that of direct interactions (Guimarães et al. 2017). These multiple interconnected networks form complex systems whose evolution is inherently chaotic and unpredictable, defying straightforward forecasts of their future trajectories.

Biodiversity also reveals how deeply interconnected living organisms are, showing that the physiology of an individual body cannot be understood solely by looking at the genetic expression of that species. This is not only because other microorganisms influence gene expression epigenetically, but also because they contribute their own genes to the organism's overall metabolism. Species should not be seen as isolated entities but as networks of relationships whose boundaries are often difficult to define. Take, for example, a multicellular organism like a human being, which hosts a vast number of bacteria and other tiny inhabitants on the skin, in the mouth, nasal cavities, gut, and even inside its own cells. It is estimated that about 90% of the cells in the human body are actually symbionts, including some arthropods like eyelash mites. Thanks to the Human Microbiome Project, conducted over five years by two research groups led by Barbara Methé at the Craig Venter Institute and Curtis Huttenhower at Harvard University, we now know that roughly 10,000 species of symbionts live on the human body. These include bacteria, fungi, protozoa, and animals such as nematodes and arthropods. Without these microorganisms, the human body could not survive. They play crucial roles not only in essential metabolic processes like digestion and nutrient absorption but also in regulating neurotransmission, hormonal balance, and immune functions, while continuously removing waste materials. This microbiome effectively turns each of us into a kind of ecosystem, full of trophic chains, symbioses, and mutualisms—producing vitamins and anti-inflammatory compounds. In this sense, we resemble a coral reef more than a pure, self-contained entity. The microbiome contributes about 8 million genes to our functioning, compared to the roughly 22,000 genes found in the human genome—over 360 times more.

The naturalist and entomologist Edward Osborn Wilson was among the first to use the term *biodiversity* and to highlight that most species are either still unknown or made up of tiny organisms. As he puts it in his book *The Diversity of Life* (1992), it is the little things that run the world. Biodiversity can be seen as a kind of safety net that cushions shocks and gives living systems the quality of resilience—the ability to withstand stress without losing functionality. As Maurizio Casiraghi explains in *Sempre più soli* (2023, 33–34), resilience does not mean preserving the status quo at all costs. Resilient systems, when disturbed, do not necessarily return to their original state; instead, they adapt and reorganize, remaining functional within a new balance. Still, Casiraghi stresses that ecosystems rich in biodiversity tend to cope better with disruptions and are therefore more stable. Human activity, however, is steadily undermining this stability and making ecosystems increasingly fragile. This happens in several interconnected ways: by reducing the available space for natural processes of renewal and resource exchange; by fragmenting habitats and limiting the flow of interactions among species; by simplifying landscapes, stripping them of complexity; by introducing pollutants that disrupt physiological systems; by altering key ecological dynamics such as food chains and mutualisms; and by spreading non-native species that destabilize local balances.

Most of the planet's biodiversity is concentrated in tropical rainforests, which host at least two-thirds of all life on Earth, even though they cover just 10% of its surface. As zoologist Stefano Mazzotti notes in *Meravigliose creature* (2024), a single hectare of these forests can contain up to 650 species of trees, and the Amazon basin alone is home to 30% of all known bird species. Several theories have been proposed to explain the extraordinary richness of life in the Amazon. One line of thought suggests that biodiversity increases in environments with abundant solar energy, a stable climate, and vast surface area—conditions that tropical rainforests typically meet. Another theory focuses on specialization: in regions with marked seasonal changes, species tend to adapt more generally, while in tropical forests, the relatively constant conditions foster highly specialized niches. A third explanation highlights the role of parasites: the warm, humid climate favors their spread, pushing species to diversify in order to break the chain of transmission and enhance their adaptive responses, as argued by researchers like Coley and Kursar (2014). Lastly, some scientists point to historical and evolutionary factors, arguing that long-term geological stability in regions like the Amazon has allowed species to accumulate over time (Jetz and Fine 2012).

In recent decades, interest in biodiversity and awareness of the immense wealth it represents have undoubtedly grown, even if full understanding of its real impact remains limited—except among naturalists. Biodiversity is not something that merely surrounds us; it concerns us at a biological level. We are interconnected with it in an anastomotic way—like an organ to the body—and it quite literally forms part of who we are. To trace the key moments that led to a greater (though still insufficient) recognition of the biodiversity crisis, it's worth recalling a few milestones: in 1948, the first major conservation organization was founded—the International Union for Conservation of Nature (IUCN)—with the mission of protecting the natural environment. In 1964 came the first edition of the Red List, classifying species by conservation status. In September 1986, the National Forum on Biodiversity helped bring the variety of life on Earth to wider attention. In May 1992, the landmark Earth Summit—the Rio Conference on Biodiversity—was held. And in 1997, the COP extended the Kyoto Protocol, committing to emissions cuts in response to climate change. The Red List recognizes five risk levels: *vulnerable*, when a species is under adaptive stress; *endangered*, when it is at risk; *critically endangered*, when it is close to extinction; *extinct in the wild*, when it is only found in captivity; and *extinct*, when no individuals remain. Over the past twenty years, the outlook has worsened dramatically: in 1998, the Red List included 3115 animal species and 3222 plant species. By 2020, those numbers had risen to 6362 animal species and 7430 plant species.

Biodiversity as a whole is the key factor for sustaining the web of life, enabling both stability and evolution. Let's begin with some figures on biodiversity. In his aforementioned essay (1992), Edward Wilson reported the number of known species at the time: a total of 1,413,000, of which 751,000 were insects—over 50%. Among these, beetles stood out with 290,000 species, followed by butterflies and moths (*Lepidoptera*) with 112,000, and wasps, bees, and ants (*Hymenoptera*) with 103,000. For all other animals, there were 281,000 species, including 123,000 other arthropods and 18,000 species of fish. Today, thanks to the work of taxonomists and advanced genetic analysis techniques, these numbers have increased. The paradox, however, is that we are driving an extraordinary number of species to extinction before we even have the chance to discover them—ignoring the self-regulating mechanisms of ecosystems. One example is the role of "umbrella species," whose protection benefits many other species within the same habitat. For instance, a study conducted by researchers at Oregon State University, based on data collected between 2001 and 2010, showed the ecological impact of reintroducing wolves to Yellowstone National Park. The presence of wolves helped vegetation recover—particularly willows along rivers, which had suffered from an overpopulation of deer—and, more broadly, led to an increase in bird and insect biodiversity (Ripple and Beschta 2011). There are also so-called keystone species, which are essential to the stability of entire ecosystems. These are often small or inconspicuous organisms, but their absence would trigger a collapse of the whole system. An example is the cleaner shrimp found in coral reefs—tiny creatures whose presence is crucial to maintaining ecological balance.

While we still don't have a precise idea of how many species exist on Earth—Wilson himself estimated the number could range from 10 to 100 million—even more elusive are the relationships between them: the intricate networks of interactions that, like the cement in a building, hold the entire structure of ecosystems together. Often, public attention focuses on species that resonate emotionally with people, such as the giant panda (*Ailuropoda melanoleuca*), which has become the iconic symbol of WWF. These are known as flagship species and can be useful for raising public awareness. However, it's clear that we cannot protect whales without also safeguarding the krill they feed on. This underscores the need for a holistic, ecosystem-wide approach rather than a fragmented view of biological communities.

The most important effort must go into cataloguing species. Current estimates reported by the *Catalogue of Life* suggest there are about 2 million known species in total. Among terrestrial vertebrates, there are approximately 11,733 known species of reptiles, 11,162 species of birds, 8474 species of amphibians, and 6399 species of mammals. In the plant kingdom, there are 374,000 known species, including 308,312 vascular plants (of which 295,383 are flowering plants and just over 1000 are gymnosperms), 44,000 species of algae, 12,700 species of mosses, and 10,560 species of ferns. Recent estimates suggest that 86% of terrestrial species and 91% of marine species have yet to be discovered. Some field studies have speculated that there may be as many as 5 million species of insects alone (Gaston 1991). Each year, scientists discover about five new species of birds and around ten new mammals—primarily microbats, rodents, and insectivores, but occasionally also small primates. The most likely candidates for future discoveries are small animals and those living in biodiversity-rich but hard-to-access environments like oceans or the underground. In total, between 15,000 and 19,000 new species are identified each year.

At the same time, some scholars have pointed out the existence of a kind of taxonomic bias in the collection of biodiversity data. This bias tends to favor charismatic or easily identifiable groups, while so-called minor organisms are often overlooked because they are hard to distinguish or invisible to the naked eye, such as prokaryotes (Larsen et al. 2017). Wilson himself noted that most bacteria remain completely unknown to us. The 4000 estimated species could in fact be underestimated by a factor of 1000, suggesting that bacteria might account for up to 90% of all species on Earth—thus challenging the idea of insect dominance. Other significantly underestimated organisms include nematodes, which may number not in the thousands (as the 15,000 currently identified suggest), but in the millions. Mites and microsporidian fungi are similarly undercounted. Fungi in general are poorly known, despite playing a crucial role in the web of life. They are essential to the symbiotic relationships they form with plants, which in turn are vital to the spread and diversification of vascular plants. According to some researchers, the number of fungal species could reach several million (Hawksworth and Lücking 2017). Finally, among the least explored ecosystems is edaphic biodiversity. The underground world hosts a dense network of interactions involving bacteria, fungi, detritivorous animals, and plant root systems—a rich and complex environment comparable in biodiversity to tropical forests and coral reefs (Monbiot 2022). Yet this underground ecosystem is at risk, particularly from deforestation and intensive farming practices (Phillips et al. 2017). As Stefano Mazzotti (2024) points out, this situation is paradoxical: soil biodiversity could actually help us understand and predict human impacts on ecosystems, especially regarding the fertility of agricultural soils.

The challenges in species identification go beyond the difficulty of accessing certain environments. A major issue is recognition itself: some species may appear morphologically distinct but are actually part of the same taxon. Conversely, there are organisms that seem identical in appearance but, when examined through DNA-based phylogenetic analysis, turn out to be genetically distinct species rather than mere subspecies. This has led to the recent reclassification and splitting of species. Notable examples include the differentiation between *Canis aureus* and *Canis anthus*, *Felis sylvestris* and *Felis lybica*, or *Vipera berus* and *Vipera walser*. The real challenge, however, lies in securing the resources to increase the number of researchers and fund large-scale observational studies. It seems paradoxical, but more money is spent on space exploration or military armament than on understanding our own planet—the only one we have—and the dynamics of its biosphere.

As mentioned earlier, most species are at risk of going extinct before they are even identified. The numbers are stark: current extinction rates are tens to hundreds of times higher than the natural background rate, particularly for certain groups of plants and animals, compared to the average over the past ten million years. Today, 37% of conifers are at risk, as well as 25% of mammals—rising to 39% for marine mammals—41% of amphibians, 19% of reptiles, 13% of birds, 27% of crustaceans, 10% of insects, and 7% of fish (Tozzi 2025). Tozzi underscores this loss with a striking metaphor: if biodiversity is a library containing billions of volumes, humanity is burning the books before learning how to read them. These terrifying figures are the result of three key historical phases. First, there was the destruction of the Pleistocene megafauna, driven partly by climate shifts but primarily by intense hunting pressure from humans. This pattern is consistent: in Africa, where the first *Homo* species appeared about a million years ago; in Eurasia around 60,000 years ago with the arrival of *Homo sapiens*; in Australia around 40,000 years ago; and in the Americas about 12,000 years ago—each wave coinciding with human colonization. Second, the development of agriculture, beginning around 8000 years ago, brought with it widespread deforestation, the use of timber for daily needs, the construction of dwellings and ships, and the progressive desertification of once-fertile lands. The spread of agriculture can be traced directly through the corresponding decline in biodiversity. Finally, the past three centuries, and especially the industrial revolution and rise of capitalist economies, have massively exploited every corner of the planet, placing all ecosystems under strain and disrupting essential ecological balances.

Biodiversity is fundamental, as Michael J. Novacek pointed out in his essay *The Biodiversity Crisis: Losing What Counts* (2001). And yet, human impact on it has been—and continues to be—devastating. The biosphere depends on the relationships made possible by biodiversity, which in turn is the result of long evolutionary processes involving the various biomes on our planet. Paradoxically, the areas with the highest levels of biodiversity are also those where self-regulating mechanisms—such as the water, carbon, and nitrogen cycles—and mutualistic interactions between species are most complex and, for that very reason, most vulnerable. A striking example of this pattern is the deforestation of the Amazon, which began in the 1940s but accelerated significantly in the 1970s. By 2017, more than 20% of the entire forested area had been cleared, amounting to 783,828 fewer square kilometers of forest than in 1970. Extensive livestock farming is the main cause—accounting for over 80%—of this deforestation, with the remainder due to agricultural exploitation. It is worth noting that the growing global consumption of meat and animal products—often seen as a marker of a more advanced lifestyle, when in reality it is more harmful to human health—is the main driver behind this trend. We know that converting plant-based protein into animal protein is a highly energy-intensive process that wastes about 90% of resources, not to mention the collateral damage in terms of water consumption, methane and CO_2 emissions, and groundwater pollution. And yet, the surging demand for meat worldwide has fueled the exploitation of the Amazon, as the need for grazing land has driven settlers to use slash-and-burn techniques to clear forest areas. Among other products contributing to this deforestation is palm oil, widely used in the food industry.

If deforestation in such hotspots represents the main threat to global biodiversity, we must not forget that many other factors are key drivers of its decline, including:

(i) Climate change, which drastically alters weather patterns, causes the tropicalization of many regions, accelerates desertification and coastal erosion, and leads to a sharp reduction in glaciers and the albedo effect;
(ii) The spread of non-native species in most ecosystems, often introduced via global transport or through synanthropic species like rats;
(iii) The steep rise in atmospheric CO_2, which is expected to soon double compared to levels just 250 years ago, leading especially to ocean acidification;
(iv) The disruption of the virosphere—the equilibrium of viral populations—caused by ecosystem imbalances and intensive livestock

farming, which amplifies the risk of spillover events (Quammen 2012);

(v) The increase in pollutants, with plastic foremost among them—estimates suggest that all plastic produced by humans now weighs twice as much as the combined mass of all living animals (Casiraghi 2023);

(vi) Industrial agriculture and livestock farming, which today occupy 40% of the planet's land surface and account for 30% of global greenhouse gas emissions, contributing significantly to the phenomenon of overexploitation (Foer 2010);

(vii) Habitat fragmentation, driven by massive land use, which not only shrinks vital habitats for many species but also drastically reduces—or even eliminates—ecological corridors, fragmenting environments and threatening genetic diversity within species;

(viii) The dramatic growth of human populations, urban expansion, and above all the increasingly energy-intensive lifestyle that has taken hold in many parts of the world since the 1960s.

I believe this erosion of biodiversity is no longer a concern only for scientists, naturalists, ecologists, or taxonomists—it has become a broader issue, a philosophical one, that should concern us all. It should prompt us to reflect on the existential meaning of human life—not to fall into a vision of self-annihilation or a pessimistic lethargy, but to critically reassess the values that have brought us to this point.

4.4 The End of the Cenozoic

If life on Earth is around four billion years old, for the first three billion years it hardly left paleontological traces. Nevertheless, it played a crucial role in preparing the planet for the conditions we know today—such as the presence of sufficient oxygen. The first living organisms to leave clear fossil traces are part of the Ediacaran Fauna, a group of multicellular organisms dating back to between 620 and 550 million years ago. These organisms resembled jellyfish (medusoids), with radial or bipolar symmetry—like *Dickinsonia costata*—or bush-like shapes, and were preserved as impressions in sediment. However, it was with the Cambrian period, the first of the Paleozoic era, that a true explosion of diverse life forms began, marking the start of the Phanerozoic era—Phanerozoic literally meaning "visible life." The Cambrian explosion was remarkable for the number of fundamental kinds of *bauplan* it introduced, and was famously described by paleontologist Stephen J. Gould in his 1990 essay *Wonderful Life*.

The Paleozoic era, also known as the Primary Era, spanned from about 540 to 252 million years ago, lasting roughly 288 million years, and included two mass extinctions. The first occurred at the end of the Ordovician period, around 443 million years ago, likely caused by intense glaciation and a drop in oceanic oxygen levels—at the time, the sea was the only habitat for life. This extinction event wiped out about 85% of all species. The second took place at the end of the Devonian period, about 360 million years ago, with around 75% of species disappearing, possibly due to the impact of an asteroid. The key point to remember, however, is that the following period—the Carboniferous—witnessed a biodiversity crisis that lasted for 20 million years. This highlights an important truth: while mass extinctions have never brought life on Earth to a complete halt, the disruptions to the biosphere they caused lasted millions of years.

Undoubtedly, when we are confronted with these time spans, it becomes difficult to fully grasp the meaning of such events. We feel disoriented because we're dealing with timeframes that clash with our inability to comprehend the immensity of biological—or geological—time. After all, humans already struggle to navigate even just a few centuries. However, using a human-scale metric is misleading, as we'll see when discussing the Anthropocene, just as it is in all situations where we place ourselves at the center of the picture—whether it's assigning a purpose to evolution by seeing the appearance of humans as its ultimate goal, or referring to the anthropic principle (Barrow and Tipler 1986). Instead, we should recognize that the history of life on Earth goes far beyond our imagination—even if we can study, measure, and understand it to some degree.

The Paleozoic era saw the development of many types of vertebrates, including amphibians and reptiles. It is precisely on reptiles that we should focus our attention, since from this lineage came the cynodonts—ancestors of mammals—and the diapsids, which eventually gave rise to dinosaurs and then birds. Dinosaurs gained enormous popularity in the second half of the twentieth century and are often cited today in metaphorical terms—as creatures that once dominated the planet only to be wiped out, a fate that could befall humanity as well. Yet even this image is inaccurate, because dinosaurs are still among us: they are birds. It's worth noting, too, that as late as the 1960s, dinosaurs were still depicted as clumsy giant lizards—and even in the film *Jurassic Park* (1993), they appeared without feathers.

Before getting to the age of the dinosaurs, we must first recall the third mass extinction, which occurred at the end of the Permian period, around 252 million years ago. It was the most catastrophic event in Earth's history—aptly named *The Great Dying*, as it led to the extinction of about 95% of all species. Life on Earth disappeared almost entirely (Benton 2003). The likely cause was a series of volcanic eruptions that released extreme amounts of CO_2, resulting in ocean acidification. The Permian extinction paved the way for the Secondary or Mesozoic era, which lasted approximately 186 million years, from 252 to 65.5 million years ago. The Mesozoic saw two major mass extinctions. The first occurred at the end of the Triassic period, around 200 million years ago, wiping out about 75% of species. Its exact cause remains uncertain. The second happened at the end of the Cretaceous period, around 65.9 million years ago, known as the extinction of the dinosaurs. It led to the disappearance of 80% of species and was caused by the impact of an asteroid, with a collision force equivalent to five billion Hiroshima atomic bombs. The molten rock from the impact was thrown into the atmosphere, reaching heights of about 10 km. The resulting seismic waves spread for thousands of kilometers, followed by a rain of fiery debris, tsunamis, and vast wildfires. Ecosystems took thousands of years to recover (Lowery et al. 2018).

With this fifth and final mass extinction, the Secondary Era came to an end and the Cenozoic Era began—the era we are currently living in—also known as the Tertiary. So far, its duration has been slightly less than a quarter of the Primary Era and a little more than a third of the Secondary. The division into three eras was officially adopted only in the twentieth century, although it had already been proposed by the English geologist John Phillips in 1841. The idea of Earth as a place where periods of biodiversity expansion are repeatedly interrupted by great mass extinctions—considering that 99.9% of all species that ever lived on our planet are now extinct—is undeniably powerful and evocative.

In 1996, researchers Richard Leakey and Roger Lewin published *The Sixth Extinction: Patterns of Life and the Future of Humankind* to describe the ongoing loss of biodiversity. They emphasized a new disaster threatening Cenozoic biodiversity: human beings and their activities. The issue of mass extinctions—now relevant again due to the disruptions humans are causing to the biosphere—first emerged in scientific debate in the nineteenth century and played a central role in the development of Lamarckian transformist theories and later Darwinian evolutionism. The first to suggest that life on Earth had been disrupted by catastrophic events was the

French naturalist Georges Cuvier, an excellent anatomist who, nonetheless, held to a fixist position. His ability to spark interest in extinct creatures helped launch a period of intense fossil collecting, with prominent figures like British paleontologist Mary Anning, Edward D. Cope, and Othniel C. Marsh.

Reflecting on the Cenozoic offers us a broader perspective on what a geological era truly is, based on the paleontological data available. These include, for example: (i) a defining event marking its beginning, such as a catastrophic process that disrupts the existing biosphere—often tied to climate shifts or major changes in the configuration of land and sea; and (ii) the emergence of a new form of biodiversity that arises from this upheaval, taking thousands—or even millions—of years to evolve. According to the current classification, an era can begin with a mass extinction, as seen with the transitions to the Secondary and Tertiary eras. However, not every extinction marks the start of a new era, as seen in the Ordovician, Devonian, and Triassic extinctions. In general, we can say that the concept of a "geological era" does not necessarily overlap with that of an extinction event, since an era is primarily defined by the type and evolution of biodiversity it encompasses. This is one of the main reasons why I reject the idea of a newly defined era called the Anthropocene.

The extinction event at the end of the Cretaceous, around 65.9 million years ago, led to the disappearance not only of non-avian dinosaurs but also of many other animal groups: the flying pterosaurs, the marine ichthyosaurs, pliosaurs and mosasaurs, ammonites, most planktonic foraminifera, and several types of birds such as the enantiornithes and hesperornithes. In short, much of Mesozoic biodiversity vanished. The Cenozoic Era—its name derived from the Greek *kainos* (new) and *zoè* (life)—marks a distinct phase in Earth's history, characterized by the development of new ecosystems and the rise of mammals and birds as dominant vertebrates. During the Eocene (56 to 38 million years ago), the Amazon rainforest began to take shape—making it an astonishing 55 million years old. In the Miocene period (23 to 5 million years ago), considered by many as the peak of Tertiary biodiversity, a lush flora thrived. Oak, alder, and plane trees flourished alongside palms, eucalyptus, sequoias, and a wide variety of herbaceous plants. Mammals experienced an extraordinary evolutionary boom, both in diversity and size, with species such as the proboscideans *Deinotherium* and *Mastodon*—as described in Steve Brusatte's *The Rise and Reign of the Mammals* (2023). This was followed by the Pliocene, during which Earth's temperature began to decline, initiating a cooling trend that eventually led to the ice ages of the Pleistocene, from 2.5 million to 12,000 years ago. These glaciations are known respectively as the Donau (ending around 1 million years ago), Günz (650,000–500,000 years ago), Mindel (400,000–300,000 years ago), Riss (200,000–120,000 years ago), and finally the Würm (75,000–10,000 years ago).

The most recent period of the Cenozoic is called the Holocene, spanning from 12,000 years ago to the present. During this time, there have been no major changes in the Earth's climate or structure. Cenozoic biodiversity has spread across the globe and is now found even in the most extreme environments: alpine choughs fly above the Himalayan peaks at altitudes over 8000 m; deep-sea fish and crustaceans swim at depths of 8000 m; and archaebacteria thrive in sulfuric hot springs, glaciers, and even in dust particles floating thousands of kilometers above Earth's surface. Despite this, the vast majority of life exists within a thin layer stretching from the treetops to the surface layers of the soil and oceans—bounded by the lithosphere (the uppermost crust of the Earth's mantle) and the troposphere (the lowest layer of the atmosphere) (Cain et al. 2011).

Today, there are still a few Cenozoic biodiversity hotspots—though many are at serious risk of decline. Some of these are islands that, due to their geographic isolation and relative protection from human exploitation, remain true sanctuaries of biodiversity, especially in terms of endemic species. One striking example is New Guinea, home to 6% of the world's known terrestrial species, about half of which are endemic. The island boasts 800 species of birds, including most species of birds-of-paradise and even the world's only known poisonous birds, the *Pitohui dichrous*, whose feathers contain a neurotoxic compound more potent than strychnine. New Guinea is covered in rainforest, hosting around 21,000 plant species and an as-yet-unknown number of invertebrates, estimated at around 200,000 species—including the world's largest butterfly, *Ornithoptera alexandrae*, with a wingspan of 30 cm.

Another biodiversity-rich island is Borneo, which supports 10,000 plant species, 350 bird species, 150 reptile and amphibian species, and 10 primate species. Borneo is also home to remarkable endemic species, such as *Phobaeticus chani*, the world's longest stick insect at 56.7 cm; flying frogs (*Rhacophorus penanorum*), which glide using webbed feet and skin flaps on their limbs; primates of the *Loris* genus, known for their toxic bite—a rare trait among mammals; and the Bornean clouded leopard (*Neofelis diardi*), found only on this island.

Madagascar is another major biodiversity hotspot, featuring a wide range of habitats—including the world's largest coral reef system and the most extensive mangrove areas in the Indian Ocean. Of the island's 13,000 plant species, nearly 90% are endemic, as are 92% of its mammals, 99% of amphibians, and 95% of reptiles. Madagascar's iconic lemurs are a standout feature of its wildlife, but their habitat is also under serious threat due to deforestation.

Let's now consider the major forest biomes, which can be categorized as follows:

(i) *Tropical rainforests*, located at low tropical latitudes, characterized by warm temperatures without seasonal variation and abundant rainfall throughout the year. Though they cover only about 11% of the Earth's surface, they are estimated to host 50% of all known species. These forests are mainly found in South America along the Amazon River, in Central Africa along the Congo River, and in Southeast Asia along the Mekong River. All are under severe threat due to extremely high rates of deforestation (Asner et al. 2009; Wright 2005).

(ii) *Seasonal tropical forests*, located north and south of the tropics, experience seasonal fluctuations in climate and rainfall, alternating between wet and dry seasons. These forests are marked by deciduous trees that shed their leaves during the dry season, lower tree density, widespread savannas, and a greater presence of herbaceous and shrubby species—also rich in biodiversity. Unfortunately, this biome is also in significant decline: only half of it remains intact today due to increasing logging pressures (Bullock et al. 1995).

(iii) *Temperate deciduous forests*, found between 30° and 50° latitude in Eurasia and western North America, are characterized by trees that lose their leaves in autumn and winter, fertile and moist soils, and distinct seasons. However, these forests have been heavily compromised by centuries of agricultural development, which depleted soil nutrients. As a result, even reforestation efforts often lead to forest compositions that differ greatly from the original ecosystems.

(iv) *Temperate evergreen forests*, typically found around 50° latitude, host lower biodiversity and are dominated by conifers. Examples exist between regions such as North and South America and southeastern Australia. These forests have also suffered extensive damage from logging for cellulose and paper production, leading to intense and widespread deforestation over time.

(v) *Boreal forests*, located above 50° latitude in the Northern Hemisphere, are defined by cold, moist soils, slow decomposition rates, and a high accumulation of organic matter. This organic buildup acts as fuel, making these forests particularly susceptible to wildfires, often triggered by lightning strikes.

Biodiversity is the living fabric of our planet—a vast, intricate web of species, ecosystems, and genetic variety that sustains life as we know it. From tropical rainforests to deep ocean trenches, every organism plays a role in maintaining the balance of Earth's systems. Yet this richness is increasingly threatened by human activity. Protecting biodiversity is not just about preserving nature's wonders—it's essential for the health, resilience, and future of our planet. If we continue on our current path of destruction, one question remains: are we facing the end of life on Earth?

4.5 The End of Life on Earth?

The ecological connections that sustain the presence of a species on Earth are many and so intricate that they would be impossible to analyze exhaustively. There are trophic, niche, epidemiological, demographic, and epigenetic connections that determine the presence or absence of a biosystemic balance involving a set of species bound by a shared destiny. Major upheavals in the biosphere do not selectively affect single species; instead, they can sweep away entire classes, such as the *Theropoda* among archosaurs or the *Temnospondyli* among amphibians. This is because each period ensures specific biocenotic conditions that do not concern individual species but rather a particular physiological blueprint and a certain overall balance (Valentine 2004; Erwin 1999). The Cenozoic, which is our geological era—beyond speculations about the so-called Anthropocene—features a specific set of living beings, including beetles, hymenoptera, birds, mammals, and angiosperms, for example. These living forms are all intertwined in the same biocenotic network currently in place.

When the foundational condition supporting a given biocenotic plan is disrupted, the entire system collapses. This may take a longer or a shorter time, but it still happens very quickly compared to earlier periods of relative stability for living forms. It is the organisms that dominated a particular epoch—such as mammals in the current geological era—that are destined to succumb in these upheavals, and the reason is straightforward. These living forms have built their entire success on specific biospheric conditions and, through specialization, have become entirely dependent on them (Gilbert and Epel 2015). Life has consistently rebounded after immense geological catastrophes in the Phanerozoic, such as the dramatic increase in volcanic activity at the end of the Permian or the asteroid impact at the end of the Mesozoic (Brusatte 2019). While this resilience should not be taken for granted, it is plausible to assume that it may still operate even in the current biodiversity crisis.

Therefore, there is good reason to believe that the fundamental nature of life on our planet is not threatened by the reckless behavior of global human policies. However, this is not the crux of the matter. The evidence we gather from severe crises in homeodynamic regimes—such as disruptions in the nitrogen, carbon, and water cycles, alterations in ocean currents, climate change, the destruction of entire biomes, progressive deforestation, the reduction of biodiversity, and compromising of the virosphere (Quammen 2012), among the most notable examples—unequivocally points to convergent processes deconstructing the current state of the biosphere (Barrau 2020; Rich 2019; Mercalli 2019). We may ponder whether human actions today can be likened to other events that have disrupted Earth's homeodynamic conditions in the past, leading to mass extinctions, or if we are essentially accelerating processes of decay already underway for reasons beyond our intervention. These questions, while intriguing, are of little significance. Indeed, even from a skeptical perspective, there is no denying that human impact on the planet holds some responsibility (Jonas 1984).

Still, believing that we are a threat to life on Earth is, in my opinion, an exaggeration. After all, the biosphere on our planet has already gone through moments of severe criticality and has always managed to recover. That said, every major upheaval has caused transitional periods lasting millions of years with extreme climatic conditions adverse to the development of a new equilibrium needed for the emergence of other forms of biosphere. Moments of crisis always involve long periods of progressive decay, accompanied by rapid implosions of the system during which a large number of species are wiped out within a short time. Catastrophes are followed by extremely long periods of readjustment—lasting millions of years—marked by challenging conditions for the rebirth of a stable biosphere. This leads me to doubt that human beings can survive during the first transition phase that would follow from a climate catastrophe. And even if they did, they would still face extinction during the implosion periods. For this reason, we must strive for the preservation of *this* world, however much this very wish may appear to be a form of anthropocentrism.

In reality, this concern goes beyond humanity—however central it may be from our point of view—and involves an entire community of living beings associated in a biospheric dimension that has built its profile over millions of years. The Cenozoic condition is nothing but the world that various mythologies and religions have envisioned as the Eden created especially for human beings—the place we commonly, albeit erroneously, call home. Our entire world is at risk, and there is no hypothetical lifeboat for the human being to be saved (Berners Lee 2021). Even if there were one, which I strongly doubt, we would still be condemned to live in a world that is out of tune with our deepest expectations—those existential a priori foundations that derive from a phylogeny shaped in our minds by this very biospheric condition (Bartalesi. 2024). Also from this perspective, which is both epistemic and aesthetic, the biosphere is not an external space that welcomes us: it is what has produced our way of being, so that we mirror it in every way. The destruction of the biosphere is the destruction of our ontological foundations.

When we speak of homeodynamic processes in the biosphere, we are referring to a condition that, although characterized by fluctuations and evolutions, is maintained by well-defined mechanisms that together allow for a certain stability. The biosphere is undoubtedly dynamic and open to transformations, but always in a long-term progression. Compromising its resilience means exposing ourselves to catastrophic trends. Therefore, what's at stake is the survival of a certain ecological balance, which is not only the place where humans live but a single system that, in transforming itself, preserves its basic configuration, as each of its parts is intimately connected.

Let me be clear: whatever dystopian image we may have in mind is far removed from the real foreshadowing of risk and can, at best, only depict the first moments of a far more devastating transition phase. To deviate from homeodynamic parameters is to set autocatalytic systems in motion, whereby the first phenomena of alteration would be compounded by the many consequences they would cause. The biosphere is a network that buffers fluctuating factors, just like an organism that constantly brings values back within the survival and self-organizing ranges. The biggest mistake we make is therefore to consider such a system as something external to us: this is not only a descriptive mistake but a philosophical fallacy.

Let's try, then, to understand whether it is reasonable to speak of a new era defined by human impact on the planet—as supported by undeniable evidence: the rapid growth of the human population, the reshaping of at least half the Earth's surface, the disruption of major natural cycles like water, carbon, and nitrogen, the loss of biodiversity, the overwhelming presence of human-made materials, and widespread pollution. The term *Anthropocene* has gained traction in recent years, after being introduced in the 1980s by biologist Eugene Stormer and later popularized by chemist Paul Crutzen at a 2000 conference. Crutzen's article "Geology of Mankind" (2002) and the essay *Welcome to the Anthropocene* helped formalize the concept, highlighting the massive influence humans have had on the planet over the past two centuries. Bringing attention to these issues—especially to humanity's role in global warming and the disruption of the biosphere—can serve a constructive purpose. For this reason, some authors, like Timothy Morton in *Dark Ecology* (2016), argue that *Anthropocene* is one of the most anti-anthropocentric terms imaginable. On the contrary, I believe that assigning a geological epoch to humans is in fact the ultimate expression of anthropocentrism. It reinforces the idea that Earth is a human-shaped entity under our control—when in truth, there has never been a more uncertain time in history for the survival of our own species.

Anthropocentrism, in fact, lies at the heart of humanity's troubled relationship with the biosphere, and the word *Anthropocene* risks reinforcing some of the worst anthropocentric biases. First, it fuels human supremacism—the belief in our power and superiority over nature—when in reality, we are increasingly vulnerable to the biosphere's shifts. Second, it encourages an anthroplastic aesthetic: the idea that turning natural entities into artificial ones is not only possible but inherently better. Third, it fosters demiurgic pride—the illusion that humans can create an alternative world apart from the biosphere. *Contra* Timothy Morton—whose critique of human responsibility I share—I believe that naming a geological epoch after humanity only deepens the arrogance of anthropocentrism, appealing not to reason but to emotion and pride. The popularity of the term *Anthropocene* likely owes not just to Paul Crutzen's authority as a Nobel laureate, but also to its flattery of our species' ego. After all, people tend to protect and glorify a sense of self that extends outward in concentric circles—from the individual to family, community, and eventually species—a collective *we* that strengthens itself by opposing a perceived *other*. And nothing plays this oppositional role better than the non-human. So, when we diminish the biosphere's meaning, we inevitably bolster human-centered thinking—even if we do so with a guilty conscience.

So what doesn't hold up in the idea of the Anthropocene? I believe there are several critical weaknesses. First, there's no consensus on when it began. Some trace its origin to the Industrial Revolution, others to the first nuclear test—Trinity, on July 16, 1945. Still others place it at the dawn of agriculture (overlapping with the Holocene some 12,000 years ago), or even further back, during the Pleistocene and the extinction of megafauna, or with the emergence of the *Homo* genus itself. Second—and to me, the most important point—a new geological era is typically defined not by extinction events, but by a distinct reconfiguration of biodiversity. By this standard, we should speak of an *Anthropocenic transition*, not an era: a temporary phase marked by sharp biodiversity decline, not yet a stable ecological state. Third, many critics argue that biodiversity loss isn't the result of humanity in general, but of specific social, economic, and cultural systems that took hold at particular moments in history. For instance, the current of Ecological Marxism popularized the term *Capitalocene*, introduced by Swedish ecologist Andreas Malm in 2009 and further developed by Jason Moore in *Anthropocene or Capitalocene? Nature, History and the Crisis of Capitalism* (2016). The term challenges the notion of collective human blame, pointing instead to capitalism as the true driver of ecological disaster. Similarly, Donna Haraway uses the term *Plantationocene* (Murphy and Schroering 2020) to highlight how plantation systems not only destroyed landscapes through simplification but also exploited human labor—a model that extended from agriculture to industry, factory farming, and even wage labor. Yet despite their differences, these terms share common ground: they all describe nature as reduced to a pool of resources, available for human use. And they all assume, in some form, the idea of a new geological era triggered by human forces—granting our species a kind of geomorphic power over the Earth.

If we were to follow this logic of attribution—which I don't think we should—it might be more accurate to trace the origins of this transformation back to the Neolithic Revolution. That was when some human groups began altering their environments through agriculture and livestock domestication, initiating a new biocenotic situation. Over time, this evolved in multiple directions with the introduction of machines and new energy sources, and with the changes they produced. From that point on, in taming rivers and digging canals, draining and ploughing land, building settlements, ports, cities, and borders, humans began reshaping the environment in ways we now perceive as an expression of humanity. Of course, the initial impact was minimal—comparable to niche construction seen in other species. Early agriculture and animal husbandry could even be read as forms of symbiosis. But from the Industrial Revolutions onward, this transformation accelerated to such an extent that human-driven changes to the biosphere became globally significant. We could then speak of a *Neolithic gradient*—a scale of environmental transformation based on the degree to which a given society has modified its surroundings. At the lower end lies the pre-Neolithic baseline, such as foraging or using caves as shelters—practices still found in some communities today. This would imply that the degree of ecological responsibility we attribute to humanity should vary depending on each society's position along this Neolithic gradient.

Even if we accept that the progressive decline in biodiversity began with the Neolithic—sowing the seeds of planetary degradation, biospheric erosion, and the human exploitation of all living and non-living systems, culminating in the damage made evident over the past two centuries—still, nothing suggests the dawn of a new geological era. Yes, I am well aware of the ecological disaster inflicted by human economies on the biosphere. But just as I would never propose an *Asteroidocene* to name the meteorite impact that disrupted the quiet world of the Mesozoic, nor a *Vulcanocene* to mark the crisis that ended the Paleozoic era, I will not do so now. The reason is simple: humanity is not ushering in a new era, but rather endangering the current one—the Cenozoic, an era that has marked the flourishing of beetles, birds, and mammals. We are not so different from the asteroid that fell 65 million years ago upon the dinosaurs' world, or the Siberian Traps that, some 250 million years ago, wiped out trilobites and two-thirds of terrestrial amphibian and reptile families. Naming a geological epoch after the human species is an absurdity born of the deeply anthropocentric lenses through which our culture views the world.

We are still within the Tertiary period, part of a biosphere that, though in constant transition and fluctuation, has defined the Cenozoic. True, we are now the driving force of a mass extinction—already underway—that will likely bring this era to an end, washing away not just the remnants of the Cenozoic biosphere but also this peculiar primate who believes itself exempt from the laws of nature. There is, of course, a vast difference between claiming authorship of a new geological era—imagining ourselves at its opening like attendees at a vernissage, playing the intellectual role of the remorseful yet unrepentant (that charming contradiction Morton so enjoys)—and acknowledging that we are, perhaps, nothing more than another accident, one of many that have befallen the Earth.

Refernces

Armstrong, David M. 1989. *Universals: An Opinionated Introduction.* London: Routledge.

Armstrong, David M. 2010. *Sketch for a Systematic Metaphysics.* Oxford: Oxford University Press.

Asner, Gregory P., et al. 2009. A Contemporary Assessment of Change in Humid Tropical Forests. *Conservation Biology* 23 (6): 1386–1395.

Barrau, Aurélien. 2020. *Le plus grand défi de l'histoire de l'Humanité.* Paris: Michel Lafon.

Barrow, John D., and Frank J. Tipler. 1986. *The Anthropic Cosmological Principle.* Oxford: Oxford University Press.

Bartalesi., Lorenzo. 2024. *Storia naturale dell'estetica.* Torino: Einaudi.

Benton, J. Michael. 2003. *When Life Nearly Died: The Greatest Mass Extinction of All Time.* London: Thames & Hudson.

Berners Lee, Mike. 2021. *There Is No Planet B.* Cambridge, MA: Cambridge University Press.

Brusatte, Steve. 2019. *The Rise and Fall of the Dinosaurs: The Untold Story of a Lost World.* London: Picador.

Brusatte, Steve. 2023. *The Rise and Reign of the Mammals: A New History, from the Shadow of the Dinosaurs to Us.* London: Picador.

Bullock, Stephen H., Harold A. Mooney, and Ernesto Medina. 1995. *Seasonally Dry Tropical Forests.* Cambridge: Cambridge University Press.

Cain, Michael L., William D. Bowman, and Sally D. Hacker. 2011. *Ecology.* Sunderland: Sinauer Associates.

Carroll, Sean B. 2005. *Endless Forms Most Beautiful.* New York: Baror International.

Casiraghi, Maurizio. 2023. *Sempre più soli. Il pianeta alle soglie della sesta estinzione.* Bologna: Il Mulino.

Clarke, Bruce, and Sébastien Dutreuil. 2022. *Writing Gaja: Correspondence of James Lovelock and Lynn Margulis.* Cambridge, UK: Cambridge University Press.

Coley, Paul D., and Tod A. Kursar. 2014. On Tropical Forests and Their Pests. *Science* 343 (6166): 35–36. https://doi.org/10.1126/science.1248110.

Crain, Deborah A., and Louis J. Guillette Jr. 1998. Reptiles as Models of Contaminant-Induced Endocrine Disruption. *Animal Reproduction Science* 53 (1–4): 77–86. https://doi.org/10.1016/s0378-4320(98)00128-6.

Crutzen, Paul J. 2002. Geology of Mankind. *Nature* 415 (6867): 23. https://doi.org/10.1038/415023a.

Dawkins, Richard. 1976. *The Selfish Gene*. Oxford: Oxford University Press.
Dawkins, Richard. 1996. *Climbing Mount Improbable*. New York: Viking.
Dennett, Daniel Clement. 1995. *Darwin's Dangerous Idea: Evolution and the Meanings of Life*. London: Penguin Books.
Erwin, Douglas H. 1999. The Origin of Bodyplans. *American Zoologist* 39 (3): 617–629.
Fleming, Theodore H. 2024. *Birds, Bats, and Blooms*. Tucson (AZ): University of Arizona Press.
Gaston, Kevin J. 1991. The Magnitude of Global Insect Species Richness. *Conservation Biology* 5 (3): 283–296.
Gattuso, Jean-Pierre, and Lina Hansson. 2011. *Ocean Acidification*. Oxford, UK: Oxford University Press.
Gilbert, Scott, and David Epel. 2015. *Ecological Developmental Biology: The Environmental Regulation of Development, Health and Evolution*. Oxford: Oxford University Press.
Gould, Stephen Jay. 1990. *Wonderful Life*. New York: W.W Norton & Company.
Gould, Stephen Jay, and Richard C. Lewontin. 1979. The Spandrels of San Marco and the Panglossian Paradigm: A Critique of the Adaptationist Programme. *Proceedings of the Royal Society of London. Series B, Biological Sciences* 205 (1161): 581–598.
Gould, Stephen Jay, and Elisabeth Vrba. 1982. Exaptation-A Missing Term in the Science of Form. *Paleobiology* 8 (1): 4–15.
Guimarães, Paulo R., Carlos J. Melián, and Stefano Allesina. 2017. Indirect Effects Drive Coevolution in Mutualistic Networks. *Nature Communications* 8:14888. https://doi.org/10.1038/ncomms14888.
Hamilton, W. D. 1964. "The Genetical Evolution of Social Behaviour, I and II." *Journal of Theoretical Biology* 7 (1): 1–52.
Haraway, Donna Jeanne. 2016. *A Cyborg Manifesto*. Minneapolis: University of Minnesota Press.
Hawksworth, David L., and Robert Lücking. 2017. Fungal Diversity Revisited: 2.2 to 3.8 Million Species. *Microbiology Spectrum* 5 (4): https://doi.org/10.1128/microbiolspec.FUNK-0052-2016.
Hinde, Robert A. 1982. *Ethology: Its Nature and Relations with Other Sciences*. Oxford: Oxford University Press.
Jetz, Walter, and Paul V. A. Fine. 2012. Global Gradients in Vertebrate Diversity Predicted by Historical Area-Productivity Dynamics and Contemporary Environment. *PLoS Biology* 10 (3): e1001292. https://doi.org/10.1371/journal.pbio.1001292.
Johnson, Pieter Thomas Jacobus, et al. 1999. The Effect of Trematode Infection on Amphibian Limb Development and Survivorship. *Science* 284 (5415): 802–804. https://doi.org/10.1126/science.284.5415.802.

Jonas, Hans. 1984. *The Imperative of Responsibility: In Search of an Ethics for the Technological Age*. Chicago: University of Chicago Press.
Kiesecker, Joseph Michael. 2002. Synergism between Trematode Infection and Pesticide Exposure: A Link to Amphibian Limb Deformities in Nature? *Proceedings of the National Academy of Sciences of the United States of America* 99 (15): 9900–9904. https://doi.org/10.1073/pnas.152098899.
Koestler, Arthur. 1990. *The Ghost in the Machine*. New York: Penguin Books.
Kolbert, Elizabeth. 2014. *The Sixth Extinction: An Unnatural History*. New York: Henry Holt and Company.
Kropotkin, Pëtr. 2011. *Mutual Aid: A Factor of Evolution*. London: Freedom Press.
Ladyman, James, and Don Ross. 2007. *Every Thing Must Go: Metaphysics Naturalized*. Oxford University Press.
Larsen, Brendan B., Elizabeth C. Miller, Matthew K. Rhodes, e John J. Wiens. 2017. "Inordinate Fondness Multiplied and Redistributed: The Number of Species on Earth and the New Pie of Life." *The Quarterly Review of Biology* 92 (3): 229–265.
Leakey, Richard E., and Roger Lewin. 1996. *The Sixth Extinction: Patterns of Life and the Future of Humankind*. New York: Anchor Books.
Lewis, David. 1986. *Philosophical Papers*. Vol. 2. Oxford: Oxford University Press.
Lewontin, Richard. 1983. The Organism as the Subject and Object of Evolution. *Scientia* 118:63–82.
Lewontin, Richard. 1991. *Biology as Ideology: The Doctrine of DNA*. New York: Harper Perennial.
Lovelock, James. 1979. *Gaia. A New Look at Life on Earth*. Oxford, UK: Oxford University Press.
Lovelock, James E., and Lynn Margulis. 1974. Atmospheric Homeostasis by and for the Biosphere: the Gaia Hypothesis. *Tellus Series A* 26 (1–2): 2–10.
Lowery, Christopher M., et al. 2018. "Rapid Recovery of Life at Ground Zero of the End-Cretaceous Mass Extinction." *Nature* 558(7709): 288–291. https://doi.org/10.1038/s41586-018-0163-6.
Marchesini, Roberto. 2022. *The Creative Animal: How Every Animal Builds Its Own Existence*. Cham, Switzerland: Palgrave Macmillan.
Margulis, Lynn. 1998. *Symbiotic Planet: A New Look at Evolution*. New York: Basic Books.
Margulis Sagan, Lynn. 1967. On the Origin of Mitosing Cells. *Journal of Theoretical Biology* 14 (3): 255–274. https://doi.org/10.1016/0022-5193(67)90079-3. (Published under the surname Sagan; the author was later known as Lynn Margulis).
Marshall, John, and R. Alan Plumb. 2007. *Atmosphere, Ocean and Climate Dynamics*. Cambridge, MA: Academic.
Mazzotti, Stefano. 2024. *Meravigliose creature*. Bologna: Il Mulino.

McCallum, Malcolm L. 2007. Amphibian Decline or Extinction? Current Declines Dwarf Background Extinction Rate. *Journal of Herpetology* 41 (3): 483–491. https://doi.org/10.1670/0022-1511(2007)41[483:ADOECD]2.0.CO;2.

Mercalli, Luca. 2019. *Il clima che cambia*. Milano: Rizzoli Bur.

Monbiot, George. 2022. *Regenesis: Feeding the World without Devouring the Planet*. London: Allen Lane.

Moore, Jason W. 2016. *Anthropocene or Capitalocene, Nature, History and the Crisis of Capitalism*. Okland: PM Press.

Morton, Timothy. 2013. *Hyperobjects. Philosophy and Ecology after the End of the World*. Minneapolis: University of Minnesota Press.

Morton, Timothy. 2016. *Dark Ecology. For a Logic of Future Coexistence*. New York: Columbia University Press.

Murphy, Michael Warren, and Caitlin Schroering. 2020. Refiguring the Plantationocene: Racial Capitalism, World-Systems Analysis, and Global Socioecological Transformation. *Journal of World-Systems Research* 26 (2): 400–415. https://doi.org/10.5195/jwsr.2020.983.

Nijhout, Frederik. 1999. Control Mechanisms of Polyphenic Development in Insects: In Polyphenic Development, Environmental Factors Alter some Aspects of Development in an Orderly and Predictable Way. *BioScience* 49 (3): 181–192.

Nilsson, Dan-Eric, and Suzanne Pelger. 1994. A Pessimistic Estimate of the Time Required for an Eye to Evolve. *Proceedings of the Royal Society of London. Series B, Biological Sciences* 256 (1345): 53–58.

Novacek, Michael J. 2001. *The Biodiversity Crisis: Losing What Counts. American Museum of Natural History*. New York: The New Press.

Odling-Smee, John. 1988. Niche Construction: The Neglected Process in Evolution. In *The Role of Behavior in Evolution*, ed. Henry C. Plotkin, 98–134. Cambridge, MA: MIT Press.

Odum, Eugen P., and Gary W. Barrett. 2005. *Fundamentals of Ecology*. Belmont: Thomson Brooks-Cole.

Ouyang, Zutao, et al. 2022. Albedo changes caused by future urbanization contribute to global warming. *Nature Communications* 13 (1): 3800.

Philip, Leanne, Suzanne Simard, and Melanie Jones. 2010. Pathways for Below-Ground Carbon Transfer between Paper Birch and Douglas-Fir Seedlings. *Plant Ecology & Diversity* 3 (3): 221–233.

Phillips, Helen, John M. Halley, J. Nicolas Urbina-Cardona, and Andy Purvis. 2017. The Effect of Fragment Area on Site-Level Biodiversity. *Natural History Museum*. https://doi.org/10.5519/0089190.

Pigliucci, Massimo, and Gerd B. Müller. 2010. *Evolution. The Extended Synthesis*. Cambridge: MIT Press.

Prigogine, Ilya, and Gregoire Nicolis. 1977. *Self-Organization in Nonequilibrium Systems: From Dissipative Structures to Order through Fluctuations*. Hoboken: Wiley-Blackwell.

Qu, Xin, and Alex Hall. 2005. *Assessing Snow Albedo Feedback in Simulated Climate Change*. Los Angeles: Department of Atmospheric and Oceanic Science, University of California.

Quammen, David. 2012. *Spillover, Animal Infections and the Next Human Pandemic*. New York: W.W. Norton &C.

Rescher, Nicholas. 1996. *Process Metaphysics: An Introduction to Process Philosophy*. Albany: State of New York Press.

Rich, Nathaniel. 2019. *Losing Earth: A Recently History*. New York: Farrar, Strauss & Giroux.

Ripple, William J., and Robert L. Beschta. 2011. Trophic Cascades in Yellowstone: The First 15 Years after Wolf Reintroduction. *Biological Conservation* 145 (1): 205–213. https://doi.org/10.1016/j.biocon.2011.09.005.

Rosenberg, Eugene, et al. 2007. The Role of Microorganisms in Coral Health, Disease and Evolution. *Nature Reviews Microbiology* 5:355–362.

Safran Foer, Jonathan. 2010. *Eating Animals*. New York: Back BayBooks.

Schrödinger, Erwin. 1940. *What Is Life?* Cambridge: Cambridge University Press.

Sessions, Stanley K., and Stephen B. Ruth. 1990. Explanation for Naturally Occurring Supernumerary Limbs in Amphibians. *The Journal of Experimental Zoology* 254:38–47. https://doi.org/10.1002/jez.1402540106.

Siedler, Gerold, Stephen M. Griffies, John Gould, and John A. Church, eds. 2013. *Ocean Circulation & Climate*. Cambridge, MA: Academic.

Smith, Robert Leo, and Thomas M. Smith. 2000. *Elements of Ecology*. New York: Longman.

Steinmann, Wolf-Dieter. 2022. *Thermal Energy Storage for Medium and High Temperatures: Concepts and Applications*. Wiesbaden: Springer.

Stewart, Philip, and Sabine Globig. 2012. *Plant Physiology*. Toronto/New York: Apple Academic Press.

Tawfik, Ahmed I., et al. 1999. Identification of the Gregarization-Associated Dark-Pigmentotropin in Locusts through an Albino Mutant. *Proceedings of the National Academy of Sciences of the United States of America* 96 (12): 7083–7087. https://doi.org/10.1073/pnas.96.12.7083.

Tozzi, Mario. 2025. Prove tecniche di estinzione. Istruzioni per salvare il salvabile. Milano: Touring Club Italiano.

Valentine, James W. 2004. *On the Origin of Phyla*. Chicago: University of Chicago Press.

Wake, David Burton, and Vance Thomas Vredenburg. 2008. Are We in the Midst of the Sixth Mass Extinction? A View from the World of Amphibians. *Proceedings of the National Academy of Sciences* 105 (Suppl 1): 11466–11473. https://doi.org/10.1073/pnas.0801921105.

Weber, Max. 2004. Science as a Vocation. In From Max Weber: Essays in Sociology, edited by H. H. Gerth and C. Wright Mills. New York: Oxford University Press.

Weisman, Alan. 2007. *The World Without Us*. Toronto: Picador Paper.

West-Eberhard, Mary Jane. 2003. *Developmental Plasticity and Evolution*. Oxford: Oxford University Press.

Wilson, Edward O. 1992. *The Diversity of Life*. Cambridge, MA: Belknap Press of Harvard University Press.

Wright, S. Joseph. 2005. Tropical Forests in a Changing Environment. *Trends in Ecology & Evolution* 20 (10): 553–560.

CHAPTER 5

Anthropocentrism

5.1 Premise

As I have argued throughout this book, one of the most significant obstacles to ecological thinking is undoubtedly anthropocentrism. This is primarily due to the epistemological challenge it presents, by isolating human beings from their context and hindering the development of relational and interconnected thinking (Stallard 2017). From a practical perspective, this translates into placing human beings and their activities at the top of the priority list (Lasch 1979). However, the challenges posed by anthropocentrism to ecological thinking extend beyond economics (Hirsch 1976) or epistemology (Bachelard 2002), as they encompass various aspects of our relationship with the biosphere. Anthropocentrism is frequently viewed either as a means of safeguarding human dignity, or as an inevitable condition tied to belonging to this species. In this chapter, we will see how both of these considerations are flawed.

In the twentieth century, discussions about anthropocentrism often centered on the relationship with other animal species and related ethical considerations, particularly regarding speciesism (Singer 1975; Regan 2004). The term "speciesism" refers to the prejudicial exclusion of animal others from the status of moral patients, and an undue overlap is often made between this form of segregation and anthropocentrism as a whole. In 1970, British psychologist Richard Ryder coined the term to describe the moral discrimination humans exercise against members of other

R. Marchesini, *Philosophical Ecology*,
https://doi.org/10.1007/978-3-032-04987-2_5

species, based solely on the fact that they are not human. The tautological—and therefore unsustainable—logic of a statement like "a human being, regardless of their physical or cognitive condition, must be considered a moral patient because they are human" is self-evident. Arbitrary discrimination—whether racist, sexist, or otherwise—always rests on circular reasoning. If we cannot identify morally relevant traits that are present in all humans and absent in all non-human animals, then any distinction becomes inevitably arbitrary and ethically indefensible.

Building on this principle, in 1975, Australian philosopher Peter Singer published *Animal Liberation*, marking a turning point in the animal rights debate. For the first time, the issue of animal interests was approached through the rigorous logic of their inherent characteristics—such as Bentham's famous question, "can they suffer?"—rather than through the lens of human sentimentality. This shift moved concern for animals beyond personal sensitivity or affection (zoophilia), toward an ethical and emancipatory stance (animalism), which would later evolve into a broader political and intersectional movement (antispeciesism). Singer's utilitarianism weighs the trivial human interest in meat consumption against the intense suffering experienced by animals in factory farms, during transport to slaughterhouses, and ultimately during slaughter itself, to argue for the ethical imperative of a vegetarian diet. The very title *Animal Liberation* frames the recognition of animal interests as part of a wider emancipation movement that calls for the affirmation of animal rights. This call became even more explicit with American philosopher Tom Regan's 1983 work *The Case for Animal Rights* (2004). Regan argued that many animals are "subjects-of-a-life," and therefore possess intrinsic value—that is, value in and of themselves, not just in relation to others. To put it in Kantian terms, they are ends, not means. Regan's natural rights perspective is even more radical than Singer's utilitarianism, concluding that any practice that treats animals as means to an end is morally unacceptable, regardless of the intended goal. The animal rights movement—later known as antispeciesism—would strongly influence the cultural landscape from the late twentieth century into the early twenty-first, leaving a lasting mark on societal habits and beliefs.

Anthropocentrism thus came to be viewed increasingly as an ethical position—one that justifies the exploitation of non-human species and subjects them to mistreatment that disregards their sentience, particularly in factory farming and modern industrial agriculture. In 1962, American biologist Rachel Carson published *Silent Spring*, a searing exposé against the mass use of pesticides. The book became a cornerstone of the environmental movement, describing the ecological damage beginning with bird populations—rendered silent and unable to herald spring—and culminating in a warning for humanity about the danger of biosphere collapse. In this sense, Carson echoed Albert Schweitzer's reflection on humankind's inability to think long term. Two years later, English activist Ruth Harrison released *Animal Machines* (1964), which exposed the appalling conditions endured by millions of animals in factory farms. Her work led to the formation of the Brambell Committee and the publication of the landmark 1965 report outlining the "Five Freedoms" of animal welfare. Then, in 1976, Swiss writer Hans Ruesch published *Naked Empress or The Great Medical Fraud* (1986), denouncing the use of animals in scientific experimentation.

Together, these contributions underscore a growing trend: the reinterpretation of anthropocentrism primarily—if not exclusively—as an ethical issue, reflecting human negligence toward other living beings. However, this is a mistake. The term *anthropocentrism* refers to a much more comprehensive conception of man's centrality to the biosphere (Marchesini 2017). There are several forms of anthropocentrism, in addition to the ethical one:

(i) Ontological anthropocentrism, the view that makes man the unit of measurement of all things and, therefore, the reference to judge otherness, as well as the subsuming principle of the world.
(ii) Ontopoietic anthropocentrism, which considers the human dimension autarchic and autopoietic and human predicates as emanative qualities of the human being.
(iii) Aesthetic anthropocentrism, the anthropometric and anthropoplastic view which places the man-made above the natural.
(iv) Epistemological anthropocentrism, considering human cognition as a universal that extends beyond the species and represents the apex of knowledge.
(v) Eschatological anthropocentrism, which places man's fate in a priority relation to the biosphere or even the universe, judging it as uniquely worthy of consideration.

Today, we must view anthropocentrism as something that affects every aspect of the human condition—a philosophical barrier that prevents us from fully grasping our own ontology, the way we develop ontopoietic processes through relationships with otherness, the very meaning of morality as the strengthening of individual empathic virtues, and the ecological risks and challenges we are facing. One could say that anthropocentrism is becoming a threat to the very survival of our species—but even that seems too cautious a statement. Anthropocentrism prevents us from understanding what it means to be human. It seeks to define the human condition—our culture, our ontopoietic drive, our relational tendencies, our constant evolution and wish to go further—by drawing a rigid boundary against the non-human and turning inward. It attempts to comprehend itself through internal reflection alone, as if it could be grounded *iuxta propria principia*.

In truth, anthropocentrism is a kind of banal egocentrism, a form of social isolation, a solipsistic drift that disregards how deeply human beings have relied on non-human entities to construct their ontopoietic reality. The glorification of this disjunction—especially within the technological sphere—as the true and only human habitat is a legacy of the humanist paradigm, which envisions man as Prometheus' heir, who carved out a protected space for himself beyond nature thanks to the titan's foresight. One of the clearest representatives of this thinking today is German philosopher Peter Sloterdijk. In the third chapter of *Not Saved: Essays After Heidegger* (2016), dedicated to "The Domestication of Being," he claims that technology enables a withdrawal from nature—a kind of adaptive detachment. Yet this is misleading: niche construction does not free humans from nature but reshapes our relationship with it. In Sloterdijk's view, the anthropogenic process unfolds solipsistically, with stone technology as its only point of reference. This view reduces humanity to an entity disconnected from nature.

This is hardly a new idea: it has venerable ancestors in figures like Giovanni Pico della Mirandola and Arnold Gehlen. This species-level egocentrism, which humanism turned into an ideology, is increasingly confining us to a mirror gallery—a world that reflects humanity back at itself in fractal form. This kaleidoscope risks numbing us with its narcissistic appeal. And yet, this childish egocentrism is precisely what threatens humanity—not only ecologically (since no Sloterdijkian sphere will shield us from biosphere collapse), but also ontologically. Because the human being is not the product of separation, but of relation. Because technology does not make us more self-sufficient—but more dependent.

5.2 General Aspects of Anthropocentrism

When talking about anthropocentrism, there is often some confusion, partly because people tend to implicitly refer to ethical anthropocentrism, forgetting that it actually stems from an ontological conception—one that places the human on a completely different plane from that of other living beings.

The result of this attitude is an overestimation of human characteristics, which leads to a denial—or at least an understatement—of the limitations and problems inherent to human nature. At the same time, it produces a total devaluation of non-human otherness. This approach severs—or at least claims to sever—the bond of encounter and exchange between the human and the non-human. It implies the idea that humans live in a separate dimension from the biosphere, disconnected from it through their technological tools and in a state of existential self-sufficiency, creating a mutual impermeability between the two supposed realms. The consequences of this mindset are far-reaching. It fosters the belief that the human being is, in ontological terms, a self-sufficient entity with no need to engage in anthropopoietic relationships with the non-human. This reinforces a vision of the biosphere not as something with which we are consubstantial, but merely as a house or a set of resources—something we inhabit and use, and from which we could, in principle, always move elsewhere. At the same time, it obscures any organismic view of the biosphere, one that includes the human as an integral part, and blinds us to the complex anastomosis that binds us to other species across every dimension of existence.

Sometimes, ideological anthropocentrism, which is the product of a specific cultural paradigm based on well-defined philosophical assumptions, is also confused with *the anthropocentric perspective*, meaning the naïve point of view that stems from our phylogenetic development. While there may be connections between the two—paradigmatic anthropocentrism can, in some ways, be seen as the absolutized form of the anthropocentric perspective—there are also important differences. Confusing these two leads to a kind of justificatory syllogism that can be summarized as follows: since the anthropocentric perspective is the natural cognitive standpoint of human beings, every form of anthropocentrism is implicitly justified. This reasoning is flawed for at least two basic reasons. First, paradigmatic anthropocentrism does not stem *ipso facto* from the anthropocentric perspective—it is an ideology shaped by a specific cultural project.

Second, it is not true that the anthropocentric perspective is an absolute cognitive limit—it is a relative one. Just as an egocentric person can make an effort to develop empathy, so too can we humans engage in forms of decentering that allow us to go beyond our default perspective. Indeed, as we will see, we have already achieved this in some respects.

For some, questioning anthropocentrism is tantamount to questioning—against all evidence—the undeniable qualities of the human animal, which place it on a level of significance above all others. This portrayal of the human as an entity of exceptional importance within the biological landscape is indeed undeniable, and it would be wrong to assume that criticism of anthropocentrism means denying the unique traits of our species. Similarly, some believe that challenging anthropocentrism amounts to blurring human prerogatives within the broader animal context, leading to a form of biological reductionism—with its related naturalistic fallacy—and opening the door to traits typically stigmatized when attributed to the natural world: amorality, the law of the strongest, the struggle for existence, aggression. In such cases, it's important to clarify that what an anthropocentric view has assigned to the "beastly condition" does not reflect the true nature of the animal world. In fact, nature displays every kind of behavior—those we admire and those we find repugnant. What's really at play here is a projection: these traits reflect certain human tendencies we would rather not acknowledge as part of our own behavioral repertoire. Indeed, ethics and pedagogy have long tried to steer humanity toward the ideal of a virtuous being. In this sense, "the animal"—as a categorical entity—is a deliberately fashioned construct, because no single species actually embodies the set of attributes assigned to it. It is meant to serve as a kind of dark mirror, a polarity invented as a counterpoint to the human. This imagined animal otherness allows humans to outline, on one side, a coherent and unified self-image, and on the other, to define an ontopoietic trajectory by contrast. In other words: humanity is a project still in the making, and its developmental path is seen as moving ever further away from the animal.

There is only one problem. Many philosophers forget that what they describe as "the animal" doesn't correspond to *any* actual entity in the zoological universe—in fact, it is more akin to Dorian Gray's hidden portrait, or Mr. Hyde lurking in the depths of Dr. Jekyll. Naturally, if we accept this biased and artificial vision of animal otherness, then any attempt to cross the disjunctive Rubicon between human and animal will be seen as a dangerous threat to human dignity, a perilous slip into the indistinct murk of bestial darkness. But the real issue is not the borders of the human cosmopolis. It is our ability to recognize that these inclinations do not belong to some imaginary animal category—they are part of human nature itself. Thus, if we hope to set limits and take preventive measures against these traits, we must take a hard look at ourselves. Sweeping our own dust under the "animal" rug will not help us deal with our problems once and for all.

As mentioned earlier, a widespread and problematic perception of the critique of anthropocentrism lies in the belief that re-establishing a connection between humanity and animality might actually serve to justify these negative tendencies. In other words: if we claim that certain objectionable behaviors are part of human inclinations, we risk giving them an ontological foundation. This is a mistake rooted in a static and deterministic view of nature—one that assumes animals are driven by instinct and that their behavior is entirely predetermined. This will be discussed at length in Chap. 6, so I won't repeat the arguments presented there to show why such an assessment is mistaken. I will only note that unfortunately, history is littered with interpretations rooted in naturalistic determinism—from the eugenic theories of Francis Galton, such as his essay *Hereditary, Character and Talent* (2018), to *The Bell Curve* (1994) by Charles Murray and Richard Herrnstein—applied to human beings. It's therefore understandable that a kind of ideological immune system has developed in response. However, this rejection of any phylogenetic basis for certain human behaviors can sometimes lead to the paradox of denying what is plainly evident. The real issue lies not in acknowledging that repugnant tendencies exist in human nature—which is essential if we hope to prevent or manage them—but rather in misinterpreting such tendencies, whether through deterministic and single-cause models or through frameworks that seek to justify them.

Anthropocentrism can be described as a kind of naïveté, similar to a child's egocentrism. First, because it assumes that the entire world revolves around the human being; second, because it tends to treat its own perspective on reality as absolute and universal; and third, because it reflects a childlike, unquestioning faith in itself and its destiny, along with an inability to engage in self-criticism. Anthropocentrism, in its boundless confidence in humanity and its abilities—as well as its implicit optimism regarding human progress and glorious manifest destiny—poses a serious obstacle: it prevents us from confronting the problematic inclinations that characterize us. As I've said before, sweeping the dust under the animal rug and projecting our flaws onto some external "other," whatever form it takes, is a grave mistake. It renders us incapable of learning from history, as we insist on attributing our own horrors—war, rape, genocide—to some other entity that has nothing to do with us. But this attitude also poses a significant barrier to taking seriously the ecological risks we are facing. In this context, questioning anthropocentrism means first, recognizing that humanity's ability to adopt a neutral perspective on its own future is limited, and that we are instead afflicted by deep myopia—unable to think in the long term—as well as pronounced astigmatism, which blurs our view of reality and distorts even what lies right beside us. Second, it means acknowledging that we suffer from a pathological optimism regarding our capabilities and future—an optimism that prevents us from confronting our own limitations and the scale of the problems before us, leading to superficiality and fatalism rather than a true assumption of responsibility.

Once again, to critique anthropocentrism does not mean to deny the peculiarities of the human being; on the contrary, it means taking into account all aspects of human nature—both the unique features that have enabled humans to reach the undeniable heights we know, and the elements of fragility, narrow-mindedness, and prejudice. It involves adopting an honest self-perception and a cautious kind of trust—far removed both from Hobbes' idea of *bellum omnium contra omnes* and from Hume's notion of natural sympathy. This also entails rethinking the construct of *animality*, with the understanding that the condition of being-animal cannot be defined through a set of common predicates to be set against those of human beings. One cannot compare a construct with a well-defined entity: animality is a *metapredicative* condition, which manifests in singular forms across different species. In this light, the human being is nothing more than a unique and specific way of being-animal.

To grasp this, we must first deconstruct the view of animality as humanity's dark mirror, and instead consider it through the metapredicative principles discussed in relation to the concept of animal agency. In other words, conceiving the human as an expression of animality does not mean reducing humans to a feral state. Rather, it means acknowledging that the emergent anthropopoietic qualities—broadly speaking, those linked to cultural development—do not directly derive from human nature, so it is incorrect to conflate nature with culture, as culture results from nonemanative processes; and yet, undeniably, they have precursors in human nature. For instance, prosocial behaviors originate in parental care and altruism, while certain cultural catalysts stem from our natural disposition—for example, the propensity to imitate other species, rooted in the marked mimetic tendencies that define humanity.

Thus, gaining a deeper understanding of animality—by deconstructing the opposition between human and animal and dismantling the idea of the latter as a dark reflection of the former—makes the human landscape come into clearer focus, in both its strengths and weaknesses. Above all, it reveals two especially valuable insights into the human condition. First, the shared traits we hold with other species, which allow us to move beyond the monadic view of humanity—the anguished solitude of man who sees himself as existentially alone, believing his perspective to be irreconcilable with that of other species. This supposed incommunicability, as we'll see, has no real basis. Second, the relational condition we share with other species: we humans have never been separate or self-sufficient. On the contrary, we have always learned from other animals, incorporating into our own cultural habits qualities that were not originally part of our species-specific repertoire, but were inspired by other forms of life.

Undoubtedly, the inability to recognize shared traits—that is, the fact that most predicates of a species (including the human) did not evolve uniquely within that animal but were already present in a common ancestor and are thus shared among multiple species—prevents us from acknowledging a common ground of belonging. For this reason, continuing to view animals as though they were aliens is fundamentally wrong. As for the relational condition, acknowledging it entails moving beyond the anthropocentric vision of human solipsism and admitting that, throughout human history, there has always been a dialogic and cultural hybridization between humans and other species.

Anthropocentrism, by contrast, constructs the animal counterpart in an arbitrary and often projective manner. It creates a disjunction and an inward closure that prevents humanity from understanding some key aspects of its own ontology—namely: (i) the *continuity* of human nature with non-human others, and thus the complex blend of similarities and differences that defines its ontic condition (i.e., its phylogenetic heritage); (ii) the *dialogue* between the human being and non-human others that has influenced the development of cultural dimensions—*ontopoiesis* (i.e., the becoming of the human condition). Anthropocentrism tends to confine the human within a self-contained space, fostering a kind of resentment and opposition toward the non-human. I believe this sentiment of preemptive defensiveness and a desire for dominance carries significant weight in how humans relate to the natural world—with the devastating consequences we are witnessing today.

Regarding the failure to grasp this shared condition—and thus the continuum between human and non-human—it is not surprising that, even today, some question the comparative method used to study certain human behaviors alongside those observed in closely related species from a phylogenetic perspective. Erich Fromm, in his book *The Anatomy of Human Destructiveness* (1992), criticized the ethological comparative method, claiming that the analogies drawn between the behavior of certain animals and that of humans are merely suggestive—appealing to animal lovers but ultimately proving nothing, given their arbitrary causal overlap. Fromm's error lies in his framing of comparison purely in terms of analogy. In ethology, a clear distinction is made between homologous structures—those stemming from common ancestry and showing anatomical overlap—and analogous structures—those that perform similar functions but do not share a common origin. Comparative analysis in ethology refers to ***homologies***, while ***analogies*** serve as useful indicators for understanding the selective pressures acting on species that share the same environment or lifestyle and are thus shaped in similar ways.

As for the pronounced tendency of the human being toward *transspecific relations*—that is, learning from other species, a topic I explored in *Over the Human* (2017)—there remains a prevailing inclination toward an autarchic interpretation. This is the idea that everything humanity has developed culturally, scientifically, and technologically is the product of its own ingenuity, as if humans were the planet's sole inhabitants and had never drawn any benefit or inspiration from the presence of other species. The only anthropogenetic perspective that is commonly acknowledged in relation to other species is an oppositional one—what Giorgio Agamben presents in *The Open: Man and Animal* (2003) as the *anthropological machine*, where relational anthropopoiesis is conceived solely in negative terms, as the subtraction of "beastly" traits. This amounts to a distancing process not unlike the notion of the dark mirror.

By contrast, the view I defended in the aforementioned 2017 essay is that, in the construction of human identity through relationships with other species, what prevailed is *positive anthropopoiesis*. This means the incorporation of heterospecific predicates into the human dimension in cultural form. Once again, anthropocentrism risks leading us astray: by refusing to acknowledge these referential debts—these models that have inspired and guided us—we not only fail to understand the origins of human culture, but, through our disinterest in animal others, we risk cutting ourselves off from our most important ontopoietic source.

5.3 The Two Faces of Anthropocentrism

Undoubtedly, anthropocentrism exhibits general characteristics rooted in the disjunction of the human from non-human others and in the tendency to consider the human perspective as absolute—giving rise to all the consequences outlined in the previous section. At the same time, several questions naturally arise: Is it possible for human beings *not* to adopt an anthropocentric approach? Is it legitimate or at least justifiable for humans to maintain an anthropocentric worldview? Given that all species adopt a self-centered criterion, isn't it perhaps excessive to expect humans to counter this tendency? If only humans question their self-centered perspective, doesn't that risk placing them in a different ontological condition from other species? Does anthropocentrism stem from a natural human condition, or is it the result of a specific philosophical stance?

Answering these questions requires identifying the different faces of anthropocentrism. Therefore, the first step must be to conduct a brief analysis in this regard. In fact, when we talk about anthropocentrism, we must necessarily distinguish between two very different expressions of the anthropocentric approach:

1. *Naïve anthropocentrism*, based simply on applying a human perspective to the world;
2. *Ideological anthropocentrism*, the product of a specific philosophical orientation and characteristic of certain cultures.

That said, I will attempt a preliminary response to the five questions posed above—statements that will become clearer after examining the two expressions of anthropocentrism: (i) Yes, it is possible for human beings to broaden their perspective—if not entirely overcoming anthropocentrism, at least mitigating it and attributing its only relative value; (ii) The anthropocentric approach is understandable, but that does not justify or legitimize doing nothing to reduce its effects—especially when it becomes a form of ideology; (iii) It is true that mitigating the anthropocentric approach requires effort—arguably a form of effort akin to, though greater than, the empathetic effort involved in social relationships. However, it is not a stretch, since such effort enhances human capacities; (iv) No, because the very characteristics of the human being have always led to decentering, through borrowing from other species. Just as humans stand out for their technical and cultural abilities, they are also marked by the capacity for empathic decentering; (v) Anthropocentrism arises and naturally exists within humans due to a disposition that is part of our nature. However, this disposition has been absolutized by a specific cultural program, which transformed it from a tendency into a philosophical paradigm.

We can therefore speak of:

1. A *natural or naïve anthropocentrism*, which stems from the inherent tendency of every species to adopt a self-centered perspective on the world. This perspective naturally results in an anthropocentric outlook, inclined to interpret reality through the lens of the species' perceptual and cognitive apparatus, projecting its own condition onto others. This underlies the tendency toward so-called naïve physics, animism in relation to natural phenomena (attributing them intentionality), and the anthropomorphic reading of animal others.
2. A *cultural anthropocentrism*, the result of a specific philosophical project that has, in some respects, taken on ideological traits. This has given rise to a *paradigmatic anthropocentrism*, which goes far beyond a self-centered perspective and turns the human being into the only entity endowed with intrinsic value—indeed, into the ultimate purpose of the universe itself.

If we do not understand the difference between these two expressions, everything becomes confused and arbitrary, as their respective consequences cannot be grasped. What is even more concerning is the tendency to justify paradigmatic anthropocentrism by appealing to the self-centered perspective—by falling into the logical fallacy of assuming that a natural human tendency directly justifies a philosophical or ideological position.

Let us therefore try to better understand the distinction between these two forms. We will see not only that there is no direct lineage from the first to the second, but—paradoxically—that the very questioning of the anthropocentric perspective has, in some ways, contributed to reinforcing paradigmatic anthropocentrism.

5.3.1 *The Anthropocentric Perspective*

The first form of anthropocentrism is the anthropocentric perspective—a concept that refers to the self-centered dimension that any animal is naturally inclined to adopt in its relationship with the world. Our tools for accessing reality, the very structure of our bodies, the emotional and cognitive traits that allow us to experience the external world—this set of species-specific intrinsic characteristics forms a particular point of view on the world. Of course, the term should not be understood literally: it is not simply about how we see reality, nor about having a particular perception of it, but rather about experiencing the world and being within it. This mode of experience depends on several factors: not just the specificities of the sensory organs, nor solely the structure of body and mind, or the temporal rhythm by which a species defines diachronic flow, but also on lifestyle, environment, and the counterparts with which one interacts.

This perspective, in turn, gives rise to two tendencies: (i) *Epistemic anthropocentrism*, where one uses one's own perceptual and cognitive systems not as a species-specific endowment, but as if they were neutral lenses; (ii) *Anthropomorphism*, where human traits are projected onto non-human others, albeit in diminished form. Both of these tendencies reflect anthropocentric biases that become epistemic obstacles to recognizing otherness.

Let's begin with epistemic anthropocentrism, often also referred to as naïve physics (Bozzi 1998). I often hear that, as human beings, we cannot help but be anthropocentric—meaning we are, in any case, destined to set ourselves up as a point of reference and interpretation of the world (Heller 2009). This is a mistake, and not a minor one. Let's try to understand why. It's undeniable that we cannot stand as neutral observers: we are encompassed within a peculiar species dimension. However, instead of generically speaking of anthropocentrism, which, as we have seen, involves several aspects, we should simply say that our *natural condition* necessarily places us within an anthropocentric perspective. Our perception and understanding of the world follow precise a priori coordinates, a legacy inherited from phylogeny.

On the other hand, while this underlying aspect cannot be ignored, it is equally incorrect to consider it an insuperable limitation. Just as a child evolves from a self-centered condition to empathetic thinking, human beings can and should broaden their horizon in their relationship with the world (Piaget 1970). The anthropocentric perspective serves as the starting point and, in some ways, defines the range of possibilities, but it is certainly not an unchangeable dimension. This insight aligns with von Uexküll's evolutionary conception of the *Umwelt* as affected by the ontogenetic processes of the individual (von Uexküll 1956). Similar to how the relationship with others causes the egocentric perspective to evolve into an empathetic one, it is possible to diminish the projective tendency.

In this sense, we can assert that technoscience has served an essential corrective function in relation to the anthropocentric perspective. It has done so by exposing the fallacies inherent in our intuitive interpretation of reality (Wolpert 2000). Take, for example, the Copernican model challenging geocentrism or Darwinian evolutionary theory dismantling the fixist and essentialist image of living things—both highly intuitive notions (Kuhn 1957). Or else, consider Einsteinian relativity or quantum mechanics, both presenting a universe that contradicts our usual experiences. Science is counterintuitive. Indeed, if phenomena were easily interpreted through intuition, we wouldn't need science in the first place. The way we perceive the world through a priori foundations, owing to their evolutionary nature of phylogenetic a posteriori ones, does not grant them universal value but, on the contrary, a specific domain of validity. Ethology has also expanded our horizons, revealing additional existential perspectives related to perception (Hughes 1999) and cognition (Griffin 2001). This new awareness has allowed encounters with heterospecific entities to become moments of revelation, leading to immersion in and experience of other worlds. Additionally, the information technology revolution has facilitated the creation of sophisticated machines capable of showcasing the numerous interpretive biases inherent in human intuition. More precisely, these machines demonstrate the adaptive nature of our epistemic apparatuses and a priori foundations.

Now, I think that claiming the inevitability of the anthropocentric viewpoint overlooks a very important fact—namely that we are already in a decentralized condition concerning the perspective coordinates inherited from our phylogenetic heritage (Plessner 2019). Contemporary human thought is influenced by the phylogenetic condition of *Homo sapiens* in some aspects, but in many others, it has evolved into a broader epistemological perspective. Human history has unfolded a perspectival dimension that extends beyond its phylogenetic inheritance, showcasing the ability to utilize the anthropocentric perspective as a starting point to explore new epistemological dimensions. So in my view, while there are certainly limits to how far we can go, we have already surpassed the anthropocentric perspective (Bachelard 2002). This *anthropo-decentralization*, the trajectory of anthropopoiesis, represents the cultural journey humanity has been on for millennia, which has been exponentially accelerated by technoscientific evolution.

The idea that human beings are confined within their own anthropocentric dimension and incapable of projecting themselves beyond their own boundaries would condemn us to an inability to fully access the existential dimension of another species. Yet, while it is true that one can never entirely abandon the anthropocentric perspective, it is undeniable that we can implement redefinitions that open us up, broadening our epistemic horizon. As mentioned, this is evident in counterintuitionist science (Wolpert 1994; Bachelard 2002), but it has also been shown by cognitive sciences, which highlight the partial nature of human thought and reasoning. If we consider the human being as the result of adaptive specialization—such is our anatomy and physiology, for instance—it becomes clear that we must also speak of cognitive specialization. It's not hard to understand that, just as there are multiple perceptual universes in the animal world, so too, in terms of intellect, the human is not a universal but a particular way of interpreting the world, to which we should assign a limited domain of validity. We could then say that the anthropocentric perspective is a form of species egocentrism which, if it cannot be completely overcome, can at least be mitigated through empathic exercises—ones that engage both cognitive and emotional aspects.

Let us now consider the tendency toward anthropomorphization and, more generally, the projective attitude—also a consequence of a self-referential perspective. Anthropomorphism is often considered the opposite of anthropocentrism because, on the surface, the human attitude toward animal otherness appears inclusive rather than exclusive. In reality, this is only an appearance: anthropomorphism, when not expressed critically (Bekoff 2004), is merely another form of anthropocentrism—projective anthropocentrism—which extends the anthropocentric view to other species via analogy. There are several types of anthropomorphism:

1. Animist anthropomorphism: attributes the same intentionality found in animals to natural phenomena.
2. A-specific anthropomorphism: ignores species-specific differences and projects the human perspective onto other species.
3. Epistemic anthropomorphism: assumes the only way to interpret otherness is through analogy, as seen in critiques of Fromm's ethology.

4. Affiliative anthropomorphism: typically used toward pets.
5. Critical anthropomorphism: begins with a continuist view and uses projection but subjects it to scrutiny.

Of all these, I at least partially endorse critical anthropomorphism, which—when applied carefully and with respect for the phylogenetic proximity between humans and another species—can be helpful in interpreting the psychological conditions and behavioral expressions of other species. However, I believe that understanding the condition of other animals is better achieved through three very specific keys of investigation: first, the *comparative analysis* of homologous traits, which shows the level of comparability between two species and reveals shared traits inherited from a common ancestor; second, *adaptive interpretation*, which considers the metapredicates common to animal life—such as sensibility, emotionality, and desire—and examines how these manifest in a species according to its specific needs; and finally *analogical evaluation*, which considers traits that may be shared because they were shaped by similar selective pressures.

In general, I remain skeptical of anthropomorphism because it continues to view the human as a model to be applied to other species, which are then inevitably seen as "less than" since they can never reach that model and remain ontologically subordinate to it. Also, by absolutizing continuity, it tends to interpret the differences between humans and other species quantitatively, whereas adaptive specialization operates very differently. Finally, it fails to recognize the qualities that another species may possess in greater or simply different measure than humans.

5.3.2 *Paradigmatic Anthropocentrism*

Let's now turn to paradigmatic anthropocentrism, whereby the human being is placed at the center of existence. This gives rise to two main tendencies:

1. Ontological anthropocentrism, which views the human as the unit of measurement, the synthesis, and the ultimate end of the universe;
2. Ethical anthropocentrism, which regards only humans as worthy of moral consideration, thereby rendering all other living beings irrelevant or assigning radically different ethical treatment to them.

Unlike the anthropocentric perspective, paradigmatic anthropocentrism stems from a cultural project that not only absolutizes the human perspective, turning it into a sort of ontological universal, but also introduces assumptions with profound consequences. This cultural project originated particularly in the Mediterranean basin and underpins Western culture. Its two primary roots are the Abrahamic tradition, which presents the human as *imago Dei*, and the Greek tradition, particularly post-Sophistic philosophy. These two cultures, eventually synthesized in Christianity, start from different ontological positions but converge in separating the human from other species, placing the former at the top and at the center of the living. Let's now have a quick look at the history of this stance.

The idea of *humanitas* is well exemplified by Publius Terentius Afer's famous line: "*Homo sum, humani nihil a me alienum puto*" ("I am human, and nothing human is alien to me"), from the comedy *Heautontimoroumenos* (165 BCE). This idea found full expression in Rome through the Scipionic Circle, linked to the Hellenistic tradition. The exaltation of *humanitas*, further developed by Stoic philosophy, then became a true cultural manifesto that elevated human virtues as the foundation of moral, educational, and political missions. This concept was later revived by fifteenth-century humanists, notably in Giovanni Pico della Mirandola's *Oration on the Dignity of Man* (1486), often considered the manifesto of the Italian Renaissance (Ries 2002). In this work, human creativity is celebrated, and—drawing on the Promethean notion of humanity lacking Epimetheus' gift—humans are seen as free and self-determining, unlike other animals. Leonardo da Vinci's *Vitruvian Man* (1490), drawn in pen and ink, visually encapsulates this notion. It inscribes the human body within two perfect shapes: the circle, symbolizing the heavens and the spiritual realm, and the square, representing the earthly dimension. In this way, the *Vitruvian Man* is no longer just one entity among others but becomes the mirror of the universe, a point of convergence between microcosm and macrocosm, synthesizing both earthly and spiritual dimensions. This same conception can be found in Descartes' philosophy, where the human is seen as the convergence of *res cogitans* and *res extensa*.

At the same time, the *Vitruvian Man*, regarding the earthly dimension, is depicted as: (i) a naked human being—an entity whose nudity symbolizes purity, self-creation (*autopoiesis*), and ultimately autonomy from anything non-human, including even technology; (ii) a universal entity—arms and legs outstretched like a compass, simultaneously measuring the world and encompassing it. If Pico's oration is the manifesto of humanist anthropocentrism, the *Vitruvian Man* is its most vivid visual representation. Where the Sophist claim "man is the measure of all things" (attributed to Protagoras by Plato in the *Protagoras*, 388 BCE) suggested the relativity of knowledge, the humanists of the fifteenth century interpreted it in the opposite sense: not as a form of relativism, but as an assertion of universality. This humanist universalism gave rise to a self-referential dimension further reinforced by the cultural byproduct of the Copernican revolution: with the fall of geocentrism, there emerged an even stronger need to anchor ourselves in anthropocentrism.

Humanist anthropocentrism thus became the foundational cultural paradigm of modernity. It marked a profound transformation in philosophy and culture, moving from medieval theocentrism to an anthropocentrism that was primarily ontological—one that sanctioned the human being's liberation from any connection to the non-human. This paradigm, based on the ontological difference between humans and animal others, produced further anthropocentric developments. For instance, the idea of human disjunction from the biosphere leads to the belief that all cultural evolution—along with the effects of culture on human nature (anthropopoiesis)—is the sole product of human ingenuity, in an autarchic reading of human ontopoiesis. This outlook assumes that the non-human plays no role in human development and is merely a backdrop—one to which humans are completely impermeable. Yet there is no question that other species have played a co-creative role in human cultural evolution (Lévi Strauss 1963; Marchesini 2017). Furthermore, this devaluation of otherness, which strips it of inherent value, leads us to see only humans as deserving of attention, both ontologically—whereby other animals are seen as instrumental beings at human disposal—and ethically, by denying other species the potential to be considered moral subjects (ethical anthropocentrism).

Technoscientific advancements have, on one hand, critically challenged species egocentrism, yet, on the other hand, they have paradoxically strengthened human narcissism. In fact, technological achievements have been wielded to assert human ontological superiority over the biosphere. To unravel this seeming paradox, it's crucial to distinguish, once more, between the anthropocentric perspective and paradigmatic anthropocentrism. The roots of this ideology, as mentioned, extend well before the seventeenth-century technoscientific revolution, as they can be traced back to Platonic philosophy, Terence and Cicero's concept of *humanitas*, and particularly the humanist revolution of the fifteenth century (Garin 1976). Renaissance humanism, in fact, played a pivotal role in shaping the subsequent scientific revolution, making it challenging to discern a linear trajectory regarding decentralization. While Western culture has embraced a disjunctive philosophy, separating humans from the biosphere, other cultures have maintained a projective vision without such a separation (Lévi Strauss 2021). It's important to underscore that the ecological thinking proposed here is neither disjunctive nor projective; it distances itself from both anthropocentrism and anthropomorphism. Ecological thinking, in fact, is rooted in a relational conception.

In general, all forms of projective thinking can be viewed as manifestations of anthropocentrism. The allure of species narcissism is undeniable and has captivated numerous authors. While it's not accurate to assert that Western thought has completely dominated all other cultural forms, it's crucial to acknowledge the impact of globalized media and commodities, which have disseminated a specific image of the relationship between humans and the biosphere. This has, in turn, eroded or marginalized alternative cultural expressions (Tarnas 1991). As mentioned, the cultural distancing between humans and nature has deep roots, starting from Plato's influence on Western philosophy—after all, Alfred North Whitehead stated that all Western philosophy is a series of footnotes to Plato (Whitehead 1979). The concept of human dignity evolved along this trajectory, initially as the distinction and separation of humanity from the earthly realm, driven by an upward aspiration. Of course there have been dissenting voices, but they weren't many and often rejected anthropocentrism only partially. The prevalent interpretation is that of man as a universal, transcending species, with no Epimethean rank (Stiegler 1998; Belliti 2004). Consequently, freedom and autopoiesis are conceived as the absence of constraints.

In short, paradigmatic anthropocentrism is an uncritical celebration of humanity, fostering an almost boundless confidence in progress. In this perspective, through a mythopoetic interpretation of human history, the task and destiny of humanity become intricately tied to taming nature. The challenge in grasping the tenets of ecological thinking arises from this distancing vision of anthropopoiesis (Remotti 1996; Fabietti 2013): rather than being perceived as a process emerging from interconnected relations with the biosphere, it is framed as opposition to it. We struggle to adopt an eco-ontological framework precisely due to the anthropocentric landmines strewn across the terrain of our philosophical reflections. This challenge extends beyond ethical considerations about our conduct, but runs deep into the core of our very thought process (Deleuze and Guattari 1997; Žukauskaitė 2023; Haraway 2016). Before anything can truly change, we need to acknowledge three key points. First, human nature is merely one of the potential variations of animality—there is nothing ontologically separate or exceptional in our biological foundation. Second, the human condition does not arise solely from the autopoietic structure of *Homo sapiens*, but takes shape through hybrid relationships with otherness. Third, the anthropopoietic process itself is not disjunctive or vertical, as traditionally conceived, but hybridizing—formed through entanglements, co-productions, and ongoing exchanges with the non-human world.

5.4 Anthropocentrism and Animality

There is a very close relationship between anthropocentrism and the way animal others are interpreted, to the point that the two aspects are intimately connected. Paradigmatic anthropocentrism, in order to sustain itself, requires a devalued counterpart, used as a contrasting term. Returning to Agamben's considerations, I believe it would be more accurate to speak of an *anthropocentric machine* rather than an *anthropological machine*. If we examine the various works that attempt to highlight the ontological uniqueness of the human being—up to the latest reflections by Sloterdijk in the aforementioned essay—we always find an unavoidable confrontation with a supposed "animal" interlocutor, constructed ad hoc (i.e., not based on any real predicative homogeneity), in order to support a theoretical structure that ultimately relies on a *petitio principii*.

Which species is being referred to is never made explicit, just as it is claimed that the human being has always been immersed in the technosphere, without referencing any specific culture or historical era. This ignores the niche construction practices of other species and, above all, overlooks the fact that technology generally enhances our relationship with the world rather than rendering us impermeable and, most importantly, it always increases ecological dependencies. One of the most common strategies in these cases is *cherry-picking*: to highlight limited reactivity, one cites the tick; to emphasize dulled sensory perception, the foraging bee; to illustrate experiential difference, the bat. Similarly, Descartes claimed that animals were machines, but then refused to explain his model—he never showed the inner mechanisms of his supposed automaton—making his proposal impossible to falsify, conveniently hiding behind the complexity of a product created by God. Paradigmatic anthropocentrism relies on logic only in appearance; in reality, it is filled with hidden rhetorical devices. More broadly, we could say that paradigmatic anthropocentrism can be easily identified as the sense of superiority and specialness that drives humans to assert their excellence whenever relating to other beings.

Sometimes such an attitude, besides being blatantly puerile, is marked by obvious methodological errors. An example of this is when one claims to compare *Homo sapiens* not with another species—the chimpanzee or the bottlenose dolphin—but with an entire category: a fictitious entity, placed as contralateral to us, which we call "animal" (Derrida 2008). This is a very serious mistake from a descriptive point of view, since we cannot identify qualities that the many non-human species have in common. When one states that only the human being is *able to* do this or that, one presumes to know the *habitus* of all other species, an arrogant statement to say the least. In ethology, one would never venture to claim to have surveyed the entire canon of expression of all known animals. What is more, naturalists themselves assert that only a fraction of the species present on Earth have been classified. Tellingly, it is precisely those who do not know about animal behavior who most readily make these sweeping claims, showing once again the importance of Socrates' principle "I know that I know nothing." It is evident that any assumption based on the principle that only human beings are able to do this or that results in a tautology, underpinned by the wish to avoid scientific analysis (Acampora 2006).

But beyond descriptive considerations, comparing human beings to other animals involves a fundamental philosophical error, which in my opinion is even more serious. Indeed, it purports to compare: (1) a predicative species profile, i.e., how that particular species expresses itself from a morphological and behavioral point of view—in this case, the human species—with (2) a metapredicative profile, i.e. one that does not concern the species-specific or adaptive traits of a species, but the general condition of animality. It is not possible, in fact, to compare the human being, defined as peculiar, with the alleged "animal" category, which inevitably can never, for implicit reasons, assume a specific character. Not to mention that the human being is also an animal, so the comparison would be groundless to begin with. It is evident that the animal condition cannot present predicates, since each species is different, but only metapredicates. The animal category must thus be re-thought through a metapredicative lens, referring to transspecific connotations which relate to *what it means to be an animal.* However, the extensive use of the animal counterpart in humanist philosophy demonstrates, and is functional to, its disjunctive project.

This not only has ethical consequences but also influences the way in which human beings think of themselves, transforming animality no longer into the basic existential dimension—into the common home that allows for sharing and osmosis—but into a contralaterality for humanity to distance itself from. The refusal to see itself as one of the many expressions of animality leads humankind to experience a sort of existential alienation from the biosphere and nature more generally, which also produces ontological solipsism. In my opinion, this is a source of suffering and deprivation. I believe that the analysis of anthropocentrism is central to understanding the current crisis in the perception of the ecological problem, starting precisely from the way in which humans represent themselves in connection with the biosphere.

Belief in their ontological difference leads human beings to see themselves as guests in the biosphere—a kind of house that welcomes them and provides resources for survival, yet remains entirely foreign to them. This sense of estrangement, induced by anthropocentrism, encourages two attitudes: on the one hand, a sense of revenge, rooted in the idea that humanity has always had to fight against a cruel, punishing nature—the opposite of Eden—which turns our species into a David battling the bestial giant Goliath, as reflected in many cultural representations: Odysseus versus Polyphemus, Theseus and the Minotaur, the Calydonian boar hunt, Saint George and the dragon, to name a few. On the other hand, an attitude of neglect, based on the belief that it doesn't really matter if the entire biosphere is wiped out. This is evident in the general indifference toward the ongoing mass extinction of species (Kolbert 2014), accompanied by absurd fantasies such as projects to bring back Pleistocene animals, and, more broadly, the idea that any human-made product is preferable to a natural one—an idea that justifies the wholesale anthropoplastic transformation of the planet.

This sense of estrangement has become widespread. When humans wonder whether other forms of intelligence exist in the universe, it hardly ever occurs to them that a non-human intelligence may already be close at hand—in the many non-human species with whom we share this planet. These animals are treated as expendable, discarded like scrap paper or consumed as mere fuel for human metabolism—which is very much like using a masterpiece to stoke a fire, as Lev Tolstoy once wrote. Despite Darwin's work, this estrangement persists to this day in humanist culture, which still holds that humans have nothing in common with non-human others. Even when the theory of evolution is not denied outright—despite overwhelming evidence—there's a tendency to relegate the idea of continuity to a distant past, far enough not to challenge the self-image of human uniqueness. The humanist interpretation of evolution is that humans evolved *from* animals, not that humans *are* animals—that is, specific rather than special—ignoring the fact that accepting the theory of a common ancestor means recognizing that most traits of a species, including humans, did not originate in it but were inherited from a long line of ancestors and are thus shared with other animals. As paleontologist Neil Shubin writes in *Your Inner Fish* (2008), even today, humans carry behavioral mechanisms and morphological precursors that evolved in fish. The sense of estrangement fostered by humanist thinking stems from the lingering influence of essentialism, as noted by Ernst Mayr in *The Growth of Biological Thought* (1982). Without acknowledging the principle of commonality—the idea that the *Umwelten* of various species overlap and are not sealed-off monads—we will never truly grasp what we share with other living beings, remaining locked, in Cartesian fashion, in the misconception that we share only the *res extensa*.

To understand this sense of estrangement, let us take as an example the famous essay *What Is It Like to Be a Bat?* (2024)—first published by Thomas Nagel in 1974. In this work, Nagel doesn't question whether a bat can think or have a mind—in short, a perspective on the world—but whether we humans are capable of understanding it. Right from the outset, he describes the bat as a fundamentally alien form of life. Unsurprisingly, he later places it on the same level as a Martian or a robot, overlooking the fact that, as a mammal, the bat shares many traits with humans—psychological traits included, such as emotional responses. Nagel's more serious mistake lies in proposing a one-to-one identity between perception and experience, arguing that echolocation necessarily leads to a radically different experience of the world. But experience cannot be reduced to raw sensory input alone; if it were, we would have to conclude that a blind or deaf person holds a perspective fundamentally incompatible with that of a sighted or hearing person—a claim that is clearly refuted by neurophysiological research. For instance, in *Picturing the Mind* (2022), Simona Ginsburg and Eva Jablonka discuss the case of Daniel Kish, a man who lost his sight to retinoblastoma and yet rides mountain bikes, plays basketball, and can easily locate a golf ball on a course. How does he do it? He produces a series of clicking sounds and extracts spatial information from the returning echoes—just like a bat. As Ginsburg and Jablonka note, studies on Kish and another echolocator show that when they register a returning echo and report what they've perceived, their visual cortex is activated, offering detailed information about the objects' properties.

The main problem with Nagel's argument is the claim that a subjective experience cannot be translated into anything objectively interpretable. Tellingly, Nagel frames the question as imagining what it would be like *for oneself* to be a bat—resorting to analogy and projective thinking. He concludes that this kind of attribution of experience is only possible between humans. I've already discussed the limitations of analogical reasoning by referencing Erich Fromm, but Nagel's reasoning touches on an even deeper issue. No one disputes the subjective nature of experience or the partiality inherent in attempting to understand another's perspective—a difficulty that applies as much to bats as to other human beings, particularly when their life experiences differ greatly from our own. This is the longstanding challenge of empathy, which, while grounded in neurophysiological mechanisms, also develops through conscious decentering. But Nagel's real question is not about ethological knowledge of the bat—it's about the subjective point of view, which, by definition, cannot be reduced to an objective report. And on that, I agree. What I do not agree with is his assumption that analogy is the only viable method for grasping another being's perspective. This methodological stance leads to an inevitable flattening of all difference, placing the bat and the Martian on equal footing. Yet while the Martian is truly alien to me, the bat is only partially different. Therefore, the proper approach should not be analogical but *homological*—based on shared qualities. Having traits in common does not make the bat's perspective fully objective, but it does make it participatory: I can partake in the bat's fear, not because of how it behaves, but because we share an emotional response to certain events. In my view, the analogical approach is anti-scientific. It overlooks both the human ability to investigate other species from an ethological and neurophysiological perspective, and the substantial shared characteristics between species—an oversight that flies in the face of the Darwinian paradigm.

Let us return, then, to Nagel's paper, which more broadly refers to the human capacity to imagine the existential condition of another species. Certainly, we cannot fully grasp what it feels like to be a bat—but we can form a reasonably accurate approximation. Other species are not alien to us; they simply inhabit a space of partial difference. In other words, every species presents a unique mix of similarity and divergence in relation to humans, proportionate to its phylogenetic proximity and the extent to which it shares a lifestyle and environment with us. Species-specific *Umwelten* are not isolated monads; they show substantial overlap—whether through inherited traits (*homologies*) or adaptive convergences (*analogies*)—which help us understand much of another species' dispositions. In this sense, we have shared foundations, a kind of ontological Rosetta Stone, which enables us to adopt an empathic approach to other species. This is especially true because animal life shares certain metapredicates: sensibility, pleasure, the search for food, instincts, and drives—all useful starting points for understanding the specific expressions of these traits in different species. Moreover, as mentioned, in the case of the bat, we are dealing with a mammal. Even though it possesses a unique perceptual system (echolocation) and the ability to fly, it still has a mammalian neurobiological structure—one that is, to a large extent, comparable to our own.

Ultimately, the work of an ethologist consists precisely in the effort to understand what it feels like to be a bat—that is, to frame the perceptual, emotional, motivational, and cognitive conditions of a subject belonging to another species. Of course one can never be fully certain, but today we do have a great number of techniques that allow us to investigate an animal's psychological states. These include: (i) specific behavioral tests, such as those for false belief, preference, differential reactivity, and violated expectation; (ii) analysis of the effects of certain neuromodulations, for instance observing behavioral changes after administering oxytocin; (iii) neural imaging techniques, such as functional MRI, to observe which brain areas are activated during a given behavior, in order to draw comparisons; (iv) ethogram charts, which document how a species carries out common functional cycles like foraging and reproduction; and (v) problem-solving tests, which reveal how an animal deals with specific challenges. Even if at first glance a bat may seem like a completely alien entity, that doesn't mean it truly is. It simply means that understanding a bat requires a counterintuitive approach—but it's the same kind of challenge we face when discussing the Copernican universe or Einsteinian relativity. Moreover, it's not necessarily the case that we are better at "decoding" animals that seem closer to us—like dogs, with whom we share daily intimacy—just because they feel more familiar.

This sense of alienness leads to two dangerous outcomes. First, the increasingly common idea that since we don't know what it's like to be a bat, any discussion about its potential status as a moral subject becomes irrelevant—lacking any moral weight. Admittedly, Nagel didn't say this himself, but many have used his argument to question the legitimacy of moral consideration for other species. Second, this perspective shuts down any possibility of connection or shared understanding between humans and other forms of life. This attitude is far more dangerous than it may appear at first, because it treats any claim about animal otherness as a projection—an anthropomorphism built on the shaky ground of analogy. And, even more damaging, it seals an unbridgeable divide between the human and other species, condemning us to ontological solipsism. This feeling of human loneliness, deeply rooted in Western culture, does nothing to help us grasp the intimate consubstantiality of the human within the biosphere—our shared experience and, most importantly, our shared fate. Humanity often believes it has a lifeboat at the ready, and that whatever dimension it may project itself into, or whatever world it may come to inhabit, it will remain always and fundamentally the same. This mindset—the very antithesis of ecological thought—leads us to believe that human predicates are the self-generated fruit of an autonomous essence, something we can carry with us like a toolkit of intrinsic qualities. It's from this idea that some of humanity's worst tendencies arise—not only neglect toward the biosphere, but even hostility toward it.

5.5 Anthropocentrism and the Human Condition

Anthropocentrism—in its disjunctive and projective facets—hinders the acknowledgment of the hybrid and copulative nature of human beings. This amounts to the mistaken interpretation of a process that has consistently characterized the entirety of human history. In essence, throughout the anthropopoietic journey, humans have consistently engaged in hybrid relationships with otherness. Simultaneously, there has been a tendency to deny these connections, attributing all credit solely to humans as the exclusive architects of cultural advancements. Never mind the arrogance implicit in such an approach—what worries me is the limitations it imposes. Presuming that one's dimension stems from an autonomous condition, an autopoietic event, and an autarchic evolution hinders the recognition of anthropopoiesis as a relational and hybrid process. This isn't solely an issue of explaining cultural phenomena but also a disregard for the significance of the ontopoietic relationship with the biosphere in the existential aspect of the human being. From a broader perspective, this translates into an inability to grasp the poietic dimension of relationships.

In ideological anthropocentrism, human beings not only perceive and depict themselves as distant from the biosphere but also envision their ontopoietic journey as a process of estrangement from nature (Sloterdijk 2011). Rejecting the criterion of specificity for that of speciality, humanity avoids all forms of sharing and rootedness, claiming to be exempt from natural constraints and free from all phylogenetic dimensioning. This propels anthropocentrism toward absolutes, exacerbating already compromised relational conditions. Human beings thus perceive themselves as immersed in a transcendent condition, contrary to the natural one, which they reject and avoid. They prefer emancipation from the biosphere over establishing a relationship with it (Loreau 2023; Kesebir and Kesebir 2017; Peck 2023). The compensatory view of culture and technology, seen as a way to make up for humanity's supposedly incomplete nature, transforms the human being into a citizen of its own cosmopolis claiming no connection to the biosphere. This legacy from the modern age poses a significant obstacle to fully understanding the bonds that connect us to all other living things.

Humanist anthropopoiesis, in fact, relies on the alleged natural incompleteness of the human being, famously depicted in Plato's *Protagoras*. This notion allows for humanity's release from any connection to animality and the celebration of its totipotence. In addition to this, there is the mythical interpretation of man as an entity who, being weak and cast out from nature, must rely solely on his wits. Consequently, an image of the human being is affirmed that strengthens this divergence from the biosphere: while other species live within their natural condition, the human being exists in the cultural dimension (Pansera 1990). Since the latter is considered the result of human ingenuity alone, the anthroposphere is transformed into a kind of separative amniotic shell, where humanity is only confronted with itself. Hence a form of ontological autarchy of humanity, which allegedly produces its own technical and cultural expressions in a state of total self-sufficiency. Ideological anthropocentrism operates as a theory of separation, standing in stark contrast to the relational perspective of ecology. Ignoring this dynamic makes it challenging to comprehend our apparent inability to recognize the shared destiny that ties us to the biosphere (Sheldrake 2021). Anthropocentrism hinges on a specific conception of the human being as a transient entity on the planet, with no actual roots in it— this is exemplified in the idea of the human soul as imprisoned in the body, or of humanity's earthly condition that aspires to an angelic dimension, often seeking purification rituals or relying on technological apparatuses.

Ideological anthropocentrism carries significant epistemological consequences, such as viewing technology merely as a tool for amplification and operational enlargement, overlooking its reorganizing effects on the human system. All these factors contribute to the choices that humans, influenced by distinct social and economic dynamics, decide to prioritize. It's crucial to recognize anthropocentrism's almost titanic effort to construct an alternative dimension (Sloterdijk 2011). This model has characterized the modern age, yielding magnificent and commendable achievements, but it has also brought about inevitable problems, which are more and more evident today (as explored in Chaps. 1 and 2). While the anthropocentric trend concerns the human being in general and is discernible in all cultures, anthropocentric extremism is undoubtedly an offshoot of the Western model (Mignolo 2011). The ongoing crisis affects all populations on the planet, but is primarily an expression of a specific culture that has elevated man to the absolute master of the biosphere, repudiating any form of relational contract with it (Fenner 2022; Weir 2023).

Now, humanism has not developed its anthropocentric perspective *ex nihilo* but by emphasizing certain connotations of the human being, such as the tendency toward technical mediation and its anthropoplastic conversion of natural entities (as highlighted by processes like agriculture and domestication). For this reason, it would be equally wrong not to take into account certain characteristics of our species that, if mythologized and incorrectly channeled, can themselves represent a problem with ecological repercussions. While the prevalence of technical intelligence in our species is undeniable, the exaltation of Prometheanism has favored the tendency to project ourselves elsewhere and to create boundaries between us and others (Jorjani 2020). Undoubtedly, this attitude has characterized certain cultural forms more than others, but it is common to all human beings. Therefore, to circumscribe the problem of the ecological crisis only to cultural aspects is incorrect.

We must not forget that even before the Neolithic revolution, our species was responsible for the destruction of the Pleistocene megafauna through a truly devastating hunting impact. We are talking about a small mass extinction wrought by humans long before the Neolithic era sanctioned the takeover of the environment from which the great civilizations arose. Still, over time, technical advancement led to the process of anthropoplastic conversion, whereby any natural entity acquires value if and to the extent that it is translated into a human product (Patel and Moore 2017). Anthropoplasty, i.e., the conversion of elements of biodiversity into entities reflecting humanity and favoring human affordances (Gibson 1979), has not only an ergonomic significance but also an aesthetic one. We could call it a process of domestication of the biosphere capable of lowering the terrifying sublime into a more reassuring beauty. It is also a conversion of value perception: the natural entity is devalued in comparison to the man-made (Foster 2022; Costa 1990). People commonly show more devotion to a computer than to an organism, despite the fact that even the most complex computer system is functionally primitive compared to a cellular organelle. The destruction of the planet, on closer inspection, is a work of conversion.

Humanism has accustomed us to view ourselves as detached from nature, starting from the notion that we belong to it only partially to begin with. This led to the understandable consequence of considering the biosphere, at most, a stage for humanity, condemned to obsolescence in the face of advancing technology (Nye 1996). The automobile replacing the horse, the tractor replacing oxen, and chemistry taking the place of manure are striking icons of the anthropoplastic conversion of the twentieth century. The anthropocentric imperative is to make our natural home more consistent with human goals. The entire biosphere then becomes a resource to be translated into a human product. If only what can be brought back to humans has value, not only is everything else a resource, but it is our task to speed up the conversion because a technology is deemed more valuable than a species, a city more than a forest, and a work of art more than an ecosystem (Marchesini 2024). It is not just a matter of handling resources sustainably but of recognizing how anthropocentric culture influences the representation of the forces at play and the values that enshrine their prevailing dynamics.

Humanist anthropocentrism generated the archetype of distancing and simultaneously constructs a hierarchical pyramid of entities based on their presumed gradient of distance from nature. Proximity to nature is thus stigmatized in the image of the wild man, in the pre-Adamic conception of Native Americans, and in the fantasies of werewolves and vampires. Humanism casts a negative light on the forest and elects the wolf as the stereotype of all evil. From an anthropocentric perspective, ecological thinking is impossible because anthropocentrism is founded on the disjunction of man from nature and the assertion of man's greater importance over the latter (Sherwood Taylor 1948). When we affirm the dignity of humanity, we inevitably discredit everything we consider non-human, deeming it worthless or even a bearer of disvalue, implicitly prompting the anthropopoietic conversion. To cut down a forest or reclaim a swamp is to transform a reality that is perceived as having no value into something that does, beyond the economic significance it may assume (Barbier 2014). If the human being always takes precedence, it is inevitable that every corner of the planet should be turned into land for cultivation or grazing livestock. The principle remains that of anthropomorphizing every recess of the world, destroying everything that resists conversion (Carson 2021).

In short, the culture of disjunction cannot comprehend *ecological thinking* because it is rooted in its very opposite—the predicative separation of entities. And this trajectory was the inevitable consequence of the seeds already present in early humanism. In fact, extracting the human being as a unique and ontologically distinct entity required a refusal to even contemplate relational qualities. Any such concession would have reintegrated man into the biosphere, whereas it was crucial to consider human predicates as emanations of the human essence alone. So can ecological thinking exist within anthropocentrism? My response is negative—it cannot. Anthropocentrism is grounded in human exceptionalism and predicative essentialism. Ecological thinking, instead, does not isolate entities; it acknowledges their primary characteristics and their impact on the whole, viewing them as beings that cannot exist outside their network of relationships.

Essentialism is the foundational principle of anthropocentrism, aiming to purify human beings so as to elevate them above all else, and serves as the primary obstacle to fully grasping ecological thinking. It interprets the predicates of a being through the lens of autonomy and emanation, and views every relationship as something external or even contaminating, akin to dirtiness and inauthenticity that calls for purification. Humanism is anthropocentric not only because it places man at the center but also because it asserts man's foundation *iuxta propria principia*, falling into predicate essentialism. This was noted by Gaston Bachelard in his essay *La formation de l'esprit scientifique* (2002), where he identified the anthropocentric perspective as an epistemological obstacle. Bachelard also recognized essentialism as a form of projective thinking that impedes the development of the scientific mindset. Unfortunately, there are still too many anthropocentric resonances that hinder a proper understanding of the relational significance of predicates. Anthropocentrism is not merely an affirmation of ontological discontinuity but, above all, a way of interpreting interactions through a logic of impermeability, claiming the self-sufficiency of different entities.

Now, after Darwin, the notion of harmony between beings has become absurd, not only due to the continuous transformations of biocenosis but also because evolution itself fosters conditions of hybridization among the living (Margulis 1998). Indeed, we could argue that the transformation of species is grounded in the very negation of concepts like perfection and purity, the prerequisites for disjunction and self-sufficiency. The biosphere, with its myriad interactions and evolutions, functions as an organism, structured on matrices of relationships that integrate metabolic functions to maintain specific internal dynamics. Indeed, the term "ecology" itself can be misleading, as it's not just about maintaining one's own home (*oikos*) but the entire body of the living, of which humans are an integral part. If we fail to grasp the implicitly connective principle of life, attempting to disjoint it in order to isolate a *primum movens*, we'll only perceive inert matter, and life will elude our field of observation.

References

Acampora, Ralph A. 2006. *Corporal Compassion: Animal Ethics and Philosophy of Body*. Pittsburgh: Pittsburgh University of Pittsburgh Press.

Agamben, Giorgio. 2003. *The Open: Man and Animal*. Stanford: Stanford University Press.

Bachelard, Gaston. 2002. *The Formation of the Scientific Mind*. Bolton/Manchester: Clinamen Press.

Barbier, Edward B. 2014. Economics: Account for Deprecation of Natural Capital. *Nature* 515:3233.

Bekoff, Marc. 2004. Wild Justice and Fair Play: Cooperation, Forgiveness, and Morality in Animals. *Biology and Philosophy* 19 (4): 489–520.

Belliti, Daniela. 2004. *Epimeteo e il Golem: Riflessioni su uomo, natura e tecnica nell'età globale*. Pisa: ETS.

Bozzi, Paolo. 1998. *Fisica ingenua. Studi di psicologia della percezione*. Milano: Garzanti.

Brambell, Frederick W. Rogers. 1965. *Report of the Technical Committee to Enquire into the Welfare of Animals Kept Under Intensive Livestock Husbandry Systems*, Cmnd. 2836. London: Her Majesty's Stationery Office.

Carson, Rachel. 1962. *Silent Spring*. Boston: Houghton Mifflin.

Carson, Rachel. 2021. *Man's War*. New York: Penguin.

Costa, Mario. 1990. *Il sublime tecnologico*. Salerno: Edisud.
Deleuze, Gilles, and Félix Guattari. 1997. *A Thousand Plateaus: Capitalism and Schizophrenia*. Minneapolis: University of Minnesota Press.
Derrida, Jacques. 2008. *The Animals that Therefore I Am*. New York: Fordham University Press.
Fabietti, Ugo. 2013. *L'identità etnica*. Roma: Carocci.
Fenner, Felicity. 2022. *Curating in a Time of Ecological Crisis*. London: Routledge.
Foster, John Bellamy. 2022. *Capitalism in Anthropocene*. New York: Monthly Review Press.
Fromm, Erich. 1992. *The Anatomy of Human Destructiveness*. New York: Holt Paperbacks.
Galton, Francis. 2018. *Hereditary Character and Talent*. Greenwood: Suzeteo Enterprises.
Garin, Eugenio. 1976. *Italian Humanism*. Westport: Praeger.
Gibson, James J. 1979. *The Ecological Approach to Visual Perception*. Boston: Houghton Mifflin.
Ginsburg, Simona, and Eva Jablonka. 2022. *Picturing the Mind: Consciousness through the Lens of Evolution*. Cambridge, MA: The MIT Press.
Griffin, Donald. 2001. *Animal Minds*. Chicago: University of Chicago Press.
Haraway, Donna Jeanne. 2016. *A Cyborg Manifesto*. Minneapolis: University of Minnesota Press.
Harrison, Ruth. 1964. *Animal Machine*. London: Vincent Stuart.
Heller, Ágnes. 2009. *Per un'antropologia della modernità*. Torino: Rosenberg & Sellier.
Herrnstein, Richard J., and Charles Murray. 1994. *The Bell Curve: Intelligence and Class Structure in American Life*. New York: Free Press.
Hirsch, Fred. 1976. *Social Limits to Growth*. London: Routledge.
Hughes, Howard C. 1999. *Sensory Exotica*. Bradford: Bradford Book.
Jorjani, Jason R. 2020. *Prometheism*. Budapest: Arktos Media.
Kesebir, Selin, and Pelin Kesebir. 2017. A Growing Disconnection From Nature Is Evident in Cultural Products. *Perspectives on Psychological Science* 12 (2): 258–269.
Kolbert, Elizabeth. 2014. *The Sixth Extinction: An Unnatural History*. New York: Henry Holt and Company.
Kuhn, Thomas. 1957. *The Copernican Revolution*. Cambridge, MA: Harvard University Press.
Lasch, Christopher. 1979. *The Culture of Narcissism*. New York: W.W. Norton & Company.
Lévi Strauss, Claude. 1963. *Totemism*. Boston: Beacon Press.
Lévi Strauss, Claude. 2021. *Wild Thought: A New Translation of "la Pensee Sauvage"*. Chicago: University of Chicago Press.

Loreau, Michel. 2023. *Nature That Makes Us Human*. Oxford, UK: Oxford University Press.
Marchesini, Roberto. 2017. *Over the Human: Post-humanism and the Concept of Animal Epiphany*. Cham: Springer International Publishing.
Marchesini, Roberto. 2024. *Posthumanist Manifesto*. New York: Lexington Books.
Margulis, Lynn. 1998. *Symbiotic Planet: A New Look at Evolution*. New York: Basic Books.
Mayr, Ernst. 1982. *The Growth of Biological Thought: Diversity, Evolution, and Inheritance*. Cambridge, MA: Belknap Press of Harvard University Press.
Mignolo, Walter D. 2011. *The Darker Side of Western Modernity: Global Futures, Decolonial Options*. Durham: Duke University Press.
Nagel, Thomas. 2024. *What Is It Like to Be a Bat?* Oxford: Oxford University Press.
Nye, David E. 1996. *American Technological Sublime*. Cambridge, MA: MIT Press.
Pansera, Maria Teresa. 1990. *L'uomo progetto della natura: L'antropologia filosofica di Arnold Gehlen*. Roma: Studium.
Patel, Raj, and Jason W. Moore. 2017. *A History of the World in Seven Cheap Things*. Oakland: University of California Press.
Peck, Steven W. 2023. Five Reason Why We Are Not Separate From Nature and Why It Matters. *Journal of Living Architecture*. https://livingarchitecturemonitor.com/articles/five-reasons-why-we-are-not-separate-from-nature-sp23.
Piaget, Jean. 1970. *Psychologie et épistémologie*. Paris: Gonthier.
Plessner, Helmuth. 2019. *Levels of Organic Life and the Human*. New York: Fordham University Press.
Regan, Tom. 2004. *The Case for Animal Rights*. Berkeley: University of California.
Remotti, Francesco. 1996. *Contro l'identità*. Bari: Laterza.
Ries, Julien. 2002. *Man and the Sense of Mystery*. New York: Chelsea House Publishers.
Ruesch, Hans. 1986. *Naked Empress or the Great Medical Fraud*. Zürich: CIVIS Publications.
Sheldrake, Merlin. 2021. *Entangled Life: How Fungi Make Our Worlds, Change Our Minds and Shape Our Futures*. New York: Random House.
Sherwood Taylor, Frank. 1948. *Man's Conquest of Nature*. London: Paul Elek.
Shubin, Neil. 2008. *Your Inner Fish: A Journey into the 3.5-Billion-Year History of the Human Body*. New York: Knopf Doubleday Publishing Group.
Singer, Peter. 1975. *Animal Liberation*. New York: New York Review/Random House.
Sloterdijk, Peter. 2011. *Bubbles: Spheres I: Microspherology*. Los Angeles: Semiotext(e).
Sloterdijk, Peter. 2016. *Not Saved: Essays After Heidegger*. Malden: Polity Press.
Stallard, Michael Lee. 2017. *Connection Culture: The Competitive Advantage of Shared Identity, Empathy, and Understanding at Work*. Alexandria: Association for Talent Development.

Stiegler, Bernard. 1998. *Technics and Time: The Fault of Epimetheus*. Stanford: Stanford University Press.

Tarnas, Richard. 1991. *The Passion of the Western Mind*. New York: Crown Publications.

Von Uexküll, Jakob. 1956. *Streifzüge durch die Umwelten von Tieren und Menschen: Ein Bilderbuch unsichtbarer Welten. Bedeutungslehre*. Hamburg: Rowohlt.

Weir, Lucy. 2023. *Philosophy as Practice in the Ecological Emergency*. Cham: Palgrave Macmillan.

Whitehead, Alfred N. 1979. *Process and Reality: An Essay in Cosmology*. New York: Free Press.

Wolpert, Lewis. 1994. *The Triumph of the Embryo*. New York: Oxford University Press.

Wolpert, Lewis. 2000. *The Unnatural Nature of Science*. Cambridge: Harvard University Press.

Žukauskaité, Audrone. 2023. *Organism-Oriented Ontology*. Edinburgh: Edinburgh University Press.

CHAPTER 6

The Phenomenology of Life

6.1 Premise

Life has often been approached through an essentialist lens, using a descriptive and explanatory dichotomy that pits a reductionist conception, which limits the biological phenomenon to physical-chemical processes, against a vitalist one that posits a mysterious "life force" to explain the peculiarities of living organisms (Driesch 2020; Bergson 1911; Teilhard de Chardin 1959). At the root of this misunderstanding lies the disjunctive way of considering life, claiming to describe and explain it synchronically and through internal examination alone. To truly understand life, on the contrary, we must consider its manifestation beyond what an entity presents internally—i.e., beyond what can be deduced from its internal description. As I have tried to argue throughout this book, in fact, life is above all a relational process that unfolds across multiple layers of connection, correlation, exchange, and ontopoiesis. This multi-relational dimension is what I call *eco-ontology*: a mode of being-through-the-world that cannot be grasped by disjunction or through a single, binary lens. Every aspect of life's phenomenology carries an intentional quality—it always points beyond itself, toward an absence or ellipsis, something at once essential and elusive.

The most obvious aspect, falling under descriptive epistemology, is that being-alive inevitably involves a continuous exchange of energy and matter with the world. Its predicative complexity can be realized (not merely

R. Marchesini, *Philosophical Ecology*,
https://doi.org/10.1007/978-3-032-04987-2_6

sustained) only insofar as it is *not* a closed system in thermodynamic equilibrium (Schrödinger 1944). Likewise, life cannot be fully understood in a synchronic way, because its being points to something that precedes it—and, in manifesting, it has already moved beyond the present. As a result, it never truly appears under our microscope, except as evanescence. Life, therefore, cannot manifest within the boundaries of a fixed definition, whether spatial or temporal, because every expression of the living refers to a transcendental dimension that defines its experience.

As mentioned, the expression of the living always refers to a prior time—a preconstitution that makes experience itself possible. This experience is understood as an intrinsic correspondence between life and the world, a feature central—though not fully resolved—in Husserl's analysis of the lifeworld, particularly in its pre-given and constitutive nature (Sowa 2008). Every instance of life's manifestation rests on a correlative predicate. We might therefore conceive of life as a chiasmatic singularity—the result of a dialectical convergence of multiple factors that integrate reciprocally. For example, the empirical gives rise to the transcendental, and vice versa (Merleau-Ponty 2013).

We cannot do without this dialectical dimensionality of life, but neither can we reduce life to the product of an encounter between ontologically separate entities—i.e., of external relations not already implied in their internal structure. Likewise, life cannot be described as something that unfolds along a single, dichotomous, and isochronous axis, as this would inevitably lead us into an infinite regress in search of some origin of life. The phenomenology of life cannot be separated from the connective process through which it manifests itself (*onto-phainresis*) and becomes (*onto-poiesis*), because its very constitution already includes, inherently, the predicative complement of an apparently external factor. Life, in fact, is based on external reference, both from an interactional perspective (external elements enabling its manifestation, or extensive heteronomy) and from a temporal one (prior information granting it a deeper, diachronic layer beyond present factors, or intensive heteronomy). To understand heteronomy—one of the most important features of the phenomenon of life—it is helpful to refer to Brentano's principle of intentionality, extending its application to capture life's relational ontology (eco-ontology): that is, its being-referred to a necessary content.

Life's intentionality does not stem from an inherent lack that must be compensated by something external. Instead, it arises from a kind of systemic redundancy—an evolutionary strategy through which life has increased in complexity by anchoring itself to the referents it finds in the world. The phenomenology of life has made dynamic disequilibrium its ontological ground; in this sense, life can be seen as driven by a desiring tension that constantly pushes it to go beyond itself (Marchesini 2018; Barbaras 2022). Seeking the essence of life—some ultimate domain that explains the living being's predicative unfolding, its beating heart—is ultimately futile. Life's very phenomenology lies in its resistance to being centered or reduced to a fixed point of origin.

We are inevitably drawn to investigate the empirical individuality of living beings as they move through the world, guided by a principle of implicit intentionality—one that proves difficult to grasp amid the plurality of life forms. This intentionality is mirrored in a plural consciousness, which we can approach only through homological gradients and levels of homology. Analogy, as Thomas Nagel (2024) warns, can be misleading; but reductionism, on the other hand, either prematurely dismisses or blurs what we seek to understand. This dynamic of conjunction—where the individuality of the living becomes entangled with the world—leads to a form of individuation that is never the realization of a fixed Self, but rather its ongoing transfiguration. Life thus appears not as a noun, but as a verbal predicate—an unfolding action.

The desiring character of life—understood as a projective tension that both grounds the subject in the here and now and simultaneously propels it beyond itself—expresses the phenomenality of the living: its predicative and creative impulse. This emergent condition, while grounded in physicochemical processes, cannot be reduced to them. This compels us to move beyond the dualism inherited from Gnostic and Cartesian traditions—a dualism that has turned the phenomenology of life into a merely extensional, and thus measurable, external object, effectively excluding it from genuine philosophical inquiry. This reduction—this uprooting of the human from life—has resulted in a condition of existential isolation. It manifests in Heidegger's notion of *being-toward-death*, a state of alienation from nature that, on one hand, brings with it a profound sense of despairing solitude, and on the other, frees the human from any perceived debt to the non-human world. What's more, it lifts moral responsibility toward the biosphere—an ethical rupture poignantly examined in Hans Jonas' *The Phenomenon of Life: Toward a Philosophical Biology* (2001).

In Jonas—as in Husserl and Bergson—we find both the urgency and the difficulty of grasping the transitive nature of life's phenomenology: the sense that, under the lens of reductive biologism, life vanishes from the biologist's slide, leaving behind only a kind of fossil—a structural trace of what once was, an echo of its former vitality. Life, in fact, reveals itself as freedom—not as the absence of constraints or an empty expanse, but as a convergence of possibilities. The space of freedom through which every living being moves is also a principle of knowledge and ontopoiesis. What we typically regard as two distinct processes—identity as something fixed, and encounter as a merely empirical phenomenon—is, in fact, the ontological dimension of life itself. Life is, by its nature, metamorphic: a transition between states, an openness to hybridization. It arises from the interweaving of different temporalities—simplified, perhaps, as past and projection. Yet even these temporal axes are plural in constitution. There is no *a priori* transcendental subject; instead, we encounter regressions into multiple moments that reach into the remote depths of life, forming an indivisible continuum. Similarly, projection into the *not-yet* is not the act of an isolated individual casting itself into the future—as though between two definable points in time—but a diachronic unfolding that reveals the potentialities life already harbors.

Morphopoiesis in living beings is thus not a mere adaptation to external conditions, but a creative appropriation of what the world offers—a way of expressing oneself. The thermodynamic disequilibrium of life is not accidental; it is intrinsic to life's phenomenology. It embodies life's inertia, its tendency to resist stability and remain in motion, escaping any final form. Life is, above all, the manifestation of a conjunctive principle. So the attributes we associate most strongly with it—sentience, affectivity, subjectivity—cannot be understood outside of a relational framework. As said, the phenomenon of life cannot be grasped through an essentialist or synchronic lens that seeks to understand its meaning by analyzing its internal content in the here and now. Doing so risks reducing it to the paradoxically amorphous polymer—first described by James Watson and Francis Crick in 1953—that resembles a library more than a librarian.

Drawing on key contributions from contemporary phenomenological philosophy (Tymieniecka 1990; Henry 2008; Barbaras 2022; Marion 2007; Verducci and Küle 2022), this chapter explores how the phenomenology of life can illuminate the ontopoietic relationships among living beings.

6.2 What Is Life?

The phenomenon of life can be described from several perspectives:

(i) From a thermodynamic standpoint, as a distinctive physical process marked by its negentropic tendencies and capacity for energy dissipation (Schrödinger 1944; Prigogine and Stengers 1977);
(ii) In terms of its morphological composition, i.e., the components that characterize it—especially nucleic acids and polypeptides—and the relationships between them (Kuriyan et al. 2013; Pasternak 1998);
(iii) Through its physiological dynamics, with reference to functions such as metabolism and reproduction, which characterize it most fundamentally (Boyd 1993; Newman 2017);
(iv) Via its evolutionary mechanisms, the processes that enable life to propagate and diversify while retaining core characteristics, yet adapting to varied environmental contexts (Pigliucci and Müller 2010);
(v) In terms of ecological systematics, that is, the relationships organisms establish with both the abiotic environment and with other living beings (Cain et al. 2011);
(vi) And finally, through the lens of biodiversity development, looking at the multitude of forms life has taken (Wilson 1986; Westheide and Rieger 2007).

This multiplicity of descriptive and explanatory levels was notably emphasized by Ernst Mayr in *The Growth of Biological Thought* (1982), through his distinction between proximate and ultimate causes. A similar multidimensional approach appears in Nikolaas Tinbergen's explanatory tetrarchy—commonly known as the *Four Questions* (1963)—which analyzes animal behavior through multiple lenses. Together, these frameworks highlight the inherent difficulty of offering a linear explanation of life, due to the dense web of causal entanglements, recursive dynamics, emergent phenomena, and singularities that define the living world.

According to Maurice Merleau-Ponty, particularly in his essay *La Nature* (1995), the concept of nature always expresses an ontology and serves as one of its privileged expressions. For Friedrich Schelling, nature is not a being-in-itself but an eternal unity—a being united by an original and absolute necessity, which the German philosopher calls *copula* (1967). This vision of entanglement is fundamental for understanding the phenomenology of life, both in its continuous state of tension (Barbaras 2022) and in its realization through conjugation (Tymieniecka 1990).

Life surrounds us and simultaneously co-constitutes us, given that our very organismic dimension is the product of a universe of symbioses. Our body is a true biome, whose physiological emergence only minimally reflects our genome. Life *is* our body, which we can experience from within because, as Edmund Husserl suggested, the *Lebenswelt* (1986) is something that concerns us intimately. Starting from this somatic condition, we can grasp the meaning of eco-ontology—that is, an ontology based on relation, emerging from relation, and impossible without it. The body itself is a complex interweaving and intersection of moments that involve not only mutual dependence but also predicative translation. Simply put, the predicate that the living manifest is never the emanation of an essence; it is always the emergent product of a crasis.

The evolution of living beings is characterized by continuous interactions among organisms, so the key principle for understanding the phenomenology of life must be found in the relational factor. This factor explains several aspects of living beings, for example their intimate primary constitution, which is based on relational dimensions that maintain an inherent microcosm organized through an internal relational system and interface with the external world via relational thresholds—such as the cytoplasmic membrane of a cell. It also accounts for the fractal formations that arise from this primary constitution by reproducing the same pattern of internal-external relationships at larger scales, such as an organism, a population, a niche, an ecosystem, or a biome, all founded on selective interfaces. The relational factor also explains the living's inclusive morphopoiesis, which occurs by translating external relations into internal ones—for example, by incorporating selective pressures into a species' conformation or by integrating the developmental environment into the organism's differential growth processes. Finally, it is the reason for life's symbiotic physiology, where a living being cannot satisfy its biological needs by simply exploiting external elements as mere resources.

In short, *the relational factor forms the basis of the ontology of life*—both in its transcendental definition, which precedes and enables experience as seen in the innate, and in its empirical expression. Relation is an ontopoietic principle that extends not only spatially but also temporally, because phylogenetic inheritance should not be understood as a simple legacy but as an unbroken continuum of presence in the world. This transcendental condition that makes living beings prior to and enabling of experience does not arise from an absolute but from the ontological continuity of life—the having-already-been-there inherent in every living being that comes to life.

Life as a process of individuation must never make us forget that the individual does not arise from nothing but is an expression of a continuum, so the search for a beginning leads only to an infinite regress. Life is like a time machine that brings the past into the present; at the very moment a human begins their existence, they carry ghostly presences of the past with them, bringing them back to life. We could say, then, that to be born is to revive fragments of the past—so that each of us, with our birth, makes present distant times scattered across four billion years of life's history on Earth. Eco-ontology, therefore, refers not only to the ontopoietic relations that define the (co)existence of living beings but also to the process—which I find both amazing and moving—by which every birth, in a sense, subsumes the history of life. It is, of course, a partial and singular recapitulation, yet remains deeply fascinating and meaningful from an ontological perspective.

As already mentioned, it is often said that science has taken poetry out of the world, but I completely disagree. If we simply compare mythological or creationist explanations with the accounts provided by evolution and paleontology, I must conclude that human imagination does not even come close to that of nature. The story of life on our planet—starting with those first cyanobacteria that enriched the Earth's atmosphere with oxygen about 2.4 billion years ago, and continuing through the evolution of all living forms across the various geological periods—is undoubtedly the most beautiful and fascinating tale imaginable. I believe we should all read this story carefully and be grateful to the great scholars who have helped us deepen our understanding of the biosphere.

The natural sciences have a long tradition dating back to Aristotle's *Historia animalium* in the fourth century BC and Pliny the Elder's *Naturalis historia* in the first century AD. This tradition continued into the modern age with important figures such as Francesco Redi, Maria Sibylla Merian, and Carl Nilsson Linnaeus. Interest in the variety of the animal world and its symbolic meanings first appears in the Greek work *The Physiologist* and then in the earliest illustrated bestiaries of medieval times, such as the *Liber monstrorum de diversis generibus* (eighth century), which focused on fantastic animals. Actual bestiaries followed from the thirteenth century onward, including the *Aberdeen* and *Ashmole* bestiaries. Authors of bestiaries that combined real species and fantastic creatures include Ulisse Aldrovandi with his *De animalibus* and Conrad Gessner with his *Historiae animalium*. Biodiversity was regarded as a quintessential collection, which led to the rise—especially between the sixteenth and eighteenth centuries—of the so-called *wunderkammer* or chambers of wonders. These were receptacles of scientific curiosities and can be seen as the precursors to modern natural science museums.

Indeed, the natural sciences have confronted humanity with the vastness of the *Lebenswelt*, the world of life, which is why I disagree with those who view science merely as reductionist, objectifying, or even eliminativist. Scientific thought has expanded our perspective on the world and helped to dismantle the anthropocentric view. It has revealed that the living world is far greater than anything we can imagine or conceive. From the eighteenth century onward, the work of explorers like Alexander von Humboldt and geologists such as Louis Agassiz and Charles Lyell significantly advanced our understanding of biodiversity. They built upon earlier contributions by scholars like Georges-Louis Leclerc de Buffon, Jean-Baptiste de Lamarck, and Georges Cuvier. The idea of an Eden designed for human beings was replaced by the view of life as a dynamic and unfolding process—a perspective that laid the groundwork for the counterintuitive notions of transformism and, later, the theory of evolution. It was Charles Darwin who, with *On the Origin of Species* (1964), proposed the first coherent theory of how biodiversity evolves. The rediscovery of Gregor Mendel's studies at the beginning of the twentieth century then illuminated the mechanisms of heredity, allowing for the development of a comprehensive evolutionary synthesis.

Yet, while the scientific method has been invaluable in revealing the counterintuitive mechanisms underlying life's diversity, it becomes limiting if treated as the only way of understanding life. Such a view risks overlooking dimensions that are instead taken up by the human sciences, which often frame them as uniquely human. This, in turn, leads to the reappearance of Cartesian dualism: not always in the same form, but always with the same underlying premise. What we need instead is a return to a phenomenology of life—one that understands the human being not as separate from, but as embedded within, the *Lebenswelt*. The phenomenological approach does not deny the validity—nor, more importantly, the value—of scientific investigation into biological phenomena. Rather, it argues that such analysis cannot be considered exhaustive. To treat it as such risks failing to capture the subjectivity of the experience of being-alive.

Phenomenology, in its role as an open-ended inquiry into the lifeworld, invites us to view the living not as a fixed object of study but as a phenomenon that offers multiple levels of interpretation and continuous reflection. This perspective stands in contrast to the reductionist and eliminativist tendencies that gained dominance in the second half of the twentieth century, following the 1953 discovery of DNA's double-helix structure and replicative mechanisms by James Watson and Francis Crick, with crucial contributions from Rosalind Franklin. That landmark discovery triggered a wave of breakthroughs and led to the rise of powerful biotechnological practices. The following decades—often referred to as the golden age of biology—were marked by a strong reductionist paradigm. During this time, the idea that genetics held the ultimate key to interpreting life gained widespread acceptance, marginalizing other explanatory frameworks. The publications of *Chance and Necessity* (1971) by Jacques Monod and *The Selfish Gene* (1976) by Richard Dawkins reflect this period's genetic determinism and the view of the living as essentially passive. However, this paradigm began to shift in the late 1980s with the emergence of new theories and approaches. The development of Evolutionary Developmental Biology (Hall 1992), niche construction theory (Odling-Smee 1988), and epigenetics (Shemer et al. 1996) opened the door to a more dynamic understanding of life. These perspectives challenged the idea of genetic primacy and pointed toward a more complex, interactive conception of evolution. This shift culminated in what is now known as the Extended Evolutionary Synthesis (Pigliucci and Müller 2010), a framework that embraces a pluralistic view of evolutionary mechanisms.

It is useful, then, to return to a reflection on the phenomenon of life—starting from the recognition that it involves far more than blind genetic replication, random mutations, and selection pressures. Living organisms not only interact with one another, but are also protagonists of their own life: they actively influence both the expression of their own traits and the selective pressures acting upon them. For this reason, we now speak of an "agency of life." This calls for a renewed philosophical inquiry into the ontology of life—one that recognizes our capacity to investigate life from within, as beings *in* life, rather than placing humanity outside of it in a dualistic framework.

In 1943, Austrian physicist Erwin Schrödinger posed a fundamental question: how can the processes occurring within living organisms be explained through the lens of physics? His first observation concerned the complexity of living systems, which, from the standpoint of thermodynamics, appear unstable, as they resist decay by absorbing energy from their surroundings—that is, by feeding on negative entropy. This idea would later be taken up and expanded by chemist Ilya Prigogine who, in the 1960s, introduced the concept of "dissipative structures": living systems as open systems that remain far from thermodynamic equilibrium by continuously exchanging energy and dissipating entropy (Prigogine and Stengers 1977).

One might expect that the discoveries of molecular biology, alongside the elucidation of biochemical and thermodynamic processes, would reinforce the idea that life is autonomous and self-contained in its explanatory power. But that is not what happened. On the contrary, Prigogine's work laid the groundwork for a relational understanding of life—an epistemology of complexity. This perspective emphasizes:

1. Open systems, which evolve through energy exchanges and dissipation;
2. Emergent phenomena, where new properties arise from combinations of elements that do not possess them individually;
3. Unpredictable processes, characteristic of systems far from equilibrium;
4. Adaptive systems, which respond to feedback loops (negative feedback below a threshold; positive feedback above), giving rise to quantum-like stability;
5. Systems at the edge of chaos, which defy linear cause-and-effect explanations and fixed, synchronous timelines.

What is emerging today is a relational understanding of the phenomenon of life—one that overcomes two traditional views: on the one hand, the essentialist conception, which posits an inner vital principle inherent to living beings; on the other, the reductionist view, which places life on the same level as purely physico-chemical phenomena, as noted by Konrad Lorenz and Franz Kreuzer (1981). For the Austrian ethologist, life is indeed composed of elements that follow the laws of physics and chemistry, but it nevertheless belongs to a different order—one marked by qualities absent in abiotic systems. If we accept that life is made of the same basic components and governed by the same laws as inanimate matter—without invoking any additional "vital principle"—then we must also recognize that life's distinctive traits arise from relational dynamics: both from internal organizational relationships and from external, interactive ones. This implies, as various scholars of the phenomenology of life have argued (each in their own way), that understanding life requires us to move beyond simplistic dichotomies—such as internal vs. external, organism vs. environment. Instead, we must seek the deep relational connections that weave together these domains into a whole.

Consider, for instance, the case of adaptive morphopoiesis—the process that shapes biodiversity through intricate chains of ecological interdependence and interconnection. What becomes immediately apparent is the impossibility of isolating a living organism from the web of relationships it maintains with its environment. Attempting to do so would be like trying to keep an organ alive after removing it from the body, while denying it access to any form of support. From this, we can infer that life unfolds according to a logic of relational recursiveness. We know, for example, that a species is shaped by the selective pressures exerted by its environment: its morphological and functional traits are a reflection of the conditions its evolutionary lineage has had to navigate. The white coats of ermines, arctic foxes, snowy owls, and polar bears, for instance, mirror the snowy landscapes and icy habitats in which they live. Likewise, the streamlined forms of dolphins, sharks, ichthyosaurs, and seals correspond to the viscosity of the aquatic medium they inhabit, with their fins molded by the resistance properties of water. Birds fly by using their wings to push against air resistance and take advantage of thermals—and so do all other flying animals. In this sense, the characteristics of air are inscribed into the internal morphology of these species by evolutionary pressures. Evolution, then, can be seen as the process through which external selective factors are internalized—embodied—via a morphopoietic, relational organization. In practical terms, this means that what was once interaction with the environment—external relation—is progressively translated into internal structure and relation.

At the same time, a living being's internal organization also transforms its external relations with the environment—converting internal relations into external ones—as clearly demonstrated by the phenomenon of niche construction. A living organism does not merely sustain itself through negentropic processes; it also modifies its environment by its very presence, thereby altering the interactions it must navigate. A niche introduces a wide range of relational adjustments. For example, it:

1. Maintains the organism in a self-regulated environment, thereby reducing the variability of interactional factors;

2. Acts as a filter, since its semi-permeable nature prevents certain external elements from entering into relation with the organism;
3. Shields the organism from potentially harmful influences;
4. Creates the conditions necessary for its survival;
5. Generates new selective pressures that drive further specialization in a given phyletic lineage (Lewontin 1983; Odling-Smee 1988; Forestiero 2009).

There is little doubt, in this sense, that Earth itself may be viewed as an extraordinary product of niche construction—as proposed by James Lovelock and microbiologist Lynn Margulis in the seminal 1979 essay *Gaia: A New Look at Life on Earth.* As explained in Chap. 3, the very features that make Earth hospitable to biodiversity, in fact, result from interactions initiated by living beings. The planet, in other words, is a product of life itself: a synergistic and self-regulating system. According to this hypothesis, the biosphere—beginning with unicellular life—has actively shaped the planet's evolutionary trajectory, modifying environmental parameters to make further life possible.

Lovelock dedicated this idea to Gaia, the Greek goddess of Earth, describing a homeodynamic system that functions, in many ways, like a living organism. This system is capable of maintaining crucial environmental variables—such as temperature, atmospheric composition, and ocean salinity—within life-sustaining gradients. It is clear that this system should not be viewed as static, but rather as governed by an evolutionary logic marked by significant discontinuities. This is because the biosphere's living landscape is in constant flux due to evolutionary dynamics, making it a metastable system.

Also, environmental parameters are themselves always shifting, as evidenced by the comparison of the last three periods of the Cenozoic—from the Pliocene (starting 5.3 million years ago), through the Pleistocene, and into the present Holocene. These reflections help us understand that while the biosphere does tend to establish dynamic equilibria—which may remain relatively stable for thousands of years—these are nonetheless evolving systems, subject to changes in their internal regulatory mechanisms that can result in radically altered scenarios. Life, in this view, is a process of continuous metamorphosis, with change present not only during major mass extinctions but also during phases of apparent environmental stability. The conclusion we must draw is that, while the Gaia system does exhibit tendencies toward self-regulation, we must not mistake this for a static condition. Any achieved balance is inherently temporary and inclined, over time, to shift or break down.

A second point to consider is that this relative stability can also be disrupted by major events, which in turn bring about profound transformations in the biosphere. These transitions often involve severe losses in biodiversity, sometimes wiping out most species from the preceding period. Today, five mass extinction events are widely recognized, though many scholars believe these are only the most visible examples—there may have been many others, less well documented. What is a mass extinction? It is simply a moment of profound disruption in the homeodynamic balances that Lovelock described—when the system shifts from self-regulation to catastrophe, and the resulting disequilibrium becomes widespread and irreversible. A catastrophic system behaves nonlinearly: its effects are neither proportional nor gradual, and certainly not predictable. This should prompt serious reflection on the risks humanity is taking by destabilizing the regulatory systems that have governed the last tens of thousands of years. For in its complexity, life is fundamentally unpredictable.

In 1947, biologist John B.S. Haldane published a book echoing Schrödinger's famous title, *What is Life?* To the question posed by the title, he responded: I am not going to answer this question. In fact, I doubt if it will ever be possible to give you a full answer (Schrödinger 1944). Haldane compared the sensation of being alive to the perception of colors—those subjective *qualia* that resist objectification. The history of life on Earth is far more complex and intricate than even the scientists of the twentieth century could have imagined when tracing seemingly linear chains of phylogenetic sequences—the most famous beginning with an ancestor resembling a chimpanzee and culminating in *Homo sapiens.* As David Quammen illustrates in his essay *The Tangled Tree* (2018), the tree of life is far more complex: its branches are interwoven, with genes flowing not only vertically—from parent to offspring—but also laterally, crossing species boundaries like a transfusion of blood between individuals. Even the human genome is a mosaic of different life forms: 8% of it consists of remnants of retroviruses that infected our ancestors. Moreover, the traditional, isolationist concept of species—as defined by Ernst Mayr in 1940 as a group of interbreeding populations reproductively isolated from others—is increasingly being challenged. Phenomena of hybridization with fertile offspring between distinct species have been observed, such as between codons and mallards, dolphins and stenellas, wolves and coyotes (Mazzotti 2024).

Life, therefore, resists all attempts to be neatly contained within our classificatory and logocentric frameworks. It reveals a richness and complexity continually affirmed by the ongoing reinterpretations of evolutionary theory. And yet, precisely through this constant refinement, the theory proves to be a powerful and still-unfinished workshop of research and reflection (Continenza and Gagliasso 1996).

6.3 The Metapredicates of Life

The living phenomenon manifests in a multitude of forms that appear radically different to our eyes—making it difficult to find any clear predicative similarity between, say, a redwood and an ant, a bacterium and a human being, or a whale and a blade of grass. Yet from a biological perspective, life is characterized by certain defining features. From the standpoint of biological dynamics, for example, we can observe the interdependent relationship between nucleic acids and proteins, which enables both the expression of specific traits and the mechanisms of heredity and performativity. Likewise, we note the presence of cellular entities enclosed by a semi-permeable membrane, capable of exchanging matter and information with the external environment while preserving an internal space defined by specific environmental gradients. Even based solely on scientific inquiry, life—though expressed in a variety of predicative forms—remains recognizable.

This plurality reflects the inevitable expression of certain fundamental metapredicates, such as conjugability and redundancy. We might say, then, that life is destined to take on a form—to be individuated—even if there are borderline cases, like viruses and prions, which are difficult to place. At the same time, life is translated into existence—that is, into a temporally present being, an expression of the telluric condition. This reminds us that life is but one possible manifestation of matter. Here, I do not wish to dwell on the biological characteristics of life, but rather on its ontological ones, by posing a question similar to Schrödinger's—but from a philosophical standpoint: *What is life*, from an ontological point of view? I am aware that I cannot rely on descriptive predicates, precisely because life appears in forms so diverse that it resists univocal classification. My approach, therefore, will be to extract qualities not through description, but through inference—assuming that the predicative features of living beings are the specific expressions of a deeper, shared condition.

A metapredicate is simply what underlies the assumption of a certain form. For example, the heteronomy of living beings—life's inherent need to rely on something external to express its qualities—is a metapredicate. As noted earlier, the first defining feature of life is its dependence on a continuous input of energy from the outside in order to sustain its intrinsic properties. This gives rise to different forms of dissipation: organisms acquire energy through various strategies, the most common being autotrophy in plants and heterotrophy in animals—though these are not the only options. Thus, negentropy manifests differently in plants, which obtain energy through chlorophyll-based photosynthesis, and in animals, which do so through feeding. This means that, from a predicative standpoint, plants and animals are quite distinct. But from a metapredicative perspective, they are similar: both are expressions of life's need to externalize energy acquisition. In this sense, plants and animals are two different ways of manifesting life.

Metapredicative evaluation begins with the insight that all living beings, while sharing certain meta-qualities, must necessarily assume different predicates. This is a crucial point. The differentiation of predicates—the emergence of distinct forms from shared metapredicates—reveals the ecological condition of living things: their fundamental interdependence. Through a process of relational morphopoiesis—a kind of mutual shaping—organisms, if they are to remain faithful to their shared meta-qualities, must differentiate from one another. For example, a predatory animal like a cheetah and an herbivore like a gazelle must each develop unique morphological and functional traits in order to fulfill the shared requirement of heterotrophic negentropy. Their predicative differences, therefore, do not obscure what it means to be an animal; rather, they perfectly illustrate it. If a carnivore had the same characteristics as an herbivore, then we would struggle to grasp what it means to be an animal at all. To sum up: the cheetah and the gazelle represent two ways of being animals, just as the animal and the plant represent two different ways of being-alive.

To establish a foundation for reflecting on the phenomenology of life, I propose outlining *eight basic metapredicates of life*. These serve to identify the predicative becoming of living beings beyond their specific expressions in existence. The metapredicates are as follows:

1. Heteronomy—the condition of being constituted in such a way as to require a specific relationship with the environment; in other

words, a living individual presupposes a certain kind of environment in order to manifest.

2. Sensibility—the reactive tendency of a living being to respond to external stimuli; that is, a sensibility to environmental variables and a propensity to react accordingly.
3. Electivity—the tendency to act as a protagonist in its interactions with the world, to orient itself and shape its own path based on inherent interests that it embodies.
4. Conjugability—the tendency to change state depending on the entities encountered, being defined by a principle of connection and relationality with the world.
5. Sympoiesis—the process of self-co-construction through relationships with other living beings; the organism can never exist in isolation or autonomy.
6. Diachrony—the impossibility of being fully explained in a single moment in time; life must be understood in reference to preceding events that inform its present manifestation.
7. Mirroring—the tendency to assume different forms or conformations in response to the environment and functional demands, such that the living being becomes a kind of distorting mirror of the world around it.
8. Redundancy—the tendency of living beings to be both more and less than what they manifest.

I will now look at these metapredicates one by one.

1. *Heteronomy*—The constitution of the living being brings the world into the present, actualizing it—not by creating it, but by extracting from the pool of possibilities a specific form, understood as a conformation, of the world. The living being, in its very makeup, expresses a particular relationship with the world; in manifesting itself, it reveals something about the world, though not objectively, but *sub specie relationis*—through the lens of relation. As previously noted, every expression of life relies on an external point of support, whether it be the energy of the sun or the release of energy through the decomposition of organic matter. Life requires a state of dynamic non-equilibrium, in which dynamism and disequilibrium mutually enable one another: movement sustains imbalance, and imbalance sustains movement. One of the fundamental traits of life, then, is its

constitutive non-autonomy—life cannot exist within the logic of a closed system.

This non-autonomy, however, should not be reduced to a purely thermodynamic consideration. In truth, the entire structure of the living being resembles a complex text that becomes legible only with the aid of extensive footnotes and bibliographical references. Life intrinsically depends on the external—it essentially leans upon what lies beyond itself. This can be described in two key ways: (i) the living being refers to something external—it is hetero-referenced, and thus exhibits a form of vital intentionality; (ii) the living being presupposes certain features of the external world—its very conformation reflects, *a priori,* the presence of specific properties of the environment in which it exists.

The heteronomic condition of living systems reveals that the achievement of a certain level of complexity—whether in an organism or an ecosystem—does not result from increasing the system's disjunction or autonomy. Rather, it depends on the ability to rely on, or anticipate, external contributions. This inherent anticipation of an external complement can be understood as the intentional character of life: being-alive always involves a reference to something beyond oneself, a prefiguration of some content present in the world. If we consider the concept of intentionality as developed in Edmund Husserl's *Logical Investigations* (2001), we see that this property—typically attributed to propositional acts such as "fear-that" or "hope-that"—is not limited to consciousness. It characterizes life as a whole, in its constant orientation toward something else.

To fully grasp heteronomy, however, one must move beyond the idea of life as incomplete. Heteronomy does not indicate a deficiency, but rather a surplus—a kind of productive tension within the system. It is as if the living strives for a higher level of expressive dynamism by leveraging an external foothold, a sort of ladder that enables access to greater complexity. This could be described as thermodynamic risk or organizational probability. Heteronomy is thus the capacity to *a priori* include something external—such as sunlight or organic matter—on the assumption that certain external elements, while not internal to the organism, are nevertheless expected to contribute to its organization. In other words, heteronomy entails the prefiguration of boundary conditions—conditions that, through evolutionary processes, have been incorporated into the very structure of the living being. Thus, the gills of a fish *presuppose* the presence of dissolved oxygen in water, just as the wings of birds presuppose the resistance of air.

Heteronomy shows us that no complex system can evolve in isolation, relying solely on its own internal contents. Evolution always involves correlation with boundary conditions, and these inevitably generate a form of dependence. Life, therefore, cannot emerge or sustain itself autonomously, nor can it be interpreted through internal criteria alone or founded *iuxta propria principia*. Every living system poses a multitude of marginal questions—not because of a deficiency, but due to redundancy. Lack and excess are intertwined: redundancy cannot exist without heteronomy, and heteronomy presupposes redundancy. The dependence of heteronomy on redundancy helps distinguish a genuine lack, which reflects the impoverishment of a system, from heteronomy, which instead marks increasing complexity and a surplus of elements capable of multiple configurations.

Ecology clearly illustrates the heteronomic nature of every organism and every life process. In evolving toward ever more improbable forms, living systems have had to rely on external organizational factors. Thinking ecologically thus requires adopting an eco-ontological perspective on the different domains of the phenomenology of life. An essentialist conception—where predicates are seen as inherent, referring to matter and form as self-contained—is not only inadequate for explaining any living entity; it is also misleading when applied to evolutionary and complex systems. Such systems always demand an explanatory note beyond the text itself, so to speak. And this note is external: the "text" of life is heteronomic because it already presupposes the presence of an annotation. Heteronomy presents itself through hierarchical degrees or levels of systemic complexity: the cells of a tissue presuppose the organism; the organism presupposes the niche it occupies; the niche in turn implies the ecosystem; and the ecosystem relies on the biome, which is itself organizationally dependent on the functioning of the biosphere.

2. *Sensibility*—One of the most relevant aspects of the life phenomenon is sensibility: the ability to perceive external conditions and respond, acting based on features present in the world. Sensibility manifests in different forms within the living world—there are many

ways organisms perceive their surroundings and behave accordingly. We can say that both consciousness and agency rely on, or result from, this fundamental capacity. Sensibility can be defined as a basic property that distinguishes abiotic phenomena from living ones. It enables living beings to detect the conditions in which they exist and to direct their internal processes in response to contextual variables. Among the metapredicates of life, sensibility perhaps most clearly expresses this being-in-relation, as it implies a constant relational state with the world.

This condition entails that the internal behavior of a living being, while grounded in inherent properties, changes in accordance with the characteristics of its context: it alters its internal relational organization—that is, the configuration of the network of relationships that define it morphologically and functionally. Conversely, if the context can influence the internal organization of a living being, then changes to this internal network also reshape the external relations the organism establishes. This reciprocal dynamic shows that internal and external relations are not two separate dimensions within the living being. Rather, they form a continuous flow of relational events that together define its organic and functional identity.

Sensibility takes on various forms in the lifeworld. In mammals, for example, it translates into a broad repertoire of dimensions: perceptual, endocrine, immune, metabolic, affective, cognitive, and up to the higher planes of consciousness. However, the focus should not be limited to any particular manifestation of sensibility, but rather on what underlies all these expressions—the fact that the living possess sensibility. Through sensibility, living organisms accomplish several feats, such as: (i) modifying their internal morpho-physiological conditions based on external factors; (ii) adapting the phenotypic translation of their genetic components in response to environmental variables; (iii) guiding their growth toward energy sources or developing organs differently based on opportunity; (iv) responding to external stimuli either through active reactions or structural changes; and (v) moving actively from one place to another.

These forms of sensibility are enabled by receptor mechanisms located on the cell membrane. In multicellular organisms, these mechanisms give rise to true sensory organs, specialized in detecting specific signals—chemical, electromagnetic, or mechanical—and converting them into signals that propagate throughout the organism via activation potential transmission systems. Examples include the vascular system in plants (phloem), the reticular nervous system found in jellyfish and cnidarians, the diffuse or ganglionic nervous system in invertebrates, and the central nervous system of vertebrates. The organism's sensory response—activation—is not isolated but propagates to other cells, creating an overall sensibility. This systemic quality exhibits emergent ones—such as the experience of desire in dopaminergic activation—giving rise to responses that vary across different life forms.

To describe mechanisms that are not based on a nervous system—which most multicellular organisms have, with the exception of sponges and placozoa—the term *tropism* is used. This refers to specific axes of orientation, such as movement toward or away from environmental gradients. For example, unicellular organisms have membrane receptors that enable appropriate physiological responses, often highly complex and involving cascades of events that reorganize the cell's internal structures. These reactions also include directional movement in response to environmental cues, known as *chemotaxis*. In bacteria, sensibility can lead to communication between cells, as internal changes trigger functional transformations across the colony. One well-known example is quorum sensing, in which bacterial cells detect the presence of others via molecular reception. When a threshold concentration—a quorum—is reached, they alter their physiological behavior. Some *Vibrio* species, for instance, become luminescent only when their population reaches a certain density, as each bacterium produces a hormone that activates luminescence only above a given concentration. A particularly notable example of orientation behavior is seen in *Amoeba proteus*, a unicellular eukaryote capable of moving along surfaces and engulfing other organisms. It does so by altering its shape to form tentacle-like protrusions, which serve both for locomotion and nutrition.

In plants, we observe various forms of tropism, such as heliotropism—the orientation toward the sun—achieved either by directing growth through the action of the plant hormone auxin or through actual movement, as in the case of the sunflower. Another example is geotropism, the response to gravity. In addition, plant physiology includes *tactisms*, which are active movements toward specific targets, and *nasties*, non-directional responses to stimuli. Examples include *tigmonasty*, as seen in the leaves of *Mimosa pudica* and *Dionaea muscipula*, and *nyctinasty*, which causes leaves or flowers to close at night. As mentioned earlier, plants have a physiological system for transmitting electrical signals—the vascular system—which functions analogously to a nervous system and is believed to contribute to plant experiences and even memory. Root systems play a central role in plant sensibility: they can orient toward specific targets, such as nutrients or water, and engage in exchanges with other plants via symbiotic networks with fungi, known as *mycorrhizae*. This underground network forms a kind of hypogeal "world wide web" (Rhodes 2017) unique to the plant kingdom, enabling global connections and communication.

As for the animal kingdom, sensibility has given rise to a vast array of specialized sensory cells, capable of responding not only to external stimuli but also to internal signals. This dual sensibility enables the establishment of homeostatic gradients and corresponding regulatory mechanisms essential for internal balance.

3. *Electivity*—Every living being has a specific orientation in the world, stemming from the fact that it possesses inherent interests—that is, concerns that affect it in a distinctive way. While a stone is carried passively by the river's current, a salmon swims upstream from the mouth to the source. Orientation is always specific and varies from species to species: the chemotropism of an amoeba, which moves in search of biological compounds to feed on, differs from the heliotropism of a cyanobacterium, which moves toward a light source. To possess inherent interests means to be *interested in the world*—a living being turns toward its surroundings through selective attention. While orientation is shaped by features of the environment, it is also guided from within, by needs or dispositions that drive the organism to follow a particular worldly path. In this sense, electivity can be defined as the mode through which each living being manifests itself underneath different variations: the amoeba and the cyanobacterium differ in the type of orientation, but both express the metapredicate of electivity.

For example, the membrane system of a cell is not merely an interface with the outside world that informs it of environmental conditions; rather, it is a selective device, equipped with specific receptors that: (i) may or may not be receptive to a given external element; and (ii) are sensitive or refractory depending on the internal condition of the organism. These receptors therefore allow the cell to activate in response to specific external and internal conditions. Within the organism, different cell populations exhibit distinct physiological responses based on membrane receptor electivity. For instance, the sensibility of pancreatic cells to blood glucose levels regulates sugar metabolism, while the response of endothelial cells to osmotic gradients produces the sensation of thirst. The same applies to neural systems: we speak of *dopaminergic* versus *serotonergic* pathways based on the presence of membrane receptors that respond differently to these neuromodulators, thereby activating specific neuronal constellations and generating particular psychological states in the subject.

These sensors reach extremely complex levels of specialization in multicellular organisms, particularly in the animal world, where we find sensory apparatuses not only capable of detecting stimuli of various kinds but also finely tuned to specific frequencies or molecules. For this reason, as several authors have rightly pointed out—exemplarily von Uexküll with his concept of *Umwelt* (2010)—life constructs its own dimension of electivity, we might say its own *world-context*, which defines its specific being-there. We should not, however, think of *Umwelten* in a Leibnizian sense, as closed and disjointed monads, because different *Umwelten* have margins of overlap, which allow living beings to communicate through a homological system—that is, through shared points of reference. Furthermore, the *Umwelt* is not properly a bubble, but a network of access to the world that constantly evolves based on the subject's experiences.

That said, it is clear that living organisms possess highly specific sensory access, such that each form of life creates its own universe: the plant world differs from the animal world, the vertebrate world from the arthropod world, the mammalian world from the avian world—and so on, down through the distinctions of order, family, genus, species, and ultimately to those peculiar to the individual. Thus, each time a species disappears, an entire world vanishes. This is deeply disheartening. Yet it is consoling to know that the legacy of a species is never entirely lost. Thanks to heredity and common descent—shared evolutionary networks—life preserves a degree of continuity and shared structure. This perspective, however, becomes incomprehensible if one adopts an essentialist and non-evolutionary view: that is, if one interprets the *Umwelt* as a sealed bubble, leading to the idea of incommunicability between living beings.

Electivity, therefore, does not reside in the *Umwelt* itself, but rather in two aspects of how living beings face the world: (i) the shared necessity of being selective in order to orient themselves—since no path through the world, no itinerary, would be possible if life did not contain within itself a principle of electivity; (ii) the interactions among living organisms, where electivity is not to be understood as disjunction, but as the capacity for selective engagement with other forms of life.

To conclude, we must avoid fixating on each entity's specific expression of electivity. In other words, to grasp the phenomenology of life, we must move beyond first appearances—the *how* of two specific entities, such as the amoeba and the cyanobacterium—and instead reveal what they share: the *what* of their relationship with the world, that is, what lies beneath or remains hidden behind the visible surface. A carnivore has a very different lifestyle than an herbivore, yet both navigate the world electively, and this capacity is part of their nature even before it manifests as a need. To emphasize the electivity of the living is to unveil it—to remove the mask of its particular form in order to reveal what is shared in the condition of being-alive.

4. *Conjugability*—Electivity transforms the beingness of the living into a particular style of being-in-the-world. In this sense, we might invoke Martin Heidegger's existential phenomenology from *Being and Time* (1978), particularly the concept of *Dasein*—that is, the being-there of the individual and their being-in-the-world (*In-der-Welt-sein*)—as a metapredicate of life itself, with two important modifications: (i) by defining this being-there as a *presencing* that goes beyond mere presence, indicating an *involvement*—meaning both an *envelopment* in the world and an *engagement* with the world; (ii) by replacing the principle of *readiness-to-hand* with that of *conjugability*, which may lead to the use of external objects, but equally to knowledge, contemplation, aesthetics, resonance, and more. For Heidegger, in fact, the entities we encounter are first and foremost ready-to-hand and usable, and as such must be understood in their spatial relation to the subject—whereby space itself becomes a declination of use, and in this way entities acquire meaning. Undoubtedly, instrumental relations do characterize aspects of the phenomenol-

ogy of life. Yet it is a stretch to claim that presence and usability are equivalent. One cannot, for example, derive the usability of a partner from reproductive interest, nor can one explain Lévinas' concept of empathy through the principle of embeddedness—that is, being embedded in a context of use.

We might say that the living presents itself as a conjugable verb, whereby, in relation to an external entity, it assumes a certain conjugative disposition—presenting itself as contemplating, knowing, embodying, etc. These attitudes toward-the-world are nothing other than different expressions of a single metapredicative quality: conjugability—the fact that life has both a root and a terminus. For this reason, while it retains a specific range of inherence—which, besides, makes it recognizable—it also always presents, within this inherence, an internal variation. This is something different from, and I would say more than, mere adaptive plasticity or deformability; rather, it is the capacity to assume the most suitable morpho-function, consistent with the characteristics of the context. Conjugability indicates that this quality is already present *a priori*—expected as a connotation of the living being—so that the internal reasons grounding its electivity never lead to fixity, but rather to flexibility.

This also means that the living exhibits a primal and absolute necessity of bonding with the external world—a necessity that, following Friedrich Schelling's formulation in *Treatise on the Relationship of the Real and the Ideal in Nature* (Snow 2015), is defined by the term *copula*: an infinite inherence of the finite to the bond (De Fazio 2020). The *copulative* nature of the living, on the one hand, makes explicit its subjectivity—its being elective, partial, and selective in its relation to the external world; on the other hand, it establishes its porosity, a quality that renders the living open to external conditions and capable of shifting its attitude toward-the-world in response to perceived opportunities or risks. The copulative principle of the living is another element that helps us grasp life's inherent intentionality—its being referable to, and therefore conjugable with, an external entity. Fear, for instance, cannot be properly understood as a self-contained psychological state, but rather as a propositional attitude: a fear-of—something. It is impossible to define such affective states—whether psychological or behavioral—without specifying the event or object toward which the attitude is directed. Thus, all affectivity—fearing, desiring, hoping, taking pleasure, rejecting, rejoicing—appears as an implicit conjugability with something present in the world.

At the same time, being a bearer of inherent interests necessarily generates a copulative tension in the living: a moving-tending toward the world that disrupts the classical dualism between causality and finality. This is because every action of the living simultaneously expresses both a necessitating causality and a projective teleology. We may examine any action through the lens of causality or finality, but when we restrict ourselves to only one, we inevitably miss something. This dual dynamic is exemplified in the animal world by the interplay between need and desire: need expresses intrinsic necessity, that which must be fulfilled to maintain life's homeostatic gradients; desire expresses an equally intrinsic and compelling projection, yet one that can hardly be aligned with necessity. The projective teleology of the living propels it into an unceasing motion beyond itself. Life is always in a state of becoming—a desiring condition, marked by a certain yearning, disequilibrium, a decentralizing tendency, a motion of transfiguration and beyond-passing.

This inherent conjugability speaks to a primal relationship with the world—as Maurice Merleau-Ponty claims in *Phenomenology of Perception*—in that the world is already presupposed by the subject who perceives it. Subjectivity, too, must therefore be understood in terms of the conjugative principle of the living: human beings can come to know themselves only *through* the world, not within the egological self-closure of the *cogito* (Zahavi and Zahavi 1999). In this regard, it is helpful to turn to another of Merleau-Ponty's essays, *The Structure of Behavior* (1976), in which he seeks to rethink the traditional opposition between the transcendental and the empirical. This metapredicate allows us to reconnect these two dimensions of the living: just as the empirical is realized through the transcendental—in the Husserlian vision, which integrates it with themes of corporeality—so too the transcendental can be seen as the product of life's evolutionary history. In the Lorentzian perspective, the *a priori* is nothing other than a phylogenetic *a posteriori*. Conjugability implies an inherent openness and interrelation, whereby—echoing Heidegger's *Basic Problems of Phenomenology* (1982)—being-there is, in its very origin, a *Mitsein* (co-being), that is, a being with and among intramundane entities.

5. *Sympoiesis*—The term *sympoiesis* was introduced by researcher Beth Dempster in 1998 to describe collective systems without clearly defined spatial or temporal boundaries. Donna Haraway later adopted and expanded the concept in *Staying with the Trouble: Making Kin in the Chthulucene* (2016), using it to describe how living beings are deeply interrelated—penetrating, entwining, and assimilating one another. Most significantly, this dynamic is symbiogenetic: creatures *co-construct* each other in a process we might describe as *co-becoming*. As Haraway puts it, organisms do not precede their relations; rather, they come into being *through* one another, challenging any essentialist conception. Often, for example, the development of a living being depends on the presence of specific symbionts, as embryogenetic processes enabling full realization are modulated by these other organisms (McFall-Ngai 2002; Douglas 2014). Today, organisms are often referred to as *holobionts*, highlighting how bodies resemble forests or coral reefs more than self-contained units (Rosenberg et al. 2007). Among the most crucial plant symbioses is that between *Rhizobium* bacteria and the roots of leguminous plants, which enables nitrogen fixation—transforming atmospheric nitrogen into a biologically accessible form. Another key example occurs in coastal ecosystems, particularly in the intertidal zone, involving marine phanerogams, nitrogen-fixing bacteria, bivalve mollusks, and sulfur-oxidizing bacteria that live inside the mollusks. This intricate web of relationships is vital for maintaining water clarity. Seagrasses, for instance, play a central role in the water cycle, oxygen production, sediment stabilization, and nitrogen fixation. Yet they would be decimated by toxic sulfides from decaying organic matter were it not for the sulfur-oxidizing bacteria hosted by bivalves, which neutralize these compounds (van der Heide et al. 2012; Roeselers and Newton 2012). A similar principle governs coral reefs, which, while formed by corals, are shaped by immense biodiversity. At their core lies a foundational symbiosis: photosynthetic algae provide the energy needed for calcium carbonate deposition, forming the reef's structure. These unicellular algae also enable essential nutrient exchanges, underscoring how the reef's existence depends on multi-species entanglement and continuous sympoietic activity.

At the root of all symbiotic forms lies *eukaryogenesis*, a process that began approximately 2.5 billion years ago. Through a symbiosis between an archaeal cell and a bacterium, this process gave rise first to mitochondria and later to chloroplasts via intracellular symbiosis. The endosymbiosis theory, which explains the origins of mitochondria and chloroplasts, was first formulated by biologist Lynn Margulis. Margulis was a vocal critic of neo-Darwinism—particularly its genocentric interpretation, exemplified by Thomas H. Morgan—which treated the genome as independent of the cytoplasm. For Margulis, symbiosis had to be recognized as a primary driver of evolution. It was during the 1960s that Margulis developed the idea that nucleated cells had evolved through symbiotic relationships among distinct bacterial lineages. Her landmark paper, *Origins of Mitosing Cells*, was initially rejected ten times before being published in the *Journal of Theoretical Biology* (1967). Today, endosymbiotic processes are widely accepted and well documented. For example, certain bacteria from the genus *Wolbachia* live in the cytoplasm of invertebrate oocytes, providing crucial developmental signals for the host organism (Krueger et al. 1996). Further studies have led to hypotheses that other cellular structures—such as cilia and sperm tails—may also have originated through endosymbiotic events. In *Symbiotic Planet: A New Look at Evolution* (1998), Margulis expresses skepticism about the role of random mutation as the sole engine of evolution. Instead, she argues that the primary source of evolutionary novelty lies in the effects of endo- and exosymbiosis. These include phenomena such as bacterial conjugation, in which genetic material is exchanged, or retroviral mechanisms, which integrate viral DNA into host genomes—both of which can initiate symbiogenetic processes.

As mentioned, the principle of sympoiesis asserts that living beings are co-constituted: they are relational entities whose interactive formation involves not only a network of internal and external relationships, but also their fundamental ontological dimension. A clear example is the lichen, which—though resulting from a symbiotic association between an autotrophic organism (such as a cyanobacterium or an alga) and a heterotrophic one (typically a fungus, like an ascomycete or basidiomycete)—must be considered a single biological entity. This relationship provides not only mutual benefit but the very conditions for survival: the fungus obtains organic compounds produced through the autotroph's photosynthesis, while the latter receives water and mineral salts from the fungus. Lichenization is an ancient process, foundational to the colonization of terrestrial environments. It created the conditions for the emergence of mosses and eventually all other plants, culminating in their proliferation during the Carboniferous period. In fact, lichens are believed to have been among the first organisms to inhabit land, breaking down rock through metabolic activity and preparing the ground for the arrival of other plant species. This progression reflects the broader process of ecological succession, where each stage prepares the conditions for the next, eventually reaching a state of climax. The principle of sympoiesis resonates with Anna-Teresa Tymieniecka's vision of a unity of all living beings, as articulated in her concept of eco-phenomenology—a unity expressed and sustained through the reciprocal contributions of all life forms. Another key idea in Tymieniecka's work is that of ontopoiesis, the continuous ontological metamorphosis that emerges through relationship. From this perspective, symbiosis is no longer simply mutual aid between distinct organisms, but, as said, a deeper process of mutual co-construction.

6. *Diachrony*—The living has its own phenomenological dimension that escapes phenomenal determination and necessarily requires synchronic evaluation. Of course, one can always carry out a synchronic analysis of the living on a processual level—for example, by explaining behavior in terms of internal and external stimuli that elicit responses in the individual. Such an analysis is certainly valid, but it remains partial. The same limitation applies if we adopt other explanatory frameworks: for instance, if we describe stimulus-response behavior in terms of its adaptive value, or refer to taxonomic classification, ontogenetic development, alignment with the pleasure principle, momentary agency, or the subject's projection toward a goal that transcends the present stimuli. These are all legitimate perspectives—but they are extremely incomplete; indeed, they fail even to offer a rough outline of what is really going on. This limitation is not merely due to the multiplicity of causal levels. It stems from a deeper issue: the living cannot be understood synchronically, because it is fundamentally a diachronic reality—an unfolding presence that moves through time.

Time is a fundamental factor in the phenomenology of life, yet it is often analyzed only through the lens of spatial or predicative expansion, neglecting the fact that being-alive is inherently a chronologically defined dimension. The Heideggerian principle of being-toward-death (*Sein-zum-Tode*), while not to be interpreted in a totalizing way, holds undeniable heuristic value: the birth of a living organism already contains within it the anticipation of its inevitable end. The death of a living being, however, should not be conflated with the inevitable dissolution that awaits all composite entities on the planet—rivers, mountains, continents—because it is not merely an accidental (though inevitable) event. Rather, it is a preprogrammed one. Death is already present within life itself, and like a partner in a *danse macabre*, it moves alongside life throughout the entirety of a living being's existence.

Death, in this sense, is not the extinguishing, consummation, or simply the cessation of life's processual unfolding. Nor is it just the annihilation of life's meaning, as in Sartre's conception. Instead, death emerges as the outcome of the very tension inherent in the living being—a perspective that deserves attention precisely due to the diachronic nature of life's phenomenology. Life does not emerge from nothing: every birth is the reawakening of an apparently inert past in the present moment of the unborn. Nor does life end in nothing: it provides for reproduction and other forms of diffusion in the *Lebenswelt*, which is extended across space and time. In this light, the death of the individual living being appears as nothing more than the passing of a torch. Here, the concept of the living being as a *projection* becomes central, returning to Heidegger and expanding on his thought to encompass the entire phenomenology of life. This includes: being *projected* into the future, where time ceases to be merely chronological and becomes *kairological*—filled with meaning and direction; and being a *project*, that is, a being shaped by something prior to one's own existence, thereby gaining an *a priori* significance.

The diachronic condition reveals the living as an entity that can never be grasped through theoretical immediacy. The here and now of the living is always, to some extent, an abstraction—because life constantly exceeds the bounds of the present. It cannot be fully contained within the conceptual "container" of the now. The phenomenological approach to life aims precisely to reveal this temporally given surplus. This does not mean we cannot formulate descriptive or explanatory accounts of the living phenomenon—we certainly can, and we can examine life from multiple perspectives—but it does mean such accounts will never be exhaustive. Reductionism, then, can be understood as a lens that sharpens certain features of life while necessarily obscuring others.

Natural science plays a crucial role in investigating the relationships that characterize living phenomena. It also serves as a powerful engine of philosophical questioning. Indeed, I believe philosophy can only proceed from the things themselves—and from the accounts that science provides. Yet even so, science is not equipped to offer a complete account of the living phenomenon (Zahavi and Zahavi 1999). One key aspect that consistently escapes the analyst's slide is precisely the diachronic condition—the temporal dimension which science must necessarily transform into a constant, into something synchronic. Let me be clear: this is not a criticism of the scientific method, but rather an acknowledgment of its domain of validity—and, by contrast, a recognition of the need for other modes of inquiry. This rooting of the living in remote time—a chronological space that precedes the emergence of individuated life—means that life always precedes its own present existence. At the same time, life flows toward the future: a kind of runoff that leads, on the one hand, toward death, and on the other, toward the emergence of new forms of life. In this, life reveals itself as a continuous movement through time.

7. *Mirroring*—One consequence of the ecological dimension of entities is their phenomenological translatability. Each entity manifests itself by bringing its internally developed content—shaped over time—into dialogue with external conditions. In doing so, it subjectively mirrors those boundary conditions—not just reflecting them, but representing them. I refer to this process as *mirroring*. Take, for example, an oak tree. Its distinctive features—the shape of its leaves, the articulation of its branches, the spread of its roots, its internal metabolism, and its requirements for rainfall, soil composition, symbiotic organisms, and light—are all the result of its evolutionary history. These characteristics arise from a longstanding dialogue between the species and its environment, a process through which external selection pressures have been internalized. In other words, the oak's inherent traits depend on the cumulative exchanges its

lineage has had with the world—heteronomously introjecting features of its phylogenetic context. Thus, the development of an oak is not simply an isolated unfolding, but a mirroring of the dialogues that occurred between the species and its environment over the course of phylogeny.

In this sense, being-an-oak reflects past events, in the way Konrad Lorenz described *a priori* structures as phylogenetic legacies. But this reflection is not a purely historical account, nor a closed and repetitive echo of events prior to the tree's individual history. The development of the oak's metabolic systems—such as shoots and roots—is shaped by interactions with external conditions: a shoot will grow when exposed to light and wither if kept in darkness. Crucially, this development does not unfold randomly; it follows the species-specific organizational blueprint. In other words, the oak, through its inherent properties, mirrors the specific boundary conditions—here, for instance, the availability of light—of the environment in which the individual tree is situated. Its root system develops in relation to the physicochemical and biological characteristics of the soil—including the influence of developmental symbionts, which must not be overlooked. In the same way, while the architectural form of the oak is defined by the species—leaf structure, dendritic patterns—each individual oak is unique. Its crown, for example, will take shape according to the light conditions of its particular setting. We can thus say that the oak's phylogenetic inheritance is mirrored in its intrinsic features—being an oak means displaying certain basic traits shaped by natural selection. At the same time, these features mirror the present environment through ontogeny. In this way, the past—embodied in the species' evolutionary history—dialogues with the present to shape the future.

The subject's present is always a recapitulation—partial and fragmentary, like a text missing a few pages. Yet it is never merely an echo of the past, sedimented events; rather, it is a true re-actualization. These ghosts of the past are revived in the present, taking on its contours and smoothing over any breaks in continuity. We might describe this as a staging of actors from the past who, time and again, step into the spotlight to meet the challenge of the here and now. In doing so, each performance becomes a layered time-portal, a crossroads of temporal dimensions. In fact, these actors arrive with their own scripts, but they cannot simply recite them unchanged.

Instead, they interpret their lines in light of the present, mirroring it. Every time a legacy takes the stage of life, it bends to the power of its context, shaped by the setting into which it is summoned. In this metaphorical theater of the subject's phenomenology, the reenacted actor must reinterpret the inherited script, adapting to the unique cadence of the audience before them. In this sense, the audience too becomes part of the performance—mirrored in the way the actor delivers the text.

The principle of mirroring describes the way an entity manifests itself—not by expressing its phylogenetic inherencies in a pure form, but by conjugating them according to the range of expressions the world enables or calls for. In manifesting itself, the subject enters into a contingent relationship with the world, such that its inherent qualities—which are not fixed predicates but open-ended potentials—are refracted through the contextual prism, outward. The landscape, in turn, becomes a play of reflections, with lights echoing and recalling one another across multiple planes. Mirroring is the natural condition of every living being that enters the realm of existence: in order to appear, it must align itself with the contingencies it encounters. What manifests, then, is never the being in itself, but the ecological dimension it mirrors through temporal and spatial interaction. If, as in the case of the oak tree, manifestation is also morphopoiesis—because every physiological process involves some form of growth induction—then mirroring necessarily involves recursive patterns. Over time, we observe an inevitable convergence among interacting entities, one that dissolves dichotomies such as identity versus otherness, or expression versus development. To think ecologically, then, is to shift from a phenomenology of life based on essence to one based on relation, allowing the entity to emerge as an expressive field shaped by its dynamic interconnection with the world.

8. *Redundancy*—Complex systems—and living systems in particular—possess a distinctive trait: redundancy of components. This redundancy grants them a degree of adaptability to external conditions by endowing the system with structural plasticity—that is, the capacity to assume different conformations depending on situational demands. Such plasticity represents a form of inherent virtuality: a property rooted in the system's internal characteristics. However, this does not imply totipotency; rather, it denotes a plurality of

possible outcomes within a defined perimeter. In other words, the system exhibits an intrinsic evolvability—one that does not determine specific outcomes but delineates a range of permitted possibilities. We must therefore conceive of this virtuality not as pure liquidity, where the absence of form allows the system to passively adapt to its container, but as a bounded plasticity: a structured flexibility that enables varied expressions within a structural plan. This leads to a compelling paradox. A living being is always more than what it expresses and, simultaneously, less—and the reasons are twofold: it is *more* because its virtual state holds many unrealized predicative possibilities, which remain latent throughout the individuation process; it is *less* because predicative translation requires the contribution—necessary or contingent—of something external.

It should be emphasized, however, that this capacity to assume different forms or functions depending on external relations is not to be understood as *complementation*, that is, the filling of some intrinsic lack in the entity. Rather, it is a process of *actualization*: the transformation of a virtual condition—endowed with many possibilities but no concrete presence—into an actual condition, in which the assumption of form entails the pruning of all other possibilities. Whereas complementation presupposes a lack in the entity, this expressive potential is grounded in redundancy. To think ecologically about the phenomenology of life is to shift from a compensatory and juxtapositional model—where a system acquires capacities in proportion to its deficiencies—toward a model of conjugability and dimensionality. In this view, the living being's inherent redundancy enables the intake of external content—whether symbiotic, informational, or objectual—not as a response to lack, but as an expression of the surplus already present within the system. Acquisition, then, is not about filling a void, but about converting a surplus. It is not a gap that makes adaptability possible, but complexity itself: the redundant architecture of the system provides the plasticity needed to adapt to changing conditions.

The process of translating a virtual starting condition into an actual outcome characterizes the phenomenology of life in many situations. We can say that being alive means existing within multiple levels of actualization—that is, translations of initial possibilities. To give a few examples: the phenotype is one among many possibilities originally present in the genotype, which is thus characterized by virtual polyphenism; ontogeny, in turn, is the construction of a specific identity from a set of possibilities present at birth; a species-specific being is a declination of a genus, which itself is the expression of a family, order, class, and further down the taxonomic tree. The principle of redundancy leads us to consider the phenomenology of the body as the privileged perspective of the living and its variation of actualization. To borrow Husserl's words from *Zur Phänomenologie der Intersubjektivität* (1973), the body reveals itself in its relation to the world. The somatic expression shows a body open to the possibilities with which it enters into relation—that is, as Merleau-Ponty suggests in *Phenomenology of Perception* (2013), the body is a node of open relations with the world.

To be a body means, from a phenomenological point of view, to consider being as an embodied entity that has in the body its zero point of experience with the world—provided, however, that we understand this experience not as perspectival remoteness (whether cognitive, contemplative, or otherwise) but as a translation of the world into the organization of the body itself. The virtual constitution of the body, in the totality of its apparatuses, transforms experience into a process of embodiment because interaction induces differential growth in the various cell populations that constitute it. To be a body is to have an ontopoietic relationship with external reality, which transforms the individual's very perception of their embodied self—not into a mere perspective, but into an actualization. The body is configured not only through its relationship with the external world but also through acts directed toward itself, of which conscious and subconscious processes are only some of the possible self-oriented relations capable of producing consequences for ontopoietic metamorphosis.

To conclude, if we aim to explore the phenomenology of life, we cannot ignore the fundamental characteristics shared by all living beings, regardless of species or other categorical distinctions. The eight metapredicates outlined above—common to life in every shape and form—are thus an essential starting point for understanding what it means to be alive.

References

Barbaras, Renaud. 2022. *Introduction to a Phenomenology of Life*. Bloomington: Indiana University Press.

Bergson, Henri. 1911. *Creative Evolution*. New York: Henry Holt and Company.

Boyd, Charles A. R. 1993. *The Logic of Life: The Challenge of Integrative Physiology*. Edited by Denis Noble. Oxford: Oxford University Press.

Cain, Michael L., William D. Bowman, and Sally D. Hacker. 2011. *Ecology*. Sunderland: Sinauer Associates.

Continenza, Barbara, and Elena Gagliasso. 1996. *Giochi aperti in biologia*. Milano: Franco Angeli.

Darwin, Charles. 1964. *On the Origin of Species: A Facsimile of the*. 1st ed. Cambridge, MA: Harvard University Press.

Dawkins, Richard. 1976. *The Selfish Gene*. Oxford: Oxford University Press.

De Fazio, Gianluca. 2020. Il chiasma tra Natura e trascendentale: "un'originaria duplicità". *Lo Sguardo. Rivista di filosofia* 30:323–342.

Douglas, Angela E. 2014. *The Symbiotic Habit: How Organisms Live Together*. Princeton: Princeton University Press.

Driesch, Hans. 2020. *The History and Theory of Vitalism*. North Charleston: Alpha Edition.

Forestiero, Saverio. 2009. Ambiente, adattamento e costruzione della nicchia. In *Life and Time: The Evolution of Life and Its History*, ed. Sandra Casellato, Pierre Burighel, and Alessandro Minelli, 253–283. Padova: Cleup.

Hall, Brian K. 1992. *Evolutionary Developmental Biology*. London: Chapman & Hall.

Haraway, Donna J. 2016. *Staying with the Trouble*. Durham: Duke University Press.

Heidegger, Martin. 1978. *Being and Time*. Oxford: Blackwell.

Heidegger, Martin. 1982. *The Basic Problems of Phenomenology*. Bloomington: Indiana University Press.

Henry, Michel. 2008. *Material Phenomenology*. New York: Fordham University Press.

Husserl, Edmund. 1973. *Zur Phänomenologie der Intersubjektivität: Texte aus dem Nachlass. Erster Teil: 1905–1920*. The Hague: Martinus Nijhoff/Springer Netherlands.

Husserl, Edmund. 1986. *Phänomenologie der Lebenswelt: Ausgewählte Texte, 2*. Ditzingen: Reclam.

Husserl, Edmund. 2001. *Logical Investigations*. Vol. 1. London: Taylor & Francis.

Jonas, Hans. 2001. *The Phenomenon of Life: Toward a Philosophical Biology*. Evanston: Northwestern University Press.

Krueger, Dana M., Richard G. Gustafson, and Colleen M. Cavanaugh. 1996. Vertical Transmission of Chemoautotrophic Symbionts in the Bivalve *Solemya velum* (Bivalvia: Protobranchia). *Biological Bulletin* 190 (2): 195–202. https://doi.org/10.2307/1542539.

Kuriyan, John, Boyana Konforti, and David Wemmer. 2013. *The Molecules of Life: Physical and Chemical Principles*. New York: Garland Science.
Lewontin, Richard. 1983. The organism as the subject and object of evolution. *Scientia* 118:63–82.
Lorenz, Konrad, and Franz Kreuzer. 1981. *Leben ist Lernen: Von Immanuel Kant zu Konrad Lorenz. Ein Gespräch über das Lebenswerk des Nobelpreisträgers*. München: Piper.
Lovelock, James E. 1979. *Gaia: A New Look at Life on Earth*. Oxford: Oxford University Press.
Marchesini, Roberto. 2018. *Beyond Anthropocentrism. Thoughts for a Post-Human Philosophy*. Sesto San Giovanni: Mimesis.
Marion, Jean-Luc. 2007. *The Erotic Phenomenon: Six Meditations*. Translated by Stephen E. Lewis. Chicago: University of Chicago Press.
Mayr, Ernst. 1982. *The Growth of Biological Thought: Diversity, Evolution, and Inheritance*. Cambridge, MA: Belknap Press of Harvard University Press.
Mazzotti, Stefano. 2024. *Meravigliose Creature*. Bologna: Il Mulino.
McFall-Ngai, Margaret J. 2002. Unseen Forces: The Influence of Bacteria on Animal Development. *Developmental Biology* 242:1–14.
Merleau-Ponty, Maurice. 1976. *Structure of Behavior*. Boston: Beacon Press.
Merleau-Ponty, Maurice. 1995. *La nature: Notes, cours du Collège de France. Suivi de Résumés de cours correspondants*. Paris: Éditions du Seuil.
Merleau-Ponty, Maurice. 2013. *Phénoménologie de la perception*. London: Routledge.
Monod, Jacques. 1971. *Chance and Necessity: An Essay on the Natural Philosophy of Modern Biology*. New York: Alfred A. Knopf.
Nagel, Thomas. 2024. *What Is It Like to Be a Bat?* Oxford: Oxford University Press.
Newman, Stuart A. 2017. "Toward a Nonidealist Evolutionary Synthesis." In Challenging the Modern Synthesis: Adaptation, Development, and Inheritance, edited by Philippe Huneman and Denis Walsh, 188–210. Oxford: Oxford University Press.
Odling-Smee, John. 1988. Niche Construction: The Neglected Process in Evolution. In *The Role of Behavior in Evolutionedit*, ed. Henry C. Plotkin, 98–134. Cambridge, MA: MIT Press.
Pasternak, Charles A. 1998. *The Molecules Within Us: Our Body in Health and Disease*. New York: Springer.
Pigliucci, Massimo, and Gerd B. Müller. 2010. *Evolution. The Extended Synthesis*. Cambridge, MA: MIT Press.
Prigogine, Ilya, and Isabelle Stengers. 1977. The New Alliance. Part One: From Dynamics to Thermodynamics: Physics, the Gradual Opening towards the World of Natural Processes. *Scientia* 112:287–318. Part Two: An Extended Dynamics: Towards a Human Science of Nature. Scientia 112:643–653.

Quammen, David. 2018. *The Tangled Tree: A Radical New History of Life*. New York: Simon & Schuster.

Rhodes, Christopher J. 2017. "The Whispering World of Plants: 'The Wood Wide Web.'" *Science Progress* 100 (3): 331–337.

Roeselers, Guus, and Irene L. G. Newton. 2012. On the Evolutionary Ecology of Symbioses between Chemosynthetic Bacteria and Bivalves. *Applied Microbiology and Biotechnology* 94 (1): 1–10. https://doi.org/10.1007/s00253-011-3819-9.

Rosenberg, Eugene, Orit Koren, Lior Reshef, Raz Zilber-Rosenberg, and Ilana-Keren Rosenberg. 2007. The Role of Micro-organisms in Coral Health, Disease and Evolution. *Nature Reviews Microbiology* 5 (5): 355–362.

Sagan, Lynn. 1967. On the Origin of Mitosing Cells. *Journal of Theoretical Biology* 14 (3): 255–274. https://doi.org/10.1016/0022-5193(67)90079-3. (Published under the surname Sagan; the author was later known as Lynn Margulis).

Schelling, Friedrich Wilhelm Joseph. 1967. *Schriften von 1799–1801. Erster Entwurf eines Systems der Naturphilosophie (1799), Einleitung zu dem Entwurf eines Systems der Naturphilosophie (1799), System des Transzendentalen Idealismus (1800), Über den wahren Begriff der Naturphilosophie und die richtige Art, ihre Probleme aufzulösen (1801)*. Darmstadt: Wissenschaftliche Buchgesellschaft.

Schrödinger, Erwin. 1944. *What Is Life?* Cambridge: Cambridge University Press.

Shemer, Roni, et al. 1996. Dynamic Methylation Adjustment and Counting as Part of Imprinting Mechanisms. *Proceedings of the National Academy of Sciences of the United States of America* 93 (13): 6371–6376. https://doi.org/10.1073/pnas.93.13.6371.

Snow, Dale. 2015. Review of: "Translation and Interpretative Introduction of 'Treatise on the Relationship of the Real and the Ideal in Nature" by F. W. J. Schelling. *International Philosophical Quarterly* 55 (2): 235–250.

Sowa, Rochus, ed. 2008. The Life-world: Explications of the Pregiven World and Its Constitution. Texts from the Estate (1916–1937). Husserliana, vol. XXXIX. Dordrecht: Springer.

Teilhard de Chardin, Pierre. 1959. *The Phenomenon of Man*. New York: Harper & Row.

Tinbergen, Niko. 1963. On Aims and Methods in Ethology. *Zeitschrift für Tierpsychologie* 20:410–433.

Tymieniecka, Anna-Teresa. 1990. *Logos and Life: The Passions of the Soul and the Elements in the Onto-Poiesis of Culture*, Book 3. Dordrecht: Kluwer Academic Publishers.

van der Heide, T., et al. 2012. A Three-Stage Symbiosis Forms the Foundation of Seagrass Ecosystems. *Science* 336 (6087): 1432–1434. https://doi.org/10.1126/science.1219973.

Verducci, Daniela, and Magdalena Küle, eds. 2022. The Development of Eco-Phenomenology as an Interpretative Paradigm of the Living World. Analecta Husserliana (The Yearbook of Phenomenological Research), vol. CXXIV. Cham: Springer.
von Uexküll, Jakob. 2010. *A Foray into the Worlds of Animals and Humans: With a Theory of Meaning*. Minneapolis: University of Minnesota Press.
Westheide, Wilfried, and Reinhard Rieger, eds. 2007. *Spezielle Zoologie. Teil 1: Einzeller und Wirbellose Tiere*. 2nd ed. München: Spektrum Akademischer Verlag.
Wilson, Edward O. 1986. *Biophilia*. Cambridge, MA: Harvard University Press.
Zahavi, Amotz, and Avishag Zahavi. 1999. *The Handicap Principle: A Missing Piece of Darwin's Puzzle*. New York: Oxford University Press.

CHAPTER 7

Animal Subjectivity and Agency

7.1 Premise

The condition of being an animal is often portrayed as an individual and self-contained dimension. The focus tends to be on the subject's ability to assert its presence in the world and to actively take part in it—not just by occupying space, but by transforming it into a field of action and becoming a peripatetic entity, one that can only be represented through four dimensions (Godfrey-Smith 2020). This view is not inherently incorrect, but it becomes problematic when it tries to explain animality solely through internal observation. As an expression of the phenomenology of life, the animal cannot be separated from the network of relationships—sympoiesis—that bring its characteristics into being (Haraway 2016). Indeed, if we isolate a single individual and try to understand the phenomenological reasons for its expressive and proactive way of being in the world, we find that something always escapes explanation (Tinbergen 1963). For instance, we cannot explain a certain behavior solely by referring to the physiological mechanisms that trigger it. We also need to account for why those mechanisms exist in the first place, which leads us to the phylogenetic processes that shaped them. One might then turn to natural selection as a basic explanatory framework, but even here we encounter a complex interplay of factors, such as the animal's active role in constructing its own niche (Odling-Smee 1988), the inevitably historical nature of evolutionary events (Gould 1990), the epigenetic transmission

R. Marchesini, *Philosophical Ecology*,
https://doi.org/10.1007/978-3-032-04987-2_7

of traits acquired by the individual (Müller 2010), the non-deterministic range of possibilities in phylogenetic processes (Lewontin 1983), and the importance of embryogenetic mechanisms in shaping the morphological and functional organization of organisms (Hall 1992).

But perhaps the most significant aspect of the relational nature of development—a point that led scholars Scott F. Gilbert and David Epel to title their work *Ecological Developmental Biology* (2015)—is the concept of *polyphenism* (West-Eberhard 2003; Suzuki and Nijhout 2006). Polyphenism shows that vastly different phenotypes can arise from the same genome, depending on the environmental conditions during development. For example, the migratory locust (*Schistocerca gregaria*) can exhibit two dramatically different phenotypes: one with a uniform green color, short wings, and solitary behavior, and another with long wings, bright coloring, and gregarious tendencies. This variation is not determined by genetics but is instead triggered by population density during development (Pener 1991). There is also seasonal polyphenism: this is found in butterflies of the *Pieridae* family, where spring morphs are darker than the much paler summer ones (Shapiro 1968). Another example involves temperature-dependent polyphenism: in certain reptiles like turtles and crocodiles, the temperature gradient determines whether the embryo develops testes or ovaries (Woodward and Murray 1993). In some gastropod mollusks and crustaceans, one can even observe predator-induced polyphenism: the presence of a predator prompts the development of larger body forms, often with thicker shells or exoskeletons and the appearance of spines. Among amphibians, predator presence causes tree frog larvae to hatch earlier (Warkentin 2005), reduces the size of frog tadpoles (Van Buskirk and Relyea 1998), and in many fish, leads to increased body thickness.

A particular form of polyphenism is linked to the type of nutrition an embryo receives during its differentiation phase. In bees, for example, the development of a queen depends entirely on the diet given to the larvae. The queen's fertility is determined by the size the larva reaches before metamorphosis, which in turn depends on the persistence of a specific hormone, neotenin. This hormone is secreted by the corpora allata, which remains functional only if the larva is fed royal jelly (Brian 1980). The jaws of cichlid fish also take on profoundly different shapes depending on their diet (Meyer 1987), and this occurs in primates as well, including humans. In these cases, the structure of the jaw adapts according to the texture of the food consumed (Corruccini and Beecher 1984), due to the mechanical pressure exerted during development. As Patrick Bateson (2017) reports, poor maternal nutrition during periods of famine can influence the pelvic development of female children, which in turn has significant effects on the development of their fetuses later on. Nutritionally induced polyphenism is a rapidly developing field of study, both in terms of gene plasticity—meaning that different phenotypes can emerge from the same genotype depending on external factors—and in terms of endogenous epigenetics, involving gene expression regulation mechanisms such as DNA methylation, which can silence specific genes.

On the other hand, it is well known that developmental conditions can significantly impact an individual's morphological, functional, and behavioral traits. For example, rat pups that receive less maternal care during their first week of life tend to be more anxious as adults. This is because early tactile stimulation regulates the production of glucocorticoid receptors in the hippocampus, which in turn moderates the intensity of the stress response (Weaver et al. 2007). In dogs, maternal deprivation—such as separating the puppy from its mother before eight weeks of age—leads to a wide range of behavioral changes in adulthood, particularly affecting arousal regulation and social behavior (Pageat 1998). Similarly, in cats, early separation from the mother not only affects social behavior but also increases sensitivity to stress in adulthood; veterinary urologist Tony Buffington (2018) coined the term "Pandora Syndrome" to describe a cluster of conditions and heightened disease susceptibility that stem from a cat's vulnerability to stress. In humans, numerous studies—following the pioneering research of Robert Hinde and John Bowlby on attachment—have shown that early relationships with caregivers have a profound impact on a child's behavioral development. These include emotional regulation and the formation of psychological profiles (Ainsworth and Ainsworth 1958), as well as the development of internal working models—mental representations of the self and of social counterparts (Cassidy and Berlin 1994).

In the early decades of the twentieth century, the study of animal behavior saw the rise of two major, and in many ways opposing, approaches: the North American behaviorist school, led by John Watson and Burrhus Skinner, which held that behavior is shaped by the environment (conditioning) and driven by external stimuli (responses); and the Central European psycho-energetic school, led by Konrad Lorenz, Nikolaas Tinbergen, and Karl von Frisch, which argued that behavior is innately programmed (instinct) and motivated by internal drives (consummatory acts). The paradigmatic differences between these two schools were so significant that it made direct comparison difficult. This was compounded by their differing methodologies: ethologists argued that behavior must be studied primarily in natural settings, where instincts had been shaped by natural selection; behaviorists, on the other hand, considered this approach flawed in terms of internal validity, since outside the controlled environment of a lab, it was nearly impossible to identify the true cause of a phenomenon due to the presence of countless confounding variables. Starting in the 1950s, the distance between these perspectives gradually began to narrow—though never completely disappear—leading to what can be defined as the First Synthesis of Behavioral Analysis.

Yet if we set aside their paradigmatic differences for a moment, we notice that both approaches share a key concern: the relational aspect of behavior. Behaviorists emphasized the functional link between environmental stimuli and the animal's responses; ethologists highlighted the importance (even necessity) of tracing the meaning of behavior back to the adaptive pressures that shaped specific instincts through phylogeny. When we consider an animal's expression in terms of its basic morphological, functional, or behavioral traits, or in the unique experiences that shape its biography, we are compelled to adopt a *relational approach* to both investigation and explanation. The animal individual is a multilayered, inseparable web of relationships and references—like a heavily footnoted text—because its ontological makeup has taken form through relationships that occurred at different times: across stages of phylogenesis, prenatal ontogenesis, various developmental stages, and lived experiences. The individual internalizes and embodies these past relationships, presenting itself to the world having already absorbed them. What appears to be intrinsic to the individual at any given moment—its internal characteristics—are not fixed essences, but the result of prior relationships that have been incorporated (Dawkins 2006). If we look at the life history of an individual from any species, we can observe the process by which external relations (between subject and world) are translated into internal configurations (of the subject). What now appears as an intrinsic trait—such as anatomical structure or behavioral repertoire—was once an external, ecosystem-level factor. After all, what is adaptation if not the incorporation of an external factor—the very one that exerts selective pressure—into an internal predicate of the species?

Indeed, every species bears on its body the marks of interactions with the environment accumulated over millions of years. Its somatic form subsumes a sequence of relational events that, if closely examined, trace all the way back to the origins of life on Earth. Therefore, we must move away from an essentialist ontology, which attempts to explain the subject by isolating it from its relations, and instead adopt an ecological ontology, where it is the principle of relation that gives form to living organisms. This relational lens not only helps us understand phylogeny, but also the relationship between individual development and evolutionary history (evo-devo). For instance, a mutation in the structure of the jaw muscles in humans may have altered the embryogenetic forces shaping the skull, thereby allowing for the expansion of the neurocranium (Carroll 2005). The relational web and the manifold connections of living beings also shed light on the complexity of ontogenesis, in which each developmental stage sets the stage for and influences the next. In conclusion, to explain an animal's behavior, we must take into account numerous relational and connective factors, such as: (i) its ties to phyletic ancestors; (ii) selective pressure factors; (iii) phylogenetic symbioses; (iv) epigenetic mechanisms starting from the zygote; (v) epigenetic developmental processes; (vi) environmental influences on the phenotype; (vii) neonatal environmental and parental conditions; (viii) its physiological state at a given moment; (ix) its situational psychological condition; and (x) the interacting elements in the here-and-now that may trigger or enable a behavior. In short, we cannot understand the expression of an animal by detaching it from all these relations.

7.2 Animal Subjectivity

Among the most significant traits of animal individuals is subjectivity—a quality that encompasses several key features: non-passivity, i.e., active engagement in behavior; unpredictability; selectivity in orientation and responses; and the possession of a unique perspective, a point of view on the world (Shettleworth 2010). Subjectivity is always present in animals—not just in moments of awareness—and reveals itself through certain recognizable traits: (i) an orientation toward the world, meaning an animal does not merely drift with external forces, but undertakes its own path, even opposing them; (ii) the possession of personal interests, which it seeks to defend and pursue in interaction with the environment, expressing its own will; (iii) expressive partiality, which gives meaning to the notion of "subjective"—a trait that combines freedom and indeterminacy, but also arbitrariness and unreliability; (iv) full presence in the here and now, characterized by the ability to inhabit the situation it finds itself in; its time is *kairological*, that is, a "time-for"; and (v) the ability to distinguish its own actions from what is happening in the world. The subject is one who chooses its conduct, not necessarily in a conscious way, and who invents behavior, going beyond learned competences (Safina 2015).

An animal is a subject because its response to events is always innovative; in order to meet the challenges posed by the singularity of reality, it cannot afford to simply repeat. This marks a clear distinction between a mechanistic conception of the animal—as a system of discrete automatisms—and the eco-ontological view of the animal as a relational entity, whose subjectivity is the *conditio* sine qua non for continuous information exchange with the environment and for its dynamic and ongoing transformation or adaptation—its intrinsic and inevitable becoming—through the incorporation of external factors into its own somatic wiring, a process that engages every system of the body. A machine performs a function based on an internal organization of mechanisms; an animal, on the other hand, becomes through relation. What constitutes it is not a pre-given function, but a field of functional virtualities that become actualized—realized—through relational engagement. Subjectivity is thus a presence: not necessarily conscious, but nonetheless capable of inhabiting the here and now—in the broader sense of participating, intervening, attending, and doing so in the first person—in order to improvise, using the resources at hand, the best possible response to a given situation. This openness to possibility, its being always situated within a field of expressive uncertainty—its condition of becoming—demands that it be there in the sense of *Dasein*, that it be present in the moment. Subjectivity, then, is the emergence of a self endowed with presence.

In relation to the nature of subjectivity, I have proposed in various works—such as *The Creative Animal* (2022)—five levels of subjectivity that help us better understand how the subjective self manifests in animal life:

1. Interface: the emergence of a self through the capacity to distinguish between effects passively experienced in interaction with external events (*sensibility*) and effects actively produced by one's own actions in the world (*agency*);
2. Affectivity: the development of a *for-Self*, based on being a bearer of personal interests that must be defended (*responsiveness*) and pur-

sued (*proactivity*), and that are regulated by the pleasure system, which serves as a compass for orientation;

3. Cognitivity: the emergence of an *in-Self*, characterized by the ability to gather information through interaction with the world and to use one's faculties as tools under personal ownership—unlike the mechanistic model based on automatisms typical of machines;
4. Mind: the possession of an internal systemic organization capable of producing a plurality of system states (psychological conditions) and of generating new representations, such that behavior always entails a bringing forth, or expression, of these internal states;
5. Consciousness: the explicit, aware articulation of internal processes—ranging across various levels of consciousness, including sentience, awareness of intentions, and self-consciousness—that characterize the psychological condition of the individual at a given moment.

These five layers, which form the basis of subjectivity, do not withdraw the individual into itself but act as relational devices—they are grounded in relationality.

Let's start with the first aspect: the interface, which shows how the sense of Self does not arise from mere self-reference, but from a relational process between subject and world. Both movement and action carried out by the individual produce sensory effects that could be attributed to changes occurring outside the subject, due to environmental phenomena. Therefore, it becomes necessary to distinguish between the changes in the environment caused by the animal's own actions and those resulting from external events. There is a difference, for instance, between touching something and being touched by something, because we know that the environment is not a static backdrop, but a place where events occur. In practice, it's a matter of being able to differentiate what happens in the world from what the individual causes in the world. In animals, a system has evolved called *reafference*, hypothesized by the physiologist Erich von Holst, through neural networks tasked with processing incoming information, distinguishing between: (i) sensibility—what happens out there in the world; and (ii) agency—what results from the animal's actions. This is why, when we move our eyes across various targets scattered along the horizon, we don't perceive the world itself as moving. This is a basic mechanism that, nonetheless, represents the earliest formation of a Self, as Peter Godfrey-Smith notes in his book *Metazoa* (2020). The phenomenon of reafference shows that the Self arises from a need for interface with external reality—that is, from a relational need, not from self-reference.

The second level of subjectivity, affectivity, shows us the animal as an entity endowed with sensibility and characterized by having intrinsic interests—that is, interests that are its own and that concern it, a *for-Self* condition. The animal has physiological needs, such as eating and drinking, but to meet these it must be driven to carry out specific actions, like foraging for food or finding water sources to quench its thirst. The receptors within the body not only initiate physiological mechanisms to counteract hypoglycemia or dehydration, but they also activate specific areas in the brain to produce the psychological sensations of hunger and thirst. Similarly, the body is filled with sensors whose activation produces feelings of pleasure or pain, together forming a general hedonic system—so we can say that the body is a network of pleasant or unpleasant sensations connected to external reality. Finally, affectivity consists of: (i) emotions, which lead the subject to respond in specific ways to surrounding events in order to defend its own interests; and (ii) motivations, which drive it to act to fulfill those interests, that is, to achieve what satisfies them. Overall, affectivity gives rise to copulative expressions toward the world, so that being a bearer of intrinsic interests means having an internal purpose that connects the individual to the world. Affective characteristics represent the driving forces of behavior, such that we understand the subject's actions not as mere reflexes to stimuli, but as a search for opportunities to fulfill its needs—primarily seeking expressive agency.

The third level of subjectivity, cognitivity, presents us with an animal in continuous evolution, as it gains experience through its relationship with the world, constantly gathering new information and thereby modifying its internal resources—*in-Self*. In the cognitive approach, knowledge endowments are viewed as tools available to the subject, in the form of processing schemas or functional representations. To think of these endowments as tools ready for use offers an alternative model to that of automatism: while an automatism controls you, a tool is something you use. Knowledge, too, should be understood as a relational endowment, meaning that: (i) it is the result of relational processes—phylogenetic in the case of innate endowments, ontogenetic for acquired ones; (ii) it is employed in all interactions with the external world, whether for evaluating different situations or for acting with competence. The individual's experiential condition means that being a subject implies being open to information from the outside. In this sense, we can see the subject as an entity that carries itself into the world through the information it possesses—and at the same time, through this engagement with the world it is continuously enriched with new information. Learning, therefore, represents a fundamental feature of animal subjectivity, as Konrad Lorenz argues in his dialogue with Franz Kreuzer in *Leben ist lernen* (1981): the continuous transformation of knowledge means that a subject can never truly repeat an experience.

The fourth level of subjectivity, mind, refers to a unified biographical condition of the subject, meaning it cannot be evaluated isochronically (i.e., in fixed or isolated moments) like other natural phenomena, because a subject always carries with it a unique and unrepeatable history. The mental condition speaks to us of a systemic dimension of the subject—one that can never be broken down into separate components, because its Being-there in different situations is always unified. This is the key difference between those who view the animal as a kind of puppet, driven by various automatisms—instincts and conditioning—and those, like myself, who consider animal behavior as an expression (from the Latin *ex-primo*, meaning "to bring forth") of an internal psychological state, an emergent result of the systematization of all its constituent elements. The subject is, therefore, a unified entity that engages with the external world by adopting a perspective in its entirety. In this sense, we can say that the mind is, in every respect, a relational structure (Siegel 1999). Furthermore, it must be emphasized that the mind itself develops from the convergence or interrelation of affective components (expressive motives) and cognitive components (expressive modalities).

The fifth level of subjectivity, consciousness, refers to the subject's capacity to take charge of—or make explicit—what it is experiencing at a given moment, and thus to have awareness of something specific. Consciousness has intentional value, in the sense given to intentionality by the German philosopher Franz Brentano, who described it as the tendency of acts of consciousness to be directed toward particular contents. In phenomenology, intentionality does not correspond to the idea of having the intention to perform a certain action—it is not related to will or decision-making, which are psychic dispositions linked to the concept of agency. Instead, intentionality is awareness-of something; conscious states are always oriented toward an object-content. Edmund Husserl expanded on this idea by distinguishing between the intentional act—thinking or being aware (noesis)—and the objective element, the content of the thought (noema). For Husserl, the defining feature of intentionality lies in this structure of thought always being directed toward something that is other-than-self. Therefore, consciousness too is a relational condition, which can be directed toward external contents—such as being aware of a fact, an experience, or a rule—but also toward internal contents, such as suffering, pleasure, emotions, motivations, thoughts, self-awareness, or theory of mind (Dennett 1996).

Thus, the animal is not a closed-off entity, but a constant movement outward—an unceasing dialogue with everything around it, an unstoppable drive toward exploration. In short, there is a deep relationship between animality and relational protagonism, between subjectivity and openness to the world. The animal always holds a relational position: (i) it moves actively toward something, (ii) it does not merely occupy space but takes its place in it, (iii) it does not merely drift in the flow of time but emerges from it and constructs its own present, (iv) it does not simply react to stimuli but perceives what happens in the world, (v) it processes the information it receives and builds a tailored response, (vi) it is not driven by pure impulses but feels sensations that predispose it to specific actions, (vii) it is not a slave to automatisms but uses its full knowledge endowments to determine the most appropriate behavior for the situation, and (viii) it is not poor in world but is world building.

The animal invents situations, turning the world into a field of opportunities, because it is driven to act upon what surrounds it and to differentiate between what comes from external reality and the perceptual changes caused by its own actions. The animal is in constant search: it does not merely orient itself toward a source, but is a protagonist in its relationship with the context, always seeking a way to act within it. The animal condition, therefore, is a projection onto the environment—a tendency to modify what it encounters and to build its own niche, rather than being at the mercy of external variables.

7.3 Agency

Among the fundamental properties in an animal's everyday life, one of the most crucial is the ability to make timely decisions that are also attuned to the particularities of the situation the individual is facing at a given moment (Griffin 2001). This faculty is referred to as *agency*, and it designates the notion that the individual must be considered an active agent of its own behavior—capable, in other words, of selecting the most appropriate response in order to face the challenges it encounters (Washburn 1916).

Agency is a competence that brings together, and derives from, more basic properties, such as: (i) having goals to achieve and interests to defend, two variables that function as the compass for the subject's orientation; (ii) the ability to assess the field of possibilities, obstacles, and risks that are specific to the current situation; (iii) access to flexible knowledge and skills that can be applied with adaptability—both in scope and use—according to the demands of the specific context; (iv) the capacity to choose based on internal simulations whether to act, when to do so, and how to execute the action, thus allowing for a fluid and non-predetermined operational configuration; (v) the ability to adjust one's behavior based on feedback received during the action itself, rather than persisting in dysfunctional or unproductive behaviors; (vi) the ability to differentiate behavior based on changing variables or specific objectives.

Agency is a relational characteristic that places the subject in constant interaction with everything around it, including through communication (Hearne 2007). It confers an intrinsic protagonism on the subject, allowing it to emerge from its surroundings, but it would be mistaken to view this property as a detachment from the relations with external reality. Agency is, in fact, a relational faculty capable of transforming the world into a network of relations that enable the subject to be present to itself through the continuous encounter with all that is other (Pouydebat 2017). We should never interpret agency as a detachment from external factors; rather, it is a bending of reality—a deformation of space-time—through one's presence, akin to the effect of a gravitational field. Agency arises from the interaction between an individual who carries echoes of the past—the set of instances that manifest within it but are, in fact, expressions of distant and ancestral phylogenetic events—and the present, while simultaneously projecting this interaction into a not-yet-present condition (Christen 2009). In this sense, agency is a *diachronic function* that links different temporal dimensions, connecting the past to the future. Thus, we might also say that, as a manifestation of the phenomenology of life, the animal, through its faculty of agency, dissipates information (Marchesini 2018)—information accumulated in the past and expended in order to actualize its own presence.

The need to make decisions undoubtedly stems from the singularity of situations: in such a richly variable context, it is inevitable that reality presents itself each time in a new form. But this need also arises from the internal fluctuations of the individual itself, which at one moment is hungry, at another is driven by mating interest, and at other times may be focused on achieving a goal or, conversely, on seeking safety. In other words, we are faced with two relational singularities: one being *appositional*, referring to a world that constantly changes its configuration in terms of opportunities, possibilities, obstacles, and risks; the other *propositional*, in relation to inherited drives—affective configurations (feeling-desiring) made up of sensations, emotions, and motivations descending from the past. The convergence of these two relational factors generates the subject's need to be present—that is, to bring forth their full capacity to produce ad hoc behaviors that are equally singular as what they are experiencing.

We could say, then, that every decision results from the convergence or interaction between the subject's proposition—meaning the synthesis of their intentions—and the situational apposition, meaning how the world presents itself to the individual at a given moment, offering a web of relationships translatable into a field of action. It is within this field of action that the subject finds the operational conditions to activate its heritage in the here and now, effectively making it present—like a kind of time machine. When the animal opens its eyes to the world, it does not find itself in a state of *thrownness* (Heidegger 1978) but rather in a state of *awakening*, because the experiences of its ancestors take the shape of hazy memories that, in the presence of the world, begin to clarify. The world becomes the key that unlocks its inner safe. The cub opening its eyes to the landscape is not gripped by anguish but by joy and excitement—like someone rediscovering something precious, who feels the thrill of coherence between inner and outer worlds. An animal's existential dimension is grounded in having-already-been-there in the world, in remembering rather than in seeing for the first time.

The animal's agency arises from this condition of the individual, whose existence is rooted in the past. Its interest and wonder before the world do not arise from a vacuum, but from impulses already present before birth—as Konrad Lorenz notes in *Behind the Mirror* (1977). To be bearers of such impulses entails—indeed, imposes—an engaged attention, a profoundly relational stance, a true rootedness or, more precisely, a being-already-configured for complementary relation by external referents—what is usually termed *heteronomy*. Earlier, I showed how heteronomy is a meta-predicative quality of the phenomenology of life, focusing on morphological structures like the gills of a fish or the wings of a bird. In this case, I am referring to heteronomy in terms of agency: all emotional and motivational predispositions of the animal, as well as its a priori knowledge, are nothing more than the negative imprint of something that exists in the world. Thus, agency is a relational faculty.

At the same time, the subject does not merely bring the past into the present—its existence is not an echo, much less a simple reiteration of inherited drives. One of the fundamental characteristics of animality, from an evolutionary standpoint, lies in the principle of *exaptation* (Gould and Vrba 1982) and *neuronal recycling* (Dehaene 2020)—both concepts traceable to François Jacob's idea of *tinkering* (Jacob 1977)—which show that any biological endowment can be coopted for new functions. Behavior is never a simple repetition of what is inherited or learned; it is rather the revitalization of the past in the present, a taking possession of one's impulses and a dealing with situational conditions through apposition. Indeed, being carriers of inner drives implies the need to transform the world into a field of agency. But to do so, one must measure that world with great care. The individual must know their action space—not in absolute terms, but in relation to the action it intends to undertake. Evaluation, therefore, becomes a sub-function of agency. Possessing inner drives necessarily implies openness to the outside. The animal does not simply open its sensory channels—it craves the surrounding world and somatizes it. An event is no longer merely an external occurrence but becomes an emotional state, a condition of the body.

Animality is one of the many possible expressions of life—a specific mode of existence with its own peculiarities. Chief among them is this specific mode of openness to the world: affectivity, and the resulting operational protagonism. Emotions and motivations define the drives—that is, the horizon of interests to be defended or pursued—and at the same time, the mode of openness to the world. The animal is open to the world precisely because it is a bearer of interests. This condition grants subjectivity to its perception—it is not neutral, impartial, and detached—and simultaneously imposes upon it the ability to make decisions. This condition compels the animal to transform the external scenario into (i) events that carry categorical value, meaning general evaluative weight in relation to the quantification of its interests—thus translating into a bodily state (emotional feeling); and (ii) a projective space that outlines the gradients of agency for expressing behavioral tendencies—thereby defining possible goals toward which to direct attention (motivational desire). One cannot understand an animal outside of its condition as an interested entity—that is, possessing an internal motive for action. Its phylogenetic legacy makes it restless, as if it had a task to accomplish and little time to do it. But precisely for this reason, its actions toward the context do not follow mechanical or repetitive patterns. Rather, they require relational capacity: the ability to extract from the singularity of each situation its specific opportunities and threats, and simultaneously to interpret one's own inner repertoire with flexibility and creativity (de Waal 2016).

Innate equipment, therefore, does not limit creativity—it is in fact the essential factor in the adaptive dance. One cannot fully grasp this *being-interested* without understanding the animal's diachronic dimension, whereby present behavior: (1) always refers to past events, since its motivational dispositions, bodily physiology, and innate cognitive tools have been shaped by phylogenesis; (2) is oriented toward something in the future, as interests—both in the motivational sense of a goal to achieve and in the emotional sense of preserving and defending one's interests—are always projections toward something yet to happen. As Nikolaas Tinbergen already understood in the 1960s through his "Four Questions" of ethology—which reformulated Aristotle's explanatory tetrarchy in evolutionary terms—an animal's action cannot be fully understood when analyzed solely in the present of its unfolding. I would argue that the convergence of these four causalities upon the subject—guided by the compass of pleasure—forms the basis for the individual's right to choose a course in the moment: in other words, agency.

The intersection of inherited tendencies, which prompt an engagement with the world to enhance one's interests, and situational singularities, in terms of present opportunities and risks, creates the conditions for a dialogical plane between subject and world. This encounter gives rise to the necessity of making informed decisions. The individual cannot escape the demand to offer singular responses—ones that are contextually appropriate, not repetitive. Experience arises from this interested openness to the world, which places the individual in a state of continuous interaction with everything around it. From this stems the need to enact flexible behaviors, to interpret the specific situation, and to choose among operative options through the faculty of decision-making. This brings us back to the notion of being interested as the foundational trait of agency. There would be no experience without emotional feeling and motivational desiring, because it is precisely from these dispositions that the yearning toward the world is born—the affectivity that somatizes it. Openness, in fact, does not merely involve reception of the surrounding world; it becomes an operative orientation, a solicitation to discover the right responses needed to carry out one's interests.

The environment is constantly changing, and this has particular importance for a living being whose survival depends on movement. It's about creating a kind of meeting point between two shifting realities: (i) the animal's own peripatetic condition—its movement across the world; (ii) the fluctuating condition of the world itself, with its continuous changes of state (Vallortigara 2021). Because the individual is an "interested being," as I've explained, it must anchor all its vital aims within the framework of space and time. Decision-making naturally follows from this, as time becomes a critical factor: opportunities can't wait, and dangers can quickly become deadly if not tackled at once. Faced with singularity, simply repeating known behaviors isn't enough. Even skills and knowledge would be useless unless they could be reshaped and adapted with flexibility (Gould and Gould 2007). Moving through the space of possible actions, then, means keeping a certain rhythm with time—living within a clear temporal dimension: being in-time as well as capable of meeting the world's ever-changing, unpredictable challenges. The animal's peripatetic condition means it must constantly engage with time, with a becoming that arises from its own subjective condition of temporality.

Animals therefore truly live in the present, but not in a static here-and-now. Their present is a personal construction: an active, attentive participation in the unfolding of events. This demands the ability to create space-time links, such as: (i) connecting events that happen simultaneously or in sequence; (ii) recognizing cause-and-effect relationships; (iii) making choices based on how events unfold over time; (iv) tying feedback to specific actions in a meaningful chronological order; (v) calculating how long actions will take to be effective; (vi) and anticipating future events by projecting how current situations might evolve. Building these temporal connections is essential for decision-making—and for truly owning one's choices—because everything happens within the flow of time.

Neurobiological studies have shown that primate brains contain special timing neurons—cells in the motor cortex that preactivate in anticipation of an event, almost like an internal countdown. Others track how much time has passed since a particular occurrence (Honing 2019). The brain's motor network, including the basal ganglia, thalamus, and motor cortex, plays a key role both in planning actions and in keeping track of their timing (Merchant et al. 2011). All the equipment an individual possesses—whether in terms of physical endowments or cognitive tools—is not designed for automatic, mechanical reactions to a changing reality. Instead, it exists to support the individual's decision-making ability, helping them keep pace with the unfolding of events over time. An animal, therefore, cannot simply repeat a learned instruction. It must monitor how events progress, adjusting its behavior according to how situations actually evolve.

The real world is made up of singularities—situations that, even if they seem similar in their sequence, each has crucial differences in their development, often decisive in the game of survival. This is why agency is essential: as Michael Tomasello explains in *The Evolution of Agency: Behavioral Organization from Lizard to Humans* (2022), agency is not just about reacting to stimuli, but about actively steering—and even planning—one's actions toward goals. To do this, an individual must pay close attention to how situations unfold during the course of action. In the biological context, the margin for error is extremely narrow; a mistake can be fatal. Therefore, an animal must not only recognize superficial patterns, but also detect the temporal qualities of a situation, analyzing any new developments in order to build action strategies that fit the specific timing of events. As Tomasello (2022) points out, an agent doesn't simply "aim and shoot" at its goals in a ballistic fashion. Rather, it flexibly controls—and even executively self-regulates—its actions, making informed decisions about what will work best at different moments as a dynamic situation unfolds.

This perspective reveals that the animal doesn't live within *kronos*—the rigid, external time—but within *kairos*: the time of relationships, rhythms, and movement. We should not imagine the present as a prison for the individual, nor think of action as a rigid, isochronic event. Instead, action resembles a flexible fabric, able to adapt to various temporal flows, creating an existential experience that is essentially diachronic. An animal moves according to its own temporal rhythm—compare, for example, the pace of a tortoise with that of a mouse. Within this kairological flow of time, it must make decisions based on connections and opportunities that demand full presence, not the kind of captivation Heidegger and other philosophers invoked when portraying animals as nothing more than preprogrammed mechanisms.

Being present in the moment is perhaps the most difficult quality of being-an-animal to explain. It forces us to confront a subjectivity that sets its own terms for how it participates in the world. We cannot think of animal agency as simply existing in an eternal present, because for an animal, a present exists only insofar as its existence is rooted in a past and projected toward a future. Agency is tied to this diachronic dimension of the subject: it means acting in the here and now, using accumulated capacities flexibly, to imagine a future condition that has not yet materialized (Bekoff and Allen 2002). If we forget this kairological dimension of the individual—the experience of the present as a space of opportunity—we lose sight of agency. We risk this, for example, by: (i) assuming behavior is entirely flexible and shaped by circumstances; (ii) thinking that the animal merely exhibits preprogrammed automatisms; or (iii) believing that the animal is incapable of projection toward a not-yet-actualized condition. The diachronic coordinates that build the present through a continuous dialogue between past and future involve the unfolding of external conditions, the individual's own actions, its history, and its anticipations.

Thinking of the animal as a machine would mean immersing it in an eternal here-and-now that cannot even be properly called a "present"—a purely reactive condition where all mechanisms are predetermined. In this view, the individual disappears, replaced by an algorithmic expression. Conversely, recognizing animal agency means acknowledging the individual's permanent state of relationality, which can only operate within singular situations (Marchesini 2022). Even the animal's physiological state is a singularity that influences its choices in the present, since internal conditions produce shifts in priorities. The body not only exhibits a psychological time that differentiates one species from another, but also varies moment to moment within the same individual (Sapolsky 2024). The body oscillates constantly—through phases, biorhythms, seasons, metabolism—all of which affect the individual's emotional state and perception of time. This shows us the necessity for the individual to face the challenge of being-in-time through ad hoc responses that are both flexible and improvisational. The combined effects of these singularities require the animal to act with creative presence, because, as Henri Bergson noted, "As soon as we go out of the encasings in which radical mechanism and radical finalism confine our thought, reality appears as a ceaseless upspringing of something new" (1911, 46–47).

An animal is a desiring entity, and to be desiring means to project oneself toward future conditions, which emerge from the dialogue between subject and world unfolding in the present. For this reason, the animal condition must be rooted in *Dasein*, being-there—that is, being at a certain time and in a certain situation; in short, being-in-relation. Agency, therefore, is an ecological condition, and we can speak of an eco-ontology of animality, whereby the individual carries within itself past (phylogenetic) relations and brings them into play in existence—*ex-sistere*, meaning to stand out, to rise from the ground, to appear—bringing them into the present through the subjective interpretation of its roots in the past. At the same time, the individual reshapes them through the contingent (ontogenetic) relationships that it establishes in its being-there, its presence in the moment, in order to project them into the future.

The desiring condition, which the humanist tradition has interpreted as originating from a lack and being fulfilled through appropriation, actually arises from an exuberance—the confluence of multiple drives in the subject's present—and is fulfilled by the subject's gift of itself to the world through action. The condition of *yearning* thus lies at the root of propositional attitudes—that is, referential states (states directed toward something) such as rejoicing-for, striving-toward, being-surprised-by—that drive the individual to complete a particular action according to a given temporal framework. At the same time, referentiality inevitably places the individual at an interface with external reality, faced with: (i) risks and opportunities, which fluctuate with environmental changes; (ii) the inherent ambiguity of situations, which never allow for a complete prediction of outcomes; and (iii) the volatility produced by the subject's own actions, since outcomes are influenced by the subject's interventions. Thus, we can say there is a direct proportionality between the gradient of uncertainty the subject faces and its need for agency.

The animal is always projected toward something yet to be realized, in an existential prospection that, even before it translates into physical movement, is already a psychological going-beyond. Yet, this movement into the future is not a predetermined condition—a mere unfolding of triggers, a cascade of mechanisms whose outcome is already set by initial conditions—but is literally the creation of a becoming. By projecting itself, the subject invents its future through the dialogue that unfolds through its encounters, that is, historically. Thinking ecologically therefore means abandoning the isochronic conception of events (Prigogine and Stengers 1984; Gould 1990) and the illusion of Laplace's demon, because the unfolding of the prospective dialogue generates historical events.

Through its agency, the animal imprints a specific direction on the course of events, bringing about a succession that cannot be inferred solely from existing variables—although, of course, it is influenced by them—because the subject always adds something that emerges in the moment. Agency draws nourishment from this occasional productivity of the individual: the ability to recycle its resources, to reconfigure them to achieve new performances, and to operate analogically to imagine something that does not yet exist. The presence of a desiring condition keeps the subject open to the world; thus, we speak of phylogenetic propensities not as self-referential determinisms, but as copulative factors. From the encounter between the subject's projections and the actual conditions of reality arises the state of uncertainty that the individual must manage, converging all its resources into a systemic orientation toward a purpose. This permanent state of uncertainty creates the need for decision-making capacity (Hinde 1966). Behavior, therefore, cannot result from mechanisms acting autonomously, because the individual needs access to all its resources in order to formulate an appropriate response to the challenges posed by the world. Returning to Tomasello's essay: "the evolution of agentive decision-making occurs mainly in response to different types of unpredictabilities, and we can have little doubt that the most unpredictable entities in the lives of most organisms are other organisms" (2022, 21).

As John Z. Young notes in *Philosophy and the Brain* (1987), the constant pursuit of goals is the essence of life preservation. This pursuit is not random; rather, it is directed toward achieving specific immediate goals that are pre-established for each being. An animal constantly faces the need to make choices in situations that always present themselves in a singular manner—bringing with them a high degree of novelty, both in the unfolding sequence of events and in the systemic organization of the phenomena encountered. In *Why Choose This Book? How We Make Decisions* (2006), Read Montague writes that the ability to assign value to different choices—and thus to exercise agency—evolved because it provided an effective solution to essential survival problems. Indeed, the most critical problem was the need for flexibility in facing uncertainties. Once again, it becomes clear that the necessity of decision-making capacity arises from the relational condition of the animal.

Subjectivity manifests itself in this protagonism in the face of uncertainty, which demands that the individual never interrupt the dialogue with external reality. The animal cannot avoid making choices and devising strategies each time, as it faces problems that always contain wide margins of unpredictability. To dwell in the world is to embark on a path that requires openness to the possibilities offered by circumstances. Ultimately, one must shape one's project by identifying the available operational space: acting on the context means thinking about the environment, not merely adapting to it (Thorpe 1979). The animal must necessarily take risks, bringing something new or untried into its behavior if it wants to carry out its intentions. Its life thus unfolds in the uncertainty of outcomes, and it must rely on a form of supervision that involves not only competence or appropriateness, but also innovation. The individual must be able to bend situations to its advantage, often through performative tinkering and, at times, the invention of entirely new strategies. The capacity to pursue goals and, therefore, to hold intrinsic values—what Henri Bergson calls *intrinsic teleology*—is a fundamental and irreducible quality. If this highlights the motive behind subjective action, it also reveals how the continuous pursuit of goals forces the individual to turn uncertainty into a condition of advantage. Indeed, while uncertainty may in some ways constitute a handicap, it undeniably also fluidifies the status quo, creating spaces for adaptive inventiveness.

An animal is constantly faced with evaluative challenges regarding the cost-benefit trade-offs of a behavior, the possibilities and obstacles of a given situation, the comparative advantages of action versus inaction, and the choice among different actions to identify the one that will work best in that context (Eibl-Eibesfeldt 1970). The subject is therefore forced to pay close attention to its surroundings—to remain interested in opportunities and alert to risks, to construct representations of the situation it finds itself in, and to recall past experiences that might help clarify ambiguous circumstances (Curio 1976). But as mentioned, this suspension of certainty, which imposes the non-interruption of the dialogue, also creates spaces for creativity. The subject can occupy these spaces thanks to its decision-making ability and its attunement to the local situation. Uncertainty, in other words, assures the individual that not everything is already predetermined. It grants a margin of possibility, which translates into a freedom of state—a prerequisite for creativity itself (Brensing 2020). It is in this condition that the animal's behavior becomes unpredictable, using all its available resources with extreme flexibility. Uncertainty demands that the subject inhibit certain responses and modulate emotional reactions based on the specific situation (Kershenbaum 2020). Self-control reveals a constant relationship between the individual and the situation, in which decision-making begins with the capacity to choose whether or not to pursue a given goal. Tomasello speaks of *global inhibition* as a basic function that allows goal-directed agents to operate through go/no-go decisions in a sequential manner—action by action—deactivating one and activating another, depending on the requirements of the goal and the situation (2022). Self-control is thus a fundamental function for sustaining the kind of purposeful, goal-directed decision-making that defines agency.

In short, agency is the ability to inhabit the present moment while also observing it from the outside—witnessing one's way of being in that moment—because only in this way can past experiences resonate with the individual and, at the same time, imagination project one into the future. The animal observes the relational world around it, but not merely as a landscape: it is the very field in which its action unfolds. It is only then that we can speak of full agency—the moment when the subject decides for itself, holding the authority to choose its actions, like a helmsman steering a ship. And it does not matter whether this decision-making faculty, which leads the subject to accept uncertainty and face the unknown, is more or less conscious; what matters is that, in some remote corner of the self, this process takes place (Libet 2004). The tendency to interpret the present moment prospectively—the moment in which the subject encounters uncertainty and makes it their own—adds another layer to the protagonism of the animal: that of being the bearer of a project or prospection, capable of grasping the moment by drawing together different temporalities into the here and now. One aspect of this self-directed control in action lies in simulation: the ability to envision not only through imagination, but also by evaluating the possible outcomes of various action plans in relation to the intended goal (Köhler 1976). Agency thus implies a kind of control room that supervises the project, establishes a range of possible operational options, simulates their course, and then decides on the action to take—continuing to monitor all stages of execution in order to apply any necessary corrections (Tolman 1932).

7.4 Animal Personality

The convergence of phylogenetic predispositions and ontogenetic experiences in the subject—combined with unpredictable circumstantial factors, moments of creativity linked to the relational reorganization of competences, and the effect of agency (i.e., decisions made in the moment based on the animal's ability to supervise the situation and operate under uncertainty)—gives rise to what is known as animal personality. The fact that every animal is a unique and unrepeatable individual—so much so that even cloning would not reproduce the same biographical structure—is almost self-evident. But here we are dealing with something even more significant: the emergence of personality refers to a condition that arises within the very concept of species, shaped by internal dynamics of animal populations and evolutionary processes, as explored in Claudio Carere and Dario Maestripieri's essay *Animal Personalities* (2013). Animal personality speaks to differing sensitivities, to internal working models that take form through social relationships, to distinct types of intelligence among individuals of the same species, to personal innovative capacities and equally individual acquisitive and transmissive predispositions, to the tendency of certain subjects to learn from other species, to episodes of transspecific alliances and cooperation, to the development of technological and cultural tools, and to inventive behaviors in niche construction that alter selective pressures on certain populations, giving rise to sympatric speciation processes.

If perception is a way of knowing external reality and not merely a disposition to react to stimuli—and if this property is the result of action rather than mere exposure—then it becomes clear that agency is intrinsically linked to the construction of a personal perspective on the world. The very interface between individual and world sets in motion a dialogical mechanism that goes beyond the mere monitoring of a landscape; it constitutes the construction of a lived experience that takes shape within all the animal's systems and, as a consequence, defines an operative dimension (Von Uexküll 1956). The echolocation of microchiropterans, for instance, not only illustrates a specific way of constructing an image of the world through ultrasonic echoes, but also reveals a particular mode of hunting, a reconfiguration of auditory faculties, and the challenges posed by prey type and the nocturnal context (Griffin 1958). Likewise, bees' ability to perceive polarized light demonstrates the importance of always knowing the position of the sun as a marker for reaching food sources (Von Frisch 1967). Perception becomes perspective, and this determines the evolution of a particular eco-ontological dimension, initiating an increasingly personal dialogue with the world. In this sense, we can say that the *Umwelt* introduced by Von Uexküll should be understood not as a static entity that imprisons the subject within a world-context—as in the mistaken Heideggerian interpretation—but rather, in agreement with the author himself, as a dimension in flux. The subject's action—or more precisely, its capacity for agency—makes the individual a bearer of a temporally prospective gaze upon the world, capable of bending becoming itself to its advantage, including the resonance of the past. The individual not only grasps the unfolding of reality with its sensory apparatus but, through this sensory and operative relationship, begins to develop a personal, perspectival dimension of the world.

Agency shouldn't be seen as the exercise of total arbitrariness, but rather as the personal interpretation of one's own behavioral script. A cat playing with a ball follows its unpredictable bounces around the furniture, building a specific sense of how things unfold and adapting its responses accordingly—but it does this by interpreting its innate predatory instincts in a unique way. Likewise, a dog that learns to touch a target with its nose is repurposing an inborn behavior through a new form of communication. This makes it clear that innate tendencies can't be excluded from the realm of agency, because the individual's ability to act is grounded in prior predispositions—traits shaped by the identity of their species. Still, we shouldn't think of these innate structures as constraints on subjectivity. Rather, they define the boundaries of what's possible. The species dimension instructs certain canons of agency—for example: the engagement behavior in wolf pups, the predatory sequence of wildcats, or food begging in estrildid chicks. However, these frameworks are rarely rigid. More often than not, they function as open patterns, offering plenty of room for flexibility, adaptation, and creative use in new situations.

We're mistaken when we assume that having innate instructions limits or obstructs the uniqueness of an individual's personal project. It's equally incorrect to think of learning as a process that simply produces behaviors to be repeated. The development of personality is, in fact, the emergence of a personal perspective on the world—one that is constantly renegotiated through the individual's interactions with external reality. From birth, animals are already equipped to bend reality through the lens of their species-specific perspective. As I pointed out in *The Creative Animal* (2022), a wasp transforms its environment into a landscape of "waspness"—a network of values filtered through its being-a-wasp—and the same applies to every animal. It becomes clear, then, that personality can only emerge *because* there is a prior bias in perspective. Innateness is what allows the individual to take the first steps toward subjective partiality. As Donald Griffin describes in *Animal Minds* (2001), beavers in captivity repurposed their usual dam-building behavior by piling materials around a pole that held unreachable food at its top. This inherited framework prompts the animal to view the world through its species' lens—but the world also offers its own constraints and possibilities. It is precisely the uniqueness of relational situations that demands and enables a personal interpretation of what is possible, allowing the real wasp to emerge from the script of virtually possible wasps. The animal is like an actor who follows only part of the script and responds only partially to the reactions of the audience: while taking both into account, it still adds a touch of personality and improvisation to the performance.

The development of personality is deeply tied to the decision-making condition I've referred to as *agency*. The affective dimension, understood as a copulative condition—that is, a way of being connected to the world—manifests in two main forms: (i) as feeling, in the sense of experiencing something directly—such as fear, unease, anger, or frustration; and (ii) as desiring, in the sense of being drawn toward a not-yet-realized state—such as longing, curiosity, exuberance, or motivation. This means the animal lives in a state of emotional becoming, shaped both by shifting physiological states and by the events it encounters. Affectivity—the ability to feel sensations and emotions—further reinforces the presence of a self that interacts with the world, generating experience that is never isolated or self-contained, but always relational and open to what surrounds it. As brains grow more complex, so does the internal "database" of stored information—what we might call the system's overall memory. At this point, personality takes another significant leap forward. Memory gives rise to an inner landscape of recollections, thoughts, mental maps, and procedures, enriching the individual's sense of subjectivity while remaining anchored to the living context.

Inner life can be understood as a kind of internal theater where ideas and concepts interact, allowing for processes of simulation, imagination, and dreaming (Peña-Guzmán 2022). The affective and cognitive evolution of animals shows us that agency is grounded in connective functions—not in isolation, autonomy or autopoiesis. Both affective states, which express this dynamic through felt experience, and cognitive states, which provide a mapped representation of it, converge within an eco-ontological dimension. Agency is often confused with consciousness, as if decision-making necessarily requires full awareness of the choices being made. This leads to the mistaken idea that experience must always be an explicit act. But consciousness, in its various intentional forms, is better understood as the act of shining a light on a process—such as an emotion, a desire, or a thought. From this perspective, affective and cognitive interiority *precedes* consciousness. From the simplest forms to the most complex, it is not consciousness that generates agency. Rather, all expressions of consciousness—whether evaluative or volitional—can be illuminated only because they are already grounded in agency (Panksepp and Biven

2012). In light of this, we must conclude that the construction of personality should be understood as the sum of all conscious and unconscious processes that allow a self to emerge—one that actively shapes its own relational becoming.

Even in its most elementary forms, moving through the world and acting upon external reality requires the construction of a personal relational dimension. The metapredicate from which all predicates of personality derive, therefore, should not be sought within the subject, but rather in how the subject positions itself in the world—what we might call a metapredicate of prospective relationality. Agency, in this sense, is expressed through the ability to build a personal trajectory of relational becoming and through the sustained attention to the external world, step by step, by comparing the consequences of one's actions with the world's responses. Yet, this moment-to-moment engagement with the uncertainty of experience is only part of the individual's existential condition. An animal's life is not a series of disconnected episodes; it more closely resembles a novel, in which each experience—thanks to the individual's memory faculties—is threaded into the fabric of its existential project. The emergence of a personal story, or an individual identity, gives rise to an existential dimension that, in miniature, mirrors that of the species (Bateson and Martin 1999). Ontogeny is indeed shaped in part by chance, but over time the individual tends to stay on the path it has begun, as if to preserve a sense of selfhood, making choices that remain coherent with that evolving personality.

Evaluation is not merely an activity that precedes action, guiding the choice that follows. Rather, it is a process intrinsically tied to everything that has come before, where each decision influences the next, generating a trajectory that is entirely singular. In this sense, we can rightly define the animal as "world building." Given the complexity involved in this process of individuation—one that traces the course of an animal's life—it is clear that simple feedback-based correction mechanisms are not enough. The individual must appropriate the unfolding of experience; a measure of creativity must necessarily accompany the process of enactment. In acting upon the world, the animal never follows a linear path, nor does it merely synthesize available variables—it imposes its own direction on becoming. The principle of individuation points to the unpredictability of choice—its arising in the very moment it is made, marking a departure from mechanical determinism. As Henri Bergson writes in *Creative Evolution*, "The essence of mechanical explanation [...] is to regard the future and the past as calculable functions of the present, and thus to claim that *all is given*" (1911, 38).

For an animal, there is never a condition where everything is predetermined, linear, or given. Action resembles a dialogue, where each argument proposed by the individual requires a counterpart with whom to engage, being ready to listen to the other's statements and to take advantage of their differences to improve one's own perspective. This counterpart with whom an animal must continually engage is the network of connections—the set of referential factors shaped by its relational perspective. For this reason, the individual cannot avoid being present in the here and now, inventing what aligns with its own project in the moment. The conditions in flux must be framed as field functions, meaning events that present a range of possibilities with varying probable outcomes (Parisi 2021). Thus, the project is never given, but unfolds, taking into account the opportunities and risks that arise during the action. Therefore, an animal cannot escape from maintaining an evaluative presence—creating, in each situation, a specific framework of possibilities, while keeping the horizon of options open. One of the fundamental aspects of agency is, therefore, the ability to translate external conditions into a personal perspective and, in doing so, to attribute value to the external world based on its own process of individuation.

Agency speaks to the individual's ability to make decisions, but we must understand that the intention to carry out a particular action and to choose between various options at a given moment is always a subjective expression. Thus, an animal is required to weigh what is presented before it, based on factors of coherence and priority. Evaluation is, first and foremost, the readiness to bring the development of the situation back to the self, following logical patterns that align with individual needs. The subject must be equipped with what Montague calls "value tags" (2006, 20). Evaluation is an essential necessity in orienting the subject and requires a set of interface conditions, the more complex the larger the range of choices available to the individual. In general terms, we can say that the capacity to assess each situation with precision and detail can be considered a meta-condition of selective pressure: indeed, only highly accurate detectors have been able to overcome the adaptive challenge throughout phylogenesis. Evaluating external conditions obviously requires specializations linked to the species' environment and lifestyle, but the evaluative capacity should be considered a fundamental presupposition of subjectivity.

Directionality, self-organization, and ontopoiesis reveal qualities that clearly differentiate biological phenomena from other physico-chemical ones—so much so that many authors, in attempting to explain this irreducible gap, have gone so far as to hypothesize the existence of a vital substance or something transcending matter. In truth, there is no need to invoke external factors to acknowledge the supervenience of new qualities: this is demonstrated by the properties of emergent systems, as outlined in Steven Johnson's *Emergence* (2001). According to emergentist theory, when matter assumes certain configurations—embedding organizational information and systemic structure—and once a critical mass and a sufficient number of interactions among its units are reached, properties arise that are not present in the individual components of the system. In living systems, agency can be considered a second-order emergence. All expressions of life are based solely on physico-chemical factors, and there is thus no need to imagine any additional elements. Yet, as Konrad Lorenz noted, while all vital phenomena are physico-chemical, it is equally undeniable—and this is what makes them original—that they are also something different and far more complex. It is therefore more consistent to regard protagonism as an emergent property, rather than the effect of some immaterial, elusive component. As Montague writes, "Biological systems possess layer upon layer of physical structure and hence layer upon layer of information-processing levels" (2006, 12). It follows, then, that we cannot overlook the qualities that reveal the living being as endowed with protagonism.

We perceive a phenomenon as objective when it passively follows the laws of physics—when it does not change trajectory on its own, lacks autonomy, and does not strive toward a goal, just to name a few traits. Today, we know from violation-of-expectation tests that young mammals and human infants are capable of distinguishing animal behavior from that of inanimate objects: they are not surprised if an animal changes course or jumps over an obstacle, but they do express surprise if an ordinary object does the same (Dehaene 2020). If we compare the dazzling displays of birds in flight or the spins of dolphins—their ability to harness physical phenomena to enact their own operational intent—we cannot help but notice a striking difference, what Lorenz calls their *originality*. And yet, identifying a clear threshold—where agency begins and where it ends—is never straightforward. Even the notion of decision-making autonomy is difficult to define, because the animal world is far more intricate and layered than our tendency to formulate universal laws can account for. What is certain is that the presence of inherent goals reveals in animals a kind of partiality in how they intersect with reality—an orientation that departs markedly from the impartiality of the objective domain and demands a field of agency in which action is possible.

To possess the faculty of agency—and thus decision-making—presupposes a compass for the peripatetic trajectory of the animal, which must indeed rely on innate or learned content—a database of references—but also on the ability to use that content non-repetitively, constantly adapting it to the encountered conditions. Orientation has received various explanations over time, such as Jacques Loeb's theory of tropisms and Charles Scott Sherrington's reflex chain. However, animal behavior cannot be explained solely as orientation by gradients or as a cascade of triggers; otherwise, we could not account for phenomena that contradict the domino model—such as the phenomenon of preactivation identified by Karl Spencer Lashley. The American neuropsychologist demonstrated that during an expressive sequence, some behaviors may occur before the usual order of execution. Therefore, while subjectivity may manifest itself in the capacity for orientation, it would be a mistake to conflate the two: directionality alone is not enough to define this attribute—an operational plan is required.

The structuring of behavior is based on two fundamental needs: (i) the need for adherence—that is, to shape action according to the situational characteristics it encounters; (ii) the maintenance of an irreducible margin of unpredictability throughout the action itself. The first stems from the necessity to adjust, step by step, one's actions to the unique features of a given situation and to the equally unique evolution that the scenario undergoes under the influence of the animal's activity. This is a process of correlated development that takes into account the consequences arising during the operation, and it is achieved through systems of correction and redefinition (Tomasello 2022). A cat playing with a ball, for example, cannot simply base its behavior on the ball's initial position, relying on pre-set behavioral patterns. Instead, it must constantly reconfigure its actions according to the consequences that unfold during play. The construction of an operational plan is therefore a process that unfolds in progress and requires a system of oversight. Moreover, in order to make action responsive to the novelty it presents—and to dynamically and precisely align it with how events unfold—one must rely on capacities that proportionally include a certain degree of virtuality, meaning the ability to take advantage of uncertainty.

In this regard, we must recognize that creativity can manifest in various forms: not only in inventing a new intervention strategy, but also in repurposing an existing resource by applying it to a new context. Clearly, this process of cooptation is possible if, and only if, it presupposes the presence of a subject—that is, a condition capable of overseeing the action and evaluating the operational field as a space of agency. We must not, however, conceive of the subject as a kind of Cartesian *homunculus*—such an explanation would trap us in an infinite regress—but rather as an emergent effect of synthesis, capable of evaluating the consequences of actions and using available resources as useful tools.

Creativity can also arise from using latent potentials within the operational toolkit that have not been previously explored. In this case, it is not a matter of applying the resource to new operational contexts, but rather activating new pathways of use within the resource itself. The faculty of agency, in this example, is not expressed through the simple choice between two options, but in the ability to carry out a kind of reconnaissance of the endowment's potentialities. This is possible only when resources—far from being automatic—are open-ended. That means: (i) they are operationally redundant, allowing for multiple applications so that no resource is tied to a single function; and (ii) they are functionally degenerative, containing a range of potential functional states. Together, these qualities allow for the emergence of multiple possible paths, opening up a true decision-making space. One can think of it like using a topographic map: it presents several routes to reach the same destination. The chosen route—the specific procedure or use of the resource—draws on the map's inherent features, which include unused possibilities that can be explored when the usual path is blocked.

When we speak of adjustment and correlation, we refer to the unpredictability of the resulting action, which stems from the need to tailor behavior to the specificity of the situation. The protagonism lies in the *becoming*—not the pre-existence—of the operational schema; the individual does not merely apply an endowment but brings forth a personal model of intervention. In this sense, agency appears as the capacity to shape one's state based on both the evolving nature of one's goals and the situational novelty—that is, the ability to remain in relation. At the same time, there is a need for singularity that cannot be explained in terms of conformity. It involves the animal's ability to face high-stakes situations with flexibility. In the contest between predator and prey, for instance, acquiring skills and behavioral strategies is essential, but equally important is keeping those operational programs open—not fully formed. This is not only to adapt to specific circumstances, but also to preserve an irreducible degree of unpredictability. In fact, any effective strategy must retain an element of operational surprise, even for the agent executing it, who might otherwise unwittingly reveal their intended moves.

The condition of agency in operational practice thus involves four fundamental aspects, which are always present—even in very simple situations that do not require particular problem-solving skills. These are:

1. The ability to assess different situations in terms of their coherence and adherence to the individual's overall project;
2. The ability to evaluate one's own actions through simulations, comparisons, and counterfactual projections of consequences;
3. The ability to adjust behavior based on the specific features of the situation and how it evolves during action;
4. The ability to maintain an irreducible margin of unpredictability within the action plan, making the individual's strategy less transparent to a competitor.

Personality, in this sense, appears as a space of decisional autonomy that the animal holds as it moves through the world—manifesting protagonism in the unfolding of its activities, and being endowed with an agency that enables it to make choices.

It is unquestionable that phylogenetic inheritance—encoded primarily in the genetic code—underpins constitutive and functional processes that determine the specific directionality of living organisms. However, it is equally undeniable that hereditary information cannot be fully explained as a mere domino cascade of pre-instructed processes. As Charles Taylor emphasized in his essay *The Explanation of Behaviour* (1964), the agency of living beings must be understood in terms of *purposefulness*—such that action is not predetermined in its development, but is shaped in real time so as to achieve its goal amid changing circumstances.

An animal explicitly demonstrates its status as an agent through its capacity for movement, which reveals the ability to approach or withdraw from a target, to deviate and choose its own trajectory, to uncover hidden opportunities, and to modify its surrounding environment. Animal agency, therefore, does not manifest solely in reactive terms—adjusting behavior in response to changing environmental conditions—but also in the ability to anticipate and often create opportunities through expectations and niche construction. While being goal-directed and possessing an intrinsic teleology are traits that help us describe living organisms—and animals in particular—these predicates can lead to problematic explanatory circularity if used to explain features that instead deserve thorough analysis. One approach to this issue was taken by Konrad Lorenz, who in the second half of the twentieth century undertook a significant revision of the innate/learned dichotomy by introducing the concept of *innate instructors*: phylogenetically adapted teaching mechanisms that allow individuals to develop a personal identity and thus a subjective style of behavioral expression through experience. Ernst Mayr (1982) revisited this topic by distinguishing between two types of phylogenetic inheritance mechanisms: *closed programs* and *open programs*. The latter enable animals to actively and selectively search for specific information in their environment, which in turn shapes the development of their identity. We might say that the innate doesn't dictate behavior, but plays a dual role: as a first guide that introduces the individual to the world, and as a student ready to grow through what that world teaches. In this sense, inheritance doesn't leave the individual passively exposed to their environment, but enables an emergent condition—one that weaves together all relational factors into a whole. As these authors suggest, agency is inseparable from the development of identity—from the capacity to bring forth what we now call personality. What most strongly signals a state of presence in an animal is its ability to draw on both inherited and environmental information to shape a unique way of being.

The animal demonstrates its capacity to be present—moment by moment—in its own existence, by actively interpreting both its phylogenetic blueprint and the surrounding environment. In the first case, it avoids merely repeating fixed instructions; in the second, it selects and reconfigures contextual variables. At the same time, it exhibits full self-ownership rooted in its condition of being interested. This goes beyond a simple break from determinism or a degree of operational freedom: it implies a principle of relational selectivity, driving the animal to maintain and expand the scope of its self-ownership.

To conclude, agency does not hinge on a single defining quality. When we consider the traits discussed above, we see that none of them—taken in isolation—guarantees the presence of decision-making authority. For example, having a sense of orientation or the ability to reflect the current situation does not, in itself, point to agency. The same is true for responsiveness, functional specialization, or the exhibition of partial states. Only when these features are integrated and expressed operationally can they give rise to the emergent effect that underpins the animal's protagonism in interpreting behavioral situations in the here and now. At the same time, it becomes increasingly clear that this decision-making subjectivity is not the exclusive property of higher animals or of systems with highly complex agency. Rather, it is a foundational condition of what it means to be an animal.

References

Ainsworth, Mary D. Salter, and Leonard H. Ainsworth. 1958. *Measuring Security in Personal Adjustment*. Toronto: University of Toronto Press.

Bateson, Patrick. 2017. *Behaviour, Development and Evolution*. Cambridge: Open Book Publishers.

Bateson, Patrick, and Paul Martin. 1999. *Design for a Life: How Behaviour Develops*. London: Jonathan Cape.

Bekoff, Marc, and Colin Allen. 2002. *The Cognitive Animal: Empirical and Theoretical Perspectives on Animal Cognition*. Cambridge, MA: The MIT Press.

Bergson, Henri. 1911. *Creative Evolution*. New York: Henry Holt and Company.

Brensing, Karsten. 2020. *What Do Animals Think and Feel? An Investigation into Emotion and Behavior*. Cambridge, MA: Pegasus Books.

Brian, Michael V. 1980. Social Control Over Sex and Caste in Bees, Wasps and Ants. *Biological Reviews* 55:379–415.

Buffington, C. A. Tony. 2018. Pandora Syndrome in Cats: Diagnosis and Treatment. *Today's Veterinary Practice* 8 (5): 31–41.

Carere, Claudio, and Dario Maestripieri, eds. 2013. *Animal Personalities: Behavior, Physiology, and Evolution*. Chicago: University of Chicago Press.

Carroll, Sean B. 2005. *Endless Forms Most Beautiful*. New York: Baror International.

Cassidy, Jude, and Lisa J. Berlin. 1994. The Insecure/Ambivalent Pattern of Attachment: Theory and Research. *Child Development* 65 (4): 971–991.

Christen, Yves. 2009. *L'animal est-il une personne ?* Paris: Flammarion.

Corruccini, Robert S., and Richard M. Beecher. 1984. Occlusofacial Morphological Integration Lowered in Baboons Raised on Soft Diet. *Journal of Craniofacial Genetics and Developmental Biology* 4 (2): 135–142.

Curio, Eberhard. 1976. *The Ethology of Predation*. Berlin/Heidelberg/New York: Springer.

Dawkins, Richard. 2006. *The Blind Watchmaker: Why the Evidence of Evolution Reveals a Universe Without Design*. London: Penguin Books Ltd.

de Waal, Frans. 2016. *Are We Smart Enough to Know How Smart Animals Are?* New York: W. W. Norton & Company.

Dehaene, Stanislas. 2020. *How We Learn: Why Brains Learn Better Than Any Machine ... for Now*. New York: Viking.

Dennett, Daniel C. 1996. *Kinds of Minds: Toward an Understanding of Consciousness*. New York: Basic Books.

Eibl-Eibesfeldt, Irenäus. 1970. *Ethology, the Biology of Behavior*. New York: Holt, Rinehart and Winston.

Gilbert, Scott, and David Epel. 2015. *Ecological Developmental Biology: The Environmental Regulation of Development, Health and Evolution*. Oxford: Oxford University Press.

Godfrey-Smith, Peter. 2020. *Metazoa. Animal Minds and the Birth of Consciousness*. Glasgow: William Collins.

Gould, Stephen Jay. 1990. *Wonderful Life*. New York: W.W Norton & Company.

Gould, James R., and Carol Grant Gould. 2007. *Animal Architects: Building and the Evolution of Intelligence*. New York: Basic Books.

Gould, Stephen Jay, and Elisabeth Vrba. 1982. Exaptation-A Missing Term in the Science of Form. *Paleobiology* 8 (1): 4–15.

Griffin, Donald R. 1958. *Listening in the Dark: The Acoustic Orientation of Bats and Men*. New Haven: Yale University Press.

Griffin, Donald. 2001. *Animal Minds*. Chicago: University of Chicago Press.

Hall, Brian K. 1992. *Evolutionary Developmental Biology*. London: Chapman & Hall.

Haraway, Donna Jeanne. 2016. *A Cyborg Manifesto*. Minneapolis: University of Minnesota Press.

Hearne, Vicki. 2007. *Adam's Task: Calling Animals by Name*. New York: Skyhorse Publishing.

Heidegger, Martin. 1978. *Being and Time*. Oxford: Blackwell.

Hinde, Robert A. 1966. *Animal Behaviour: A Synthesis of Ethology and Comparative Psychology*. New York: McGraw-Hill.

Honing, Henkjan. 2019. *The Evolving Animal Orchestra: In Search of What Makes Us Musical*. Cambridge, MA: The MIT Press.

Jacob, François. 1977. Evolution and Tinkering. *Science* 196 (4295): 1161–1166.

Johnson, Steven. 2001. *Emergence: The Connected Lives of Ants, Brains, Cities, and Software*. New York: Scribner.

Kershenbaum, Arik. 2020. *The Zoologist's Guide to the Galaxy: What Animals on Earth Reveal About Aliens – And Ourselves*. New York: Viking.

Köhler, Wolfgang. 1976. *The Mentality of Apes*. New York: Liveright.

Lewontin, Richard. 1983. The Organism as the Subject and Object of Evolution. *Scientia* 118:63–82.

Libet, Benjamin. 2004. *Mind Time: The Temporal Factor in Consciousness*. Cambridge: Harvard University Press.

Lorenz, Konrad. 1977. *Behind the Mirror: A Search for a Natural History of Human Knowledge*. Londra: Methuen.

Marchesini, Roberto. 2018. *Beyond Anthropocentrism. Thoughts for a Post-Human Philosophy*. Sesto San Giovanni: Mimesis International.

Marchesini, Roberto. 2022. *The Creative Animal: How Every Animal Builds Its Own Existence*. Cham: Palgrave Macmillan.

Mayr, Ernst. 1982. *The Growth of Biological Thought: Diversity, Evolution, and Inheritance*. Cambridge, MA: Belknap Press of Harvard University Press.

Merchant, Hugo, et al. 2011. Measuring Time with Different Neural Chronometers during a Synchronization-Continuation Task. *Proceedings of the National Academy of Sciences* 108 (49): 19784–19789. https://doi.org/10.1073/pnas.1112933108.

Meyer, Axel. 1987. Phenotypic Plasticity and Heterochrony in *Cichlasoma managuense* (Pisces, Cichlidae) and Their Implications for Speciation in Cichlid Fishes. *Evolution* 41 (6): 1357–1369.

Montague, Read. 2006. *Why Choose This Book? How We Make Decisions*. New York: Dutton.

Müller, Gerd B. 2010. "Epigenetic Innovation." In Evolution: The Extended Synthesis, edited by Massimo Pigliucci and Gerd B. Müller, 179–207. Cambridge, MA: MIT Press.

Odling-Smee, John. 1988. Niche Construction: The Neglected Process in Evolution. In *The Role of Behavior in Evolution*, ed. Henry C. Plotkin, 98–134. Cambridge, MA: MIT Press.

Pageat, Patrick. 1998. *Pathologie du comportement du chien*. Maisons-Alfort: Éditions du Point vétérinaire.

Panksepp, Jaak, and Lucy Biven. 2012. *The Archaeology of Mind: Neuroevolutionary Origins of Human Emotions*. New York: W.W. Norton & Company.

Parisi, Giorgio. 2021. *In un volo di storni. Le meraviglie dei sistemi complessi*. Milano: Mondadori.

Peña-Guzmán, David M. 2022. *When Animals Dream*. Princeton: Princeton University Press.

Pener, Moshe P. 1991. Locust Phase Polymorphism and Its Endocrine Relations. *Advances in Insect Physiology* 23:1–79.

Pouydebat, Emmanuelle. 2017. *L'intelligence animale: cervelle d'oiseau et mémoire d'éléphant*. Paris: Éditions Odile Jacob.

Prigogine, Ilya, and Isabelle Stengers. 1984. *Order Out of Chaos: Man's New Dialogue with Nature*. Toronto: Bantam Books.

Safina, Carl. 2015. *Beyond Words: What Animals Think and Feel*. New York: Henry Holt and Co.

Sapolsky, Robert M. 2024. *Determined: The Science of Life Without Free Will*. New York: Vintage.

Shapiro, Arthur M. 1968. Photoperiodic Induction of Vernal Phenotype in *Pieris protodice* Boisduval and LeConte. *Wasmann Journal of Biology* 26:137–149.

Shettleworth, Sara J. 2010. *Cognition, Evolution, and Behavior*. Oxford: Oxford University Press.

Siegel, Daniel J. 1999. *The Developing Mind: Toward a Neurobiology of Interpersonal Experience*. New York: Guilford Press.

Suzuki, Yuikiro, and H. Frederich Nijhout. 2006. Evolution of a Polyphenism by Genetic Accommodation. *Science* 311 (5761): 650–652. https://doi.org/10.1126/science.1118888.

Taylor, Charles. 1964. *The Explanation of Behaviour*. London: Routledge & Kegan Paul.

Thorpe, William H. 1979. *The Origins and Rise of Ethology*. London: Heinemann Educational Books.

Tinbergen, Niko. 1963. On Aims and Methods in Ethology. *Zeitschrift für Tierpsychologie* 20:410–433.

Tolman, Edward. 1932. *Purposive Behavior in Animals and Men*. New York: Century Company.

Tomasello, Michael. 2022. *The Evolution of Agency: Behavioral Organization from Lizards to Humans*. Cambridge, MA: The MIT Press.

Vallortigara, Giorgio. 2021. *Pensieri della mosca con la testa storta*. Milano: Adelphi.

Van Buskirk, Josh, and Rick A. Relyea. 1998. Natural Selection for Phenotypic Plasticity: Predator-Induced Morphological Responses in Tadpoles. *Biological Journal of the Linnean Society* 65:301–328.

Von Frisch, Karl. 1967. *The Dance Language and Orientation of Bees*. Cambridge, MA: Harvard University Press.

Von Uexküll, Jakob. 1956. *Streifzüge durch die Umwelten von Tieren und Menschen: Ein Bilderbuch unsichtbarer Welten. Bedeutungslehre*. Hamburg: Rowohlt.

Warkentin, Karen M. 2005. How Do Embryos Assess Risk? Vibrational Cues in Predator-Induced Hatching of Red-Eyed Treefrogs. *Animal Behaviour* 70 (1): 59–71.

Washburn, Margaret F. 1916. *Movement and Mental Process*. Boston: Houghton Mifflin.

Weaver, Ian, et al. 2007. The Transcription Factor Nerve Growth Factor-Inducible Protein A Mediates Epigenetic Programming: Altering Epigenetic Marks by Immediate-Early Genes. *The Journal of Neuroscience* 27 (7): 1756–1768.

West-Eberhard, Mary Jane. 2003. *Developmental Plasticity and Evolution*. Oxford: Oxford University Press.

Woodward, David E., and James Dickson Murray. 1993. On the Effect of Temperature-Dependent Sex Determination on Sex Ratio and Survivorship in Crocodilians. *Proceedings of the Royal Society of London. Series B, Biological Sciences* 252:149–155.

Young, John Zachary. 1987. *Philosophy and the Brain*. Oxford: Oxford University Press.

CHAPTER 8

Environmental Ethics

8.1 Premise

The relationship between human beings and the biosphere plays out on multiple fronts: through social, political, and cultural choices that affect individuals, local communities, and global systems alike. These interactions form a complex web that creates a systemic condition. This inevitably involves individuals in their lifestyles; local communities in the organization of services and territorial governance; national institutions in their development policies and sustainability decisions; and the supranational level in shaping shared agendas and agreements.

If we were to identify two key levers for transforming today's widespread ecological neglect and disregard for biospheric disruption, I believe they would be: (1) Changes at the individual level, but especially within local communities (a bottom-up approach); and (2) Supranational policies capable of exerting binding influence over the environmental directions of individual states (a top-down approach). This means activating a dual mechanism of intervention, first conceptualized in 1980s Japan as *dochakuka*, and translated by sociologist Roland Robertson into the concept of *glocalism*—a process that operates simultaneously at both the global and local levels (Mihr 2022). The principle was later taken up by scholars such as Zygmunt Bauman in his essay *On Glocalization* (1998), and promoted by the Glocal Forum, an international cooperation organization founded in Zurich in 2001 by Uri Savir and Jan Stenbeck. Today, however, we can

R. Marchesini, *Philosophical Ecology*,
https://doi.org/10.1007/978-3-032-04987-2_8

observe that the practical application of this principle—"think globally, act locally" and vice versa—has largely failed to meet its objectives. Instead, it has often paved the way for new forms of disguised mercantile colonialism and de-territorialization.

That's why we need to understand the limitations that have emerged over the past thirty years of critique against globalization. First and foremost, influencing individual choices is neither easy to achieve nor particularly effective in terms of impact. Most people act according to lifestyles that are shaped less by critical reflection and more by influence processes that operate through media and cultural conformity. To have real grassroots impact, we would need to establish supra-individual processes—such as small communities capable of developing informed and shared lifestyles, much like the Benedictine and Franciscan monastic traditions (Rozzoni and Limata 2022). These would disrupt mass media influence and instead promote a new form of community-based conformity. Unfortunately, if we examine the evolution of media over the past century—from the rise of radio to today's social media, from advertising campaigns to the influencer phenomenon—it becomes clear that conformity has been encouraged in different ways, eroding community-based realities. These communities once thrived on collective habits rooted in conviviality and shared daily life.

Today, we see the horizontal spread of certain shared lifestyles—such as the overuse of digital technologies, the constant drive to upgrade hardware and software, diets centered around animal protein, and the rising consumption of electricity and fossil fuels. Alongside this, there is a fragmentation into micro-communities united by specific shared interests. However, these micro-communities cannot fundamentally alter dominant horizontal lifestyles—often, they are themselves rooted in them (as in the use of social media)—and at most, they serve as niche markets. At the same time, institutions that should be shaping development policies and sustainability regulations find themselves increasingly at the mercy of large capital interests. These interests are powerful enough to influence not only national governments but also supranational organizations. Ethics—traditionally operating on this dual front of personal conduct and political decision-making—now finds itself powerless to make a meaningful impact.

Among the factors that have played a major role in undermining or rendering ecologically sustainable choices irrelevant, several stand out:

(i) The anthropocentric belief that immediate human interests have greater value—or at least priority—over the long-term well-being of the biosphere (Stengers 2015);
(ii) A lifestyle increasingly marked by nihilism toward collective values, with the individual pursuit of hedonism and self-fulfillment (Lasch 1979);
(iii) The inability to think medium or long term, with a fixation on the present that often escalates into a kind of hyper-present or obsession with immediacy (Rushkoff 2013);
(iv) The unavoidable clash between biodiversity conservation policies and the capitalist economy that currently drives global decision-making (Chakrabarty 2014);
(v) The tendency to underestimate ongoing environmental changes—or even to deny them outright. This includes influencers like Jordan Peterson and others who mock or discredit green policies;
(vi) The fundamental tension between caring for the biosphere and the pursuit of scientific, technological, and overall human development.

Major fossil fuel companies invest millions of dollars each year to sow doubt about climate change. As Steve Coll reports in *Private Empire* (2013), these companies even pay intellectuals, popular communicators, and researchers to push their narrative. A 2024 study by the Center for Countering Digital Hate (CCDH), titled *The New Climate Denial*, highlights a massive disinformation campaign across major digital platforms. These campaigns claim that climate change either isn't happening or isn't caused by human behavior. The effects of this disinformation are already visible, especially among teenagers, 70% of whom spend at least eight hours a day on digital platforms. According to the study, one-third of young people in the UK believe that the scientific narrative around the climate crisis is exaggerated—if not outright false. The same study notes a rise in climate-denial content, particularly on YouTube, where videos promote the idea that climate science is unreliable and that climate change is irrelevant, harmless, or even beneficial. This has fostered skepticism in 37% of young people.

One of the factors that has weakened the environmentalist movement is the emergence of a critical stance toward technology—at times almost neo-Luddite—and occasionally tinged with anti-scientific attitudes, including New Age tones and spiritual syncretism. These tendencies have not helped promote a scientific form of environmentalism, even though the latter has produced influential figures such as Rachel Carson, author of *Silent Spring* (1964), and Max Nicholson, who wrote *The Environmental Revolution* (1970). Also worth mentioning is economist Kenneth Boulding's essay *The Economics of the Coming Spaceship Earth* (1966), which criticized GDP as an inadequate measure of a society's value—an idea that later resonated in Robert Kennedy's speech at the University of Kansas during his 1968 presidential campaign.

One reason skepticism persists around the degradation of the biosphere or the human impact on it is the low predictability of medium-term consequences, especially when compared to the immediate benefits of exploiting natural resources. To understand this caution toward risk, we could borrow the framework of insurance sciences, which rests on three key principles: the principle of prudence, which urges us to act carefully in complex systems, given that there is never a direct, linear relationship between actions and their consequences; the precautionary principle, which holds that a lack of definitive proof does not mean the absence of a problem or the justification for recklessness; and the principle of prevention, summed up in the adage "prevention is better than cure." Today, relying on prudence alone is no longer sufficient. The scientific community has reached broad consensus: climate change is real and primarily caused by human activity. A study published in *Environmental Research Letters* reviewed 11,944 scientific papers on climate change and found that 97% agreed with these conclusions (Cook et al. 2016). It is therefore clear that if we continue deforesting at the current pace, emitting greenhouse gases at current levels, adopting Western-style diets on a global scale, and accelerating the erosion of biodiversity, endangering the balance of ecosystems, our planet will look entirely different within a few decades.

The precautionary principle, developed by Hans Jonas in his essay *The Imperative of Responsibility* (1984), is a moral imperative requiring us to act in ways that ensure human life can continue on Earth. It demands that we take full responsibility for the future, behaving as if catastrophe is not just the most probable outcome, but the most real. In this context, Jonas introduces the concept of the *heuristics of fear*—a mode of thinking that uses fear as a guide in decision-making, especially when dealing with long-term risks whose effects are not immediately apparent. The precautionary principle rests on two pillars: first, the understanding that human activity has already had a considerable impact on the planet; second, the ongoing work of climate scientists and ecologists, whose annual data show systemic trends and disruptions. Broadly speaking, the precautionary principle should guide concrete courses of action in the face of potential risks—even when scientific predictions remain uncertain. Uncertainty, in other words, should not be used as a justification for inaction. This approach received explicit recognition in environmental policy following the 1992 Rio de Janeiro Earth Summit. Principle 15 of the Rio Declaration states:

> In order to protect the environment, the precautionary approach shall be widely applied by States according to their capabilities. Where there are threats of serious or irreversible damage, lack of full scientific certainty shall not be used as a reason for postponing cost-effective measures to prevent environmental degradation.

Today, more than thirty years after the Rio Declaration, the degradation of the biosphere is far more evident and backed by solid scientific evidence. When it comes to environmental protection, we should therefore speak not merely of a *precautionary* principle—which relies on preliminary indications—but of a *preventive* one, which applies to objective risks supported by scientific data. The impacts of human activity on the biosphere are now plainly visible: climate change, biodiversity loss, melting glaciers, ocean acidification, and the accumulation of pollutants are undeniable phenomena. It's clear that, beyond any debate about the measures we should take, what we will leave to future generations is a world far less aligned with their expectations. Our children will inherit a planet that is more fragile in its self-regulating systems, significantly poorer in resources, and burdened with complex, hard-to-solve problems. So, what judgment will they pass on us? And, more importantly, does that future judgment—inevitably formed in hindsight—carry moral weight for us today? It's a question that should unsettle us, especially when we look back at history and reflect on how we now judge the actions of our ancestors. We ought to understand that while moral action takes place in the here and now, its true significance must also be assessed prospectively—because its consequences always extend into the future.

8.2 Different Approaches in Environmental Ethics

The emergence of environmental ethics occurred around the same time as the development of bioethics—particularly with reference to the essay *Bioethics: Bridge to the Future* (1971) by biochemist Van Rensselaer Potter. Environmental bioethics developed as a form of applied ethics and interdisciplinary research, aimed at examining the social problems and political choices stemming from the interaction between human action—such as economic, social, and technological development—and the environment. Among its central concerns is the responsibility toward future generations (Weiss 1989). The key distinction between bioethics and earlier approaches to environmental ethics—such as those of Aldo Leopold, Rachel Carson, and Lynn White Jr.—lies in how it addresses the impact of new technologies on the realm of *bios*, posing a fundamental question: Is everything that is technically possible also ethically acceptable?

Strong links between bioethics and environmental ethics quickly became evident. To highlight this connection, Potter himself introduced the concept of *global bioethics* in his 1988 essay *Global Bioethics: Building on the Leopold Legacy*. Environmental bioethics has since expanded to include themes such as intergenerational justice (Vanderheiden 2001) and social justice in the context of climate change (Hiskes 2008). It regards the precautionary principle as an essential compass for evaluating the potential impacts of new technologies, especially in light of their energy demands. The theme of limits emerged in the 1970s, thanks in part to the work of the Club of Rome, which commissioned the Massachusetts Institute of Technology (MIT) to conduct an analysis of five key variables: global population growth, agricultural production, non-renewable resources, industrial output, and pollution. This led to the publication of *The Limits to Growth* (Meadows et al. 1972), a report that predicted long-term collapse scenarios unless more sustainable behaviors were adopted. A few years earlier, in 1968, Paul R. Ehrlich and Anne H. Ehrlich had published *The Population Bomb*, raising alarm about the planet's ability to feed a rapidly growing global population. However, it soon became clear that the real issue lay less with population numbers and more with energy-intensive lifestyles—such as diets rich in animal protein and the heavy use of resources in capitalist economies.

It was in this context that the idea of *degrowth* began to take shape in the late 1990s. This school of thought advocates for the controlled, selective, and voluntary reduction of economic production and consumption in order to maintain ecological balance between humans and the biosphere (Schneider et al. 2010). A leading voice in this movement is Serge Latouche, who, in his book *Farewell to Growth* (2009), presents degrowth as a true paradigm shift—an alternative to the dominant model of overproduction and consumerism. This is not just an economic rethink, but a call to envision a different kind of society, as illustrated by economist

Kohei Saito in *Capital in the Anthropocene* (2020). Degrowth-oriented analysis also looks at technologies that can reduce or even resolve environmental impacts, including sponge cities that retain water, renewable energy, bio-architecture, sustainable agriculture, and biotechnologies for waste degradation. On the other hand, many emerging technologies—particularly digital ones—are far from sustainable. Consider, for example, the energy demands of artificial intelligence: by 2027, it is estimated that AI could consume 100 terawatt-hours of electricity annually, an amount comparable to the energy use of an entire country. As Liu and Zhai (2025) note, CO_2 emissions increase in step with computational power in AI-driven high-performance computing systems. Water consumption is another major concern with AI—companies operating data centers use vast quantities, prompting researcher Shaolei Ren from UC Riverside to call for awareness of AI's water footprint (Reece 2024).

Reflecting on the development of environmental ethics helps us understand that ecology—as a science capable not only of describing biospheric phenomena but also of sparking philosophical debate—has not always significantly influenced ethical frameworks. Specifically, it has not always succeeded in pushing ethical thinking beyond the human-centered model that separates the moral subject from their connections, nor in overcoming dualistic logic. What we've primarily seen is the recognition of new morally relevant subjects—such as future generations, non-human animals, various life forms, and ecosystems—without a corresponding shift in the core principles that guide moral orientation. This doesn't mean that ecology, as a descriptive science of biospheric relationships, can or should directly prescribe normative rules or ethical conduct. Rather, what ecology offers is a lens that highlights aspects often overlooked by traditional humanistic ethics, such as: the diversity of interests involved, which challenges the idea of a universalist ethical vision or the simple extension of universal rights to more beings; the importance of connections that sustain life, making it impossible to maintain an atomistic view of beings; and the collapse of traditional dichotomies, such as the divide between science and ethics—not because one can generate the norms of the other directly, but because they exist in a relationship of mutual dependence, where each raises problems that call for answers from the other.

The question of the moral relevance of our actions that affect biodiversity lies at the heart of philosophical ecology. It begins with a systemic understanding of moral posture: any ethical concern that focuses solely on individuals misses the point. In environmental ethics, the moral subject can no longer be considered as an isolated, atomistic entity—one that deserves attention independently of its connections, or that is evaluated solely based on self-referential interests. From the perspective of philosophical ecology, our ethical gaze should also encompass the relational dimension of each individual being. Descriptive ecology shows us that every living organism exists in a constant web of interactions with other entities—its very existence depends on being embedded in a network of exchanges from which it cannot be separated. Also, its manifestation—its way of expressing itself—depends on the type of relationship it is part of at any given moment.

Moreover, descriptive ecology teaches us that any action directed at an individual inevitably affects the entire community of beings—not uniformly or equally, but according to the specific relational gradient established by that being. This gradient is not based on physical, ontological, or aesthetic proximity—such as family members, species similarity, pets, or charismatic animals like pandas—but on relational exchange: the degree of connection, dependence, and the ripple effects in the event of disruption. These links often go unnoticed by those unaccustomed to ecological thinking—thinking that is connective, built on networks of dependency, cascading consequences, and feedback loops. To think ecologically means to see the world through the relationships that beings establish with one another. It means adopting a perspective that does not isolate an entity but recognizes it through the web of connections that make it what it is. In short, we must acknowledge the trans-individual nature of all living beings.

The essentialist view—which sees an entity's qualities as internal and intrinsic attributes, and regards the entity as self-referential (rigidly bounded, closed in its autopoiesis, and emanating inherent properties, even if reliant on external resources)—has implications that extend beyond ontology and epistemology. It also shapes the moral posture we tend to adopt. To think ecologically means to shift perspective. As Gregory Bateson suggests in *Steps to an Ecology of Mind* (1972), it means paying attention first and foremost to the *patterns that connect*. This too requires a counterintuitive mindset, because—as gatherers, which we still are by virtue of the foraging behaviors that characterize our species—we tend to focus on icons: the flower, the seashell, the mineral. In other words, we highlight closed structures and draw them out from their context.

Our syllegic nature—our tendency to seek out discrete items (the berry, the fruit, the larva, the egg), collect objects, and store them—leads us to focus on the entity as such, on its *idea* (from the Greek *idein*, "to see"), that is, on its isolated and recognizable appearance. This undoubtedly limits our ability to trace the evident or hidden connections needed for ecological thinking. It prevents us not only from thinking systemically—which is essential—but also from seeing the entity as anything other than a closed-off, self-sufficient unit of meaning. We seem to have a species-specific tendency to reason in terms of disjunctions and dichotomies, to seek categorical order based on separations and meaningful differences—a phylogenetic legacy that influences our epistemological structures (as discussed by Popper 1972; Lorenz 1977; Campbell 1974). On the other hand, as argued by American biologist Daniel Simberloff in his essay *A Succession of Paradigms in Ecology* (1980), our tendency toward typological thinking—seeking order and similarity where there is disorder and entanglement—is a legacy of Platonic idealism. One thing is for certain: whether evolutionary or metaphysical in origin, these influences have shaped not only our ways of knowing, but also our approaches to environmental ethics.

To summarize the various moral postures observable today in relation to broader behavioral and decision-making frameworks—economic, political, social, and cultural—that clearly stem from specific ethical stances toward biodiversity (particularly in terms of how priorities are set), we can identify four main approaches:

1. *The anthropocentric approach*: This view holds that only human interests—especially those that can be asserted in the here and now—are morally relevant. Anything outside the human sphere lacks moral standing, including future generations, who are not considered entitled to any rights.
2. *The anthropo-referential approach*: This perspective recognizes intrinsic (i.e., non-instrumental) value to non-human entities and future generations, but ties their moral relevance to the meaning assigned by human beings—the only agents capable of translating intrinsic value into moral consideration.
3. *The Individual-centric approach*: This assigns intrinsic value and moral relevance to both human and non-human individuals, understood as distinct entities characterized by specific qualities. These qualities influence the degree and nature of moral attention they receive.
4. *The ecocentric approach*: This expands intrinsic value and moral relevance not only to individuals, but also to supra-individual entities—

such as species, ecosystems, parks—as well as abiotic elements with environmental significance, like rivers, mountains, wetlands, or even developmental potential.

8.2.1 *The Anthropocentric Approach*

The anthropocentric approach is undoubtedly the most dominant from a political, social, and economic standpoint. This holds true both at the level of individual choices—for instance, the trend toward overconsumption has shown no real decline over the past five decades, despite the environmental issue gaining prominence in the mid-1970s—and at the level of society as a whole, which has become increasingly organized around values that emphasize self-referential individuals in competition with everything around them, freed from any constraints regarding the instrumental use of biodiversity. The economic model itself is based on consumption—a parameter used to determine a nation's level of well-being—and on a capitalist framework that, by exacerbating social inequalities, supports the entire self-centered and competitive structure of society.

The anthropocentric approach highlights the incomparable value of the human being over all other entities, the absolute priority of immediate human well-being, an unreserved faith in humanity's ability to make wise and far-sighted decisions, and the separation of humans from the rest of nature. It rests on the notion that humans now belong to a different realm: the artificial and the technological. This approach assumes that the biosphere has no value in itself but only instrumental value, insofar as it contributes to human well-being. This can be described as human chauvinism: a worldview that places the absolute moral primacy of the human species as an axiom and views the biosphere only in terms of its utility (Bartolommei 1995). This primacy is justified using criteria of merit—such as technical, deliberative, or planning capabilities—or ontological relevance, such as judgment, choice, or human superiority, to infer the moral irrelevance of all things non-human.

The anthropocentric approach can be broken down into six specific subcategories:

1.1 The Frontier Ethic, which is based on the principle of full human freedom to use natural resources, to pursue personal fulfillment, and to free oneself from all constraints—even to the point of establishing a moral duty to maximize one's life in all respects, from economic achievement to hedonic or experiential fulfillment. This ethic treats everything as a resource to be used for immediate profit (Udall 1963).

1.2 The Stewardship Ethic, or responsible management, which advocates for the enlightened use of natural resources. In this view, the human being plays the role of administrator of the biosphere—still seen in instrumental terms—but framed as prudent economic management, guided by a “fatherly” attitude (Passmore 1974).

1.3 The Social Justice Ethic, which prioritizes human needs. However, unlike the previous subcategories, this ethic is not aimed at individual fulfillment or even moderate and responsible use of natural resources. Instead, it encourages exploitation as a means to achieve greater social equity (Hardin 1973).

1.4 The Substitutability Ethic, which rests on the idea that prospective evaluations should not overlook the technological solutions that humans can develop. From this perspective, environmental alarmism is a form of shortsightedness that fails to consider humanity’s capacity to resolve ecological issues through technology (Nordhaus and Shellenberger 2009).

1.5 The Transhumanist Ethic, centered on the idea of singularity, envisions the human being as destined to break away from nature entirely. This ethic signals a radical discontinuity from the biosphere through technology and reflects a detached, even dismissive, attitude toward nature—seen as a now-irrelevant past in light of humanity’s tech-centric destiny (Kurzweil 2005).

1.6 The Accelerationist Ethic, which considers the ethic of responsible resource management to be shortsighted. It argues that exploitation processes should actually be intensified, as they drive the very changes needed to transition from an era of human dependence on nature to the full realization of the Anthropocene (Noys 2008).

8.2.2 *The Anthropo-Referential Approach*

The anthropo-referential approach does not deny that non-human entities have intrinsic value. However, it holds that moral value—that is, the attribution of moral dignity to non-human beings—is still something that depends on, or is at least related to, human judgment. In other words, even if biodiversity possesses its own intrinsic value, this does not automatically translate into a binding moral obligation for humans. Instead, it may or may not acquire moral relevance depending on whether humans assign such value to it. In this view, the biosphere or future generations do not, in themselves, impose a moral claim on us. They are not considered full moral patients by default. Rather, they are entities that are recognized *by humans* as worthy of moral attention. According to this approach, their ethical relevance depends on the moral recognition granted by human beings. This position makes human attribution a *conditio* sine qua non for moral relevance.

It's important not to confuse this view—which sees value as something that concerns humans (i.e., anthropo-referential)—with the broader idea that any prescriptive ethical direction is necessarily the result of human judgment. In fact, even non-anthropocentric ethics—numbers 3 and 4—typically acknowledge this second point: there is no shared denominator among the diverse forms of life from which a universal normative system can be derived. This is why the Stoic motto "follow nature," expressed by Cicero as *"Natura sequi volumus"* and echoed in Barry Commoner's 1971 work *The Closing Circle* in the phrase "Nature knows best," is open to considerable misinterpretation. Chief among the reasons is the ambiguity of the concept of "nature" itself. Are we referring to human nature or to nature in general? If we mean the former, we can only agree with John Stuart Mill, who, in *Utilitarianism* (1998), argues that conformity with human nature has nothing to do with what is just or unjust. If we mean the latter, we must ask whether it even makes sense to speak of a "law of nature," given that what we actually observe is a plurality of natural expressions.

The anthropo-referential approach acknowledges that humans may have duties—albeit indirect ones—toward the biosphere. This implies a human responsibility toward nature, no longer seen merely as a resource to be exploited or managed moderately, but as a good in itself, worthy of preservation and protection. Even within this approach, there are multiple subcategories:

2.1 The ethics of conservation regards biodiversity not as a mere resource, but as a value in itself—one that must be protected and maintained for its intrinsic biological and ecological significance. This value does not depend on human utility, such as environmental defense or the discovery of potential medicinal compounds. However, protecting it requires subordinating human interests to those of the biosphere as a whole.

2.2 The spaceship ethics or non-disjunction approach sees human survival and the survival of the biosphere as deeply intertwined and inseparable. The biosphere, like a spaceship, binds us to a shared destiny that must be preserved. Just as there are no purely private interests on a spaceship, this perspective recognizes no strict boundary between human and non-human interests (Boulding 1966).
2.3 The ethics of protection considers the biosphere an anthropological value that transcends the economy. It touches on fundamental aspects of human existence, positioning nature as a supreme good—for example, as a living laboratory that offers not only scientific knowledge but also philosophical insight. It also has educational value, offering meaningful experiences that contribute to personal development (Derr 1974), and aesthetic value, as in Edward Wilson's concept of *biophilia* (1986), comparable to the value of a work of art (Hargrove 1989).
2.4 The ethics of non-human reference treats nature as an anthropopoietic entity (Marchesini 2017), meaning one that contributes to human cultural and ethnic development. This is due to its symbolic content, its role in ritual construction, and its place in the traditions of various peoples (Sagoff 1985).
2.5 The ethics of protecting universal human values sees nature as a source of experiences that foster moral character. These include the development of empathy and the capacity for a more objective—less anthropocentric—perception of reality (Norton 1988).
2.6 The ethics of value incorporation emphasizes the role of nature in shaping and reinforcing human communities. Natural entities should never be seen as mere objects or resources. Rather, they serve as references capable of embodying models, paradigms, symbols, or moral qualities. They can transmit or exemplify values—or inspire virtue and the pursuit of the good life.

8.2.3 *The Individual-centric Approach*

The individual-centered approach diverges from the anthropocentric view in two key ways: first, it assigns intrinsic value to non-human entities; second, it holds that moral relevance is not something granted by humans, but rather something rooted in the inherent characteristics of the individual as a bearer interests, which intersect with those of humans—a relationship described as *dialogical inherence*.

In this view, both the duties humans have toward these entities and the limitations on what can be done to them—or the rights they are granted—are based on their intrinsic value. Non-human beings are thus considered as individuals who, by virtue of having inherent interests, deserve moral consideration independently of how humans happen to treat them. This perspective emphasizes individuals—each seen through the lens of their own life experience, rather than as elements within a broader ecological system. Hence the label anti-anthropocentric individualism.

I've already discussed the importance of moving beyond anthropocentrism, not just as a form of human chauvinism, but as an epistemological and ontological obstacle to the development of human potential. Here, I will focus specifically on ethical anthropocentrism, understood as speciesism—a form of segregation against other living species, based on a preconceived bias: namely, whether a being does or does not belong to the species *Homo sapiens*. It is considered a bias because it rests on a tautology, which can be expressed as: *humans have rights because they are human*. By contrast, Type 3 ethics begins from the assumption that there are no morally relevant differences between humans and non-human biological entities that would justify unequal moral consideration.

The individual-centered approach includes several subcategories:

3.1 Sensio-centric ethics: This approach focuses on the capacity of a biological entity to feel pain (Bentham 1948) or to be vulnerable to suffering (Derrida 2008), as well as to experience pleasure in its various forms. It is grounded in utilitarian philosophy, which evaluates the moral weight of actions based on their consequences—specifically, the duty to choose actions that maximize overall good. It rejects distinctions between humans and other animals (Singer 1975).

3.2 Subjectivity-based ethics: This perspective centers on the individual's perspectival quality—that is, being a *subject-of-a-life*, with a unique viewpoint on the world. This is expressed through both affective capacities (emotions and motivations) and cognitive capacities (a biography and a sense of intentionality). Such qualities entitle the individual to inalienable rights that must never be infringed and to always be respected as an end, never as a means (Regan 2004).

3.3 Biocentric ethics: Here, moral consideration is extended to any living being, regardless of specific traits. This includes not only animals but also, for instance, plants. Each living entity has its own inherent purpose, or intrinsic teleology, which presents a moral challenge for humans. The very fact of being alive—marked by individuality, uniqueness, and unrepeatability—reveals an intrinsic interest that calls for moral attention (Taylor 1986; Schweitzer 1953).
3.4 Empathy-based ethics: Building on Hume's idea of moral sympathy, this approach cautions against simply extending universal human rights to other beings in a way that risks anthropomorphizing their interests. Instead, it stresses the importance of recognizing the diversity of interests expressed by different living beings. Human rights and duties should thus be grounded in an appreciation for diversity itself (Marchesini 2025).
3.5 Pluralist ethics: This view holds that ethical issues concerning other living beings should be approached pragmatically. Rather than applying one fixed ethical framework, it calls for a flexible evaluation based on the specifics of each situation—seeking the most appropriate ethical lens to assess each case (Callicott 1989).

8.2.4 The Ecocentric Approach

The ecocentric approach extends moral value not only to individuals but also to *supra-individual* entities. These include: symbiotic and holobiont systems (e.g., multi-species organisms); categorical entities (e.g., species or other taxonomic groups); systemic entities (e.g., ecosystems or biomes); global entities (e.g., the biosphere); and even abiotic entities (e.g., a mountain or a river). Two foundational texts represent this approach: *A Sand County Almanac and Sketches Here and There* by Aldo Leopold (1949a), and *Ecology, Community and Lifestyle* by Arne Naess (1989). One key challenge of the ecocentric view lies in defining moral prescriptions, which are more straightforward in individual-centered approaches (e.g., Ross 1930; Attfield 2018). Ecosystems are by nature dynamic and constantly evolving, lacking a stable, well-defined identity. This makes preservationist practices inherently problematic. Yet, it is equally evident that the moral interests of individuals cannot be meaningfully safeguarded without acknowledging the network of relationships that connect them to other life forms.

A forest, for example, deserves moral attention that extends beyond the welfare of any single organism. Faced with ecosystems like coral reefs or tropical rainforests, one cannot help but recognize that their complex interdependent systems deserve moral consideration. Their interdependence relationships, which are of primary importance, often conflict with the preservation of individual organisms (Johnson 1992). This global approach holds that not only individual beings warrant moral consideration and impose ethical constraints on human behavior, but also trans-individual and even abiotic entities—since these often provide the conditions for life itself. It is based on the idea that systemic well-being can outweigh the welfare of isolated parts, and that the tendency to fragment reality into individuals reflects an anthropocentric bias (Rolston III 1988; Birch and Cobb 1991; Westra 1993).

The ecocentric approach encompasses several positions:

4.1 Holistic Ethics, which does not focus on individual beings but rather on life as a relational flow. According to this view, life cannot exist in isolation; organisms are always embedded in interdependent relationships that sustain them. It is therefore impossible to separate and preserve individual entities in isolation (Goldsmith 1992; Vernadsky 1998; Capra 2004).

4.2 Earth Ethics. Building on Leopold's original proposition, Earth ethics emphasizes the development of cooperative practices rooted in Darwinian conceptions of morality—as a prosocial extension from familial and emotional bonds within a group. Moral action, in this view, should be expansive, progressively extending the moral sphere to include the entire biosphere (Baird Callicott 2014).

4.3 Biotic Community Ethics. This perspective sees the biotic community as an indivisible whole that must be safeguarded in its entirety. It discourages focusing on individual entities and instead emphasizes the organismic interconnectedness of the biosphere (Latour 2017; Bunyard and Goldsmith 1988).

4.4 Ecosophy. First introduced by philosopher Arne Naess in the 1960s, ecosophy repositions the human being within the ecosphere. It advocates a normative transformation of our relationship with the whole, starting from personal experience and extending to ethical conduct and lifestyle. This approach was also adopted by philosopher Felix Guattari (2000), who identified three core relationships—with nature, society, and subjectivity—and by theologian Raimon Panikkar (1995).

4.5 Bioempathetic Ethics. This approach explores the deep emotional bond between humans and nature in all its forms. It extends beyond sympathy for individual beings to embrace an affective involvement with nature in its ongoing becoming and ontopoietic process (Carlson 2010; Giraldo and Toro 2020).

8.3 Philosophical Criteria Applied to the Non-human

Kinds of environmental ethics also differ in terms of the specific philosophical approaches they adopt, reaching different conclusions and proposing markedly diverse criteria for conduct—an important aspect that does not always make the most effective prescription or the decision-making process immediately evident. These different moral frameworks, which stem from divergent conceptions of the good or the good life, of the values to be pursued, and of the moral obligations that follow, lead to contrasting answers. For some, this is a limitation; for others, it can instead represent a strength—what John Baird Callicott calls *ethical triangulation*. According to the American philosopher, environmental ethics should not be conceived as a single doctrine adhering to one exclusive framework, but should develop through a dialogue among diverse positions and voices (1989).

Some scholars support this pluralistic approach, which combines different philosophical perspectives and values; others argue that, although commendable, such attempts risk generating contradictions (Attfield 2018). Nevertheless, it is important to note that—beyond the differences that will be detailed—the various strands of environmental ethics share common ground, much like other applied ethics that confront problems requiring pragmatic solutions. These are typically sought through compatible practical strategies and negotiated logic, as seen in fields like labor ethics or medical bioethics, where the theoretical dimension—while relevant—must ultimately engage with the practical reality of specific situations.

To begin with, some general considerations are needed—considerations that, in the case of environmental ethics, are far from self-evident:

(i) The distinction between the descriptive and the normative realms, according to which one cannot derive a normative claim from a factual statement, as per Hume's well-known law. Still, when moral reflection is extended to non-human entities, it becomes inevitable to refer to descriptive parameters;
(ii) The dichotomy between the moral agent—who is called to act morally—and the moral patient worthy of moral consideration. In the human domain, a sort of symmetry exists between these two roles, where those recognized as moral patients are also typically moral agents. In the case of non-humans, however, this symmetry collapses: the moral patient is not a moral agent;
(iii) The standard criteria of justice—such as retributive, distributive, and compensatory justice—which, in the field of environmental ethics, lead to extremely complex and often ambiguous situations, given the multiplicity of interests and the interwoven relationships and consequences present within ecosystems;
(iv) The distinction between narrow morality—understood as abstentions, such as recognizing others' rights as limits to one's own behavior or relinquishing some rights for oneself—and expanded morality, meaning the duties one has toward otherness. This dualism tends to blur in environmental matters, as the problems of the biosphere are closely tied to human action, raising questions of systemic character.

Among the different families of environmental ethics, we can identify:

1. Utilitarian ethics, which evaluates conduct based on informed choices the subject is called to make, aiming to assess the consequences of each option and select the one that maximizes overall good—thus also called *consequentialist ethics*;
2. Deontological ethics, where moral behavior is grounded in basic rules derived from principles of rationality and coherence, independent of consequences. These moral rules should therefore be universally valid;
3. Contractarian ethics, which holds that moral principles and judgments must arise from statements made by rational, self-interested individuals—as in a contract—acting impartially. This impartiality should be guaranteed by a hypothetical condition of ignorance regarding one's current or future status;

4. The ethics of care, which emphasizes certain prosocial human qualities—such as caregiving and responsibility—viewed in Darwinian terms as precursors or prerequisites for the evolution of moral behavior, and thus essential elements in our relationship with the biosphere;
5. Virtue ethics, which considers the pursuit of a virtuous life by the individual as the only true guarantee of reliable, engaged, and consistent moral action, particularly in the context of expanded or proactive morality—an aspect of paramount importance in environmental ethics;
6. Capability ethics, which links moral behavior to the opportunity for each being to develop its own inherent potential—central when considering the specific characteristics of other living beings;
7. The ethics of sympathy, in its two forms: as sympathy ethics, which highlights analogies between human beings and non-human entities, and as the broader sentiment of affiliation that humans may feel toward non-human others, such as animals or plants.

According to (1) utilitarian thought, an action is considered right or wrong based on the consequences it produces. Moral action, in this view, involves choosing behaviors that maximize overall good—hence, it represents a consequentialist (or end-state-oriented) concept of justice. For Jeremy Bentham, an action is good if it aligns with its utility in producing the greatest amount of pleasure and the least amount of suffering for the greatest number of individuals, as outlined in his main work, *An Introduction to the Principles of Morals and Legislation*. When it comes to extending this moral reasoning to non-human patients, we can recall Bentham's well-known words:

> It may come one day to be recognized, that the number of legs, the villosity of the skin, or the termination of the os sacrum, are reasons equally insufficient for abandoning a sensitive being to the same fate. What else is it that should trace the insuperable line? Is it the faculty of reason, or perhaps, the faculty for discourse? [...] The question is not, Can they reason? nor, Can they talk? but, Can they suffer? (1948, 311)

This approach to environmental ethics is known as *sentiocentrism*—that is, the view that the capacity to experience pleasure or pain is the key criterion for moral consideration. The dividing line between those who deserve moral regard and those who do not, then, is not based on language or cognitive abilities, but on sentience (Bartolommei, 1995). Bentham also links utility, pleasure, and happiness, suggesting that these values are achieved through sensitivity to suffering and an impartial point of view. However, it's important to note that these three terms—utility, pleasure, and happiness—are not synonymous. In particular, the use of the word *happiness* in reference to other species can be misleading. A more accurate way to put it might be *expressive satisfaction*, which, by its very nature, is species-specific and thus cannot be considered a universal quality, since it reflects differing interests (L. Battaglia, 2011).

The Australian philosopher Peter Singer developed a utilitarian theory focused on non-human animals in his essay *Animal Liberation* (1975). He argues that excluding animals from moral consideration lacks any real justification and is ultimately the result of circular reasoning. In 1970, British psychologist Richard Ryder coined the term *speciesism* to describe the moral discrimination humans apply to members of other species simply because they do not belong to the human one (Ryder 1971). Indeed, the circular logic of a statement such as "a human being, regardless of physical or cognitive condition, deserves moral consideration because they are human" is clear. Like racism or sexism, speciesism is based on a *petitio principii*—a logical fallacy where the conclusion is assumed in the premise. If we cannot identify morally relevant traits that are present in all human beings and absent in all other animals, then any such discrimination is arbitrary and ethically indefensible.

Several criticisms have been raised against the utilitarian approach. For example, it is difficult—sometimes impossible—to predict the consequences of specific actions in nature. This unpredictability undermines the practicality of a consequentialist ethic. On top of that, in nature, the well-being of a group often conflicts with that of the individual, but prioritizing the former over the latter is not necessarily the ethically correct choice. Moreover, consequentialism does not account for intentions, overlooking the ethical significance of the difference between voluntary and involuntary actions (Attfield 2018). A further risk is that utilitarianism may establish a new form of universalism without a solid foundation—extending humanist values to animals while simultaneously rejecting the premises of humanism itself (Braidotti 2013). Finally, utilitarianism fails to consider distributive and compensatory justice—that is, the fair allocation of benefits.

The deontological or post-Kantian perspective (2) remains centered on the individual, attributing value either to intentional subjectivity or simply to being-alive, viewing this condition as carrying an inalienable moral right. In this view, a being must always be treated as an end in itself, never merely as a means. Moral relevance derives from a being's intrinsic value. This goes beyond merely assigned value: it is the capacity of a being to confront the human agent with a moral question. In deontological ethics, this translates into a non-negotiable categorical principle. Inherent value, in fact, is not subject to personal evaluation, cannot be measured, separated from the being, or compared across entities—it is incommensurable. This clearly reveals the non-consequentialist nature of the approach: it is grounded in an a priori ought that must be respected regardless of the outcomes of a given action. The mere presence of a being imposes an ethical obligation on the moral agent—not to disregard its interests, nor to instrumentalize it for purposes unrelated to its own nature. This moral relevance can apply to specific entities, such as animals endowed with a subjective perspective and intrinsic intentionality; to living beings lacking a nervous system or consciousness but still considered individual life forms; and even to supra-individual systems such as ecosystems, which confront humanity with their moral standing.

In the context of a deontological approach to non-human animals, the work of American philosopher Tom Regan, *The Case for Animal Rights* (2004), stands out. In this book, Regan develops the idea of animal rights based on the notion that animals are *subjects-of-a-life*. His argument is that if we accept the principle of assigning value to human life regardless of rationality or cognitive ability, then we must also assign value to the lives of other animals. According to Regan, a being qualifies as a subject-of-a-life if it possesses beliefs and desires, a perspective on the world, memory that enables autonoetic or biographical experience, an emotional life marked by interests and preferences, and some awareness of its own future. In his view, the subject-of-a-life is an entity that displays propositional attitudes—not necessarily conscious ones—and a sense of self shaped by goals and subjective interests.

A different strand of deontological thought emphasizes respect for life-in-itself, particularly evident in recent debates about plant ethics (Kallhoff et al. 2019). While the moral value of plant life has traditionally been discussed through instrumental or anthropocentric lenses, some authors argue for recognizing an inherent value in plants. Doing so, however, requires humans to move beyond an animal-centered imagination, since the rhizomatic nature of plant life resists intuitive ethical categories—yet this very resistance can offer fertile ground for new moral perspectives (Pouteau 2018).

The contractarian view (3) can be traced back to Thomas Hobbes, particularly his work *Leviathan* (1651), where he argued that political and moral systems arise from a kind of social agreement designed to avoid the *bellum omnium contra omnes*—the "war of all against all"—which he considered the natural state of humankind. Within this framework, rational agents would be motivated to establish a political system to prevent mutual harm. Contemporary thinkers, however, argue that while Hobbes' model may be useful for regulating individual behavior, it does not meet the requirements of a true contract. For a contract to be legitimate, its rules must be accepted under conditions of impartiality (Dworkin 1977).

The leading figure in modern contractarian ethics is John Rawls, who in *A Theory of Justice* (1971) asks us to imagine choosing the rules of society from behind a "veil of ignorance"—without knowing what position we will occupy within that society. According to some contractarian theorists, only rational agents—motivated by self-interest but acting impartially behind the veil—can be included within the moral scope of contractarianism (Carruthers 1992). This effectively limits the approach to humans, making it anthropocentric. Restricting the contract to currently existing rational agents, even if impartial, risks excluding not only non-human subjects and environmental ethics but also the interests of future generations. However, Rawls himself later addressed this concern in *Political Liberalism* (1993), where he argues that contracting parties would also care about their descendants. As such, they would have reason to adopt the just savings principle—practices aimed at conservation and equitable distribution—for the benefit of future generations.

The contractarian approach, however, does not appear to account for non-human beings, nor is it easily extendable to them. Even when it comes to future generations, it presents several unclear and contradictory aspects (Benhabib 1992). After all, only individuals who genuinely care about the fate of future generations—those who possess a degree of altruism, an element not envisioned in contract theory—can ensure fairness between present and future generations (Attfield 2018). According to some scholars, applying contractarian principles exclusively to human beings is itself a form of bias, as the theory should, in order to maintain the impartiality it claims, be extendable beyond any preconceived conditions of privilege. In his essay *On Beasts, Persons and the Original Position*

(1979), Donald VanDeVeer argues that the veil of ignorance should be extended to include the kind of sentient being one might be—whether a human or, say, a pig—so that our moral judgments would become truly impartial. Similarly, the American philosopher Mark Rowlands, in *Animal Rights: Moral Theory and Practice* (2009), adopts a contractarian framework to argue for moral consideration toward non-human animals. All this notwithstanding, I believe these efforts suffer from a number of conceptual stretches—chief among them the attempt to establish shared positions on future-oriented choices, which assume a universality of interests that simply does not exist. Moreover, as already noted in the discussion of responsibilities toward future generations, it is highly unlikely that if humans alone were to decide, they would place the interests of other species among their priorities, especially given our limited and unreliable understanding of those species' actual needs.

The ethics of care (4) is grounded, on the one hand, in a natural human tendency—namely, parental motivation (*epimelesis*)—and, on the other, in the recognition of needs expressed by the other. A foundational text for understanding this ethical approach is Carol Gilligan's *In a Different Voice* (1982), in which the emphasis on "difference" highlights how the ethics of care has long been overlooked by moral philosophy in defining what constitutes a good life. In contrast, care should be seen as a fundamental moral attitude, forming the basis of relationships and the very foundation of human coexistence. The ethics of care can be understood in several ways: as the recognition of the vulnerability inherent in being alive, which calls for an attitude of benevolence toward all living beings, human and non-human alike; as the necessity of self-care, understood as the cultivation of personal well-being, which underpins many anthropocentric ethical approaches; and as a principle of responsibility, particularly as formulated by Hans Jonas in *The Imperative of Responsibility* (1984), where he links humanity's technological capabilities to the moral duty that follows from them.

Care for the external world is one of the most distinctive features of the human condition, according to Martin Heidegger (1978), who sees care as an essential ontological trait that defines our being in constant relation to the world. As Luigina Mortari writes in *Filosofia della cura* (2022, 35), "for human beings, living is always co-living, since no one can fully realize the project of existence alone." It also is worth recalling John Passmore's *Man's Responsibility for Nature* (1974), in which the Australian philosopher portrays human beings as stewards or trustees of the natural world, and therefore responsible for its care. According to Passmore, humanity's task is to enable nature to realize its full potential by acting in synergy with it. This view highlights the proactive meaning of care—*epimeleia* as active life—echoing the thought of Hannah Arendt (2018). Care, then, is a movement beyond the self: it entails assuming responsibility toward others and, at the same time, represents a form of self-education through the act of giving. It becomes a universal principle of sharing, one that places relationality at the heart of Being. This is evident in philosophical approaches that value fragility and the needs expressed by any form of otherness, as seen in the works of Judith Butler (2004), Joan Tronto (1993), and Jean-Luc Nancy (2000).

As Elena Pulcini notes in *La cura del mondo* (2013), *epimeleia* is strongly linked to the principle of responsibility, understood in terms of "pre-occupation"—a forward-looking and anticipatory concern for the possible futures of the beings around us. Michael Slote, in *The Ethics of Care and Empathy* (2007), emphasizes the importance of moral emotions in the development of ethical thinking. While the ethics of care may initially seem confined to human relationships, this narrow view has been effectively challenged by Josephine Donovan and Carol J. Adams (2007), who argue that care is deeply connected to recognizing the interests of otherness. In fact, the very concept of care implies going beyond self-interest—and, by extension, beyond the interests of those with whom we share personal bonds. If we were to care only for ourselves or our loved ones, we would have to deem it ethically acceptable to neglect the vast majority of humanity.

Virtue ethics (5) offers another compelling framework for the development of environmental ethics. According to Dale Jamieson (2008), virtues benefit both the moral agent and the moral patient, in the present as well as the future—making them worthy of cultivation. This ethical approach rests on the assumption that moral character must be formed through the development of certain virtues—such as generosity, altruism, fairness, prudence, courage, and sensitivity—which are essential for moral action. Virtue ethics draws on Aristotle's conception of character and virtue, as developed in the *Nicomachean Ethics*. In practice, virtues equip the moral agent to navigate the complexity of real-life moral challenges with greater clarity and confidence. This approach emphasizes the importance of building a stable moral character, allowing one—as Daniel Dombrowski (1984) and Nathan Nobis (2002) argue—to act virtuously regardless of context or the particular subject of one's moral action.

In *After Virtue* (1985), Alasdair MacIntyre argues that human beings must fulfill their *telos*—their intrinsic purpose—through the cultivation of virtue, starting from their own human nature. This potential, the ability to realize one's intrinsic ends, can only be fully achieved by allowing others to pursue their own interests as well. Starting from this evolutionary view of moral capacity, several authors—including Stephen Clark (1977), Bernard Rollin (1981), and Rosalind Hursthouse (2001)—highlight the close connection between virtue ethics and respect for the interests of non-human others.

Virtue ethics originates with the work of British philosopher Elizabeth Anscombe, who, in her article *Modern Moral Philosophy* (1958), criticizes deontology for relying on precepts that have lost their metaphysical foundation, and consequentialism for focusing solely on outcomes, which, in her view, leads to a kind of moral corruption. Her most important work is the book *Intention* (2000), in which she emphasizes the importance of highlighting the role of will in moral action. Building on this, Iris Murdoch argues in *The Sovereignty of Good* (1970) that moral action is a matter of subjective choices and preferences, and that achieving the good requires personal growth. She speaks of a purification of the gaze, a process that frees the individual from selfishness and encourages a movement away from self-centeredness—what she calls *unselfing*.

Philosopher Philippa Foot, in *Natural Goodness* (2003), proposes a naturalistic conception of morality that in some ways echoes Frans de Waal's views in *Good Natured* (2001) and *Mama's Last Hug* (2019). Foot starts from a strong concept of human nature as inclined toward the good and capable of cultivating virtues—understood as qualities that help fulfill the intrinsic tendencies of the human being. Along similar lines, Alasdair MacIntyre sees virtues as qualities to be developed from human nature itself, enabling it to reach its potential. Thus, attaining virtue represents the expression of a fully realized human nature. Later we will see how this perspective aligns well with the ethics of empathy, understood as both a dimension inherent to human nature and a capacity that requires active cultivation—what we might describe as a sentimental education of the individual.

Capability ethics (6) is an approach developed by Indian philosopher Amartya Sen and American philosopher Martha Nussbaum, made known through works such as *The Quality of Life* (1993). This perspective places at its core the capabilities of individuals and their specific ability to act morally, aiming to ensure that every entity has the opportunity to live a life worth living. In other words, this kind of ethics sees the individual as a bearer of unique traits that must be given the best possible conditions to be realized. The question it poses is not so much what a person's state of well-being is, but rather what opportunities are available to them. This makes it an approach centered on individual choice and freedom, based on the belief that the fundamental good of a society lies in promoting a range of opportunities that people may choose to exercise—or not. As is clear, this is a markedly pluralistic perspective when it comes to interests and values, making it especially well suited to emphasizing the diversity of interests that deserve recognition and respect. Capability ethics rests on the assumption that the core capabilities of beings differ in quality, not quantity: herein lies an ontological recognition of otherness, grounded in the affirmation of the inherent diversity of interests.

For both authors, it is not possible to reduce this singularity of condition without falling into distortions that inevitably overlook specific traits and, above all, fail to offer the ontological recognition of otherness. Capabilities do not concern only the intrinsic qualities of the individual, but also include the context in which the individual operates. While the authors stress the importance of considering the political, social, and ecological environment, nothing prevents us from also taking into account the ecological dimension in which a species functions. Capabilities emerge from the possibilities available to individual; these include being able to live according to one's inclinations and emotions, to access education and develop critical awareness, to engage with others in a relational context suited to one's needs for integration, to live in connection with other species and in harmony with the natural world, to enjoy leisure and recreational activities, and at the same time to participate in political life with guarantees of freedom of speech and association. For Nussbaum, well-being is measured more by the ability to carry out one's own life plans than by any general condition of welfare or wealth. See, in this regard, her essay *Not for Profit* (2010).

The ethics of sympathy (7) represents another line of reflection that views the feeling of sympathy as a fundamental driver of ethical attention toward otherness. Sympathy generates a sense of shared experience, allowing the moral agent to recognize and identify with the moral patient—thus prompting a genuine concern for the interests of the other. The ethics of sympathy has always played a pivotal role in philosophical thought, as exemplified by thinkers like David Hume, Adam Smith, and Max Scheler. Among the most important contributors to the ethics of sympathy is

Francis Hutcheson, whose *An Inquiry into the Original of Our Ideas of Beauty and Virtue* (2004) sees sympathy as a public emotional sense—a kind of contagion—rooted in human nature. This idea would later be echoed by Charles Darwin in his account of the development of prosocial behavior in humans. Instead, for David Hume, in the *Treatise of Human Nature*, moral behavior arises from a dual register: first, an affective element, and second, the imaginative capacity to project oneself into the experiences of others (Mercer 1972). By contrast, for Adam Smith—particularly in *The Theory of Moral Sentiments* (2010)—sympathy is a principle grounded in our ability to approve or disapprove of others' actions. It functions as a form of concordance and involves the presence of an impartial and disinterested imaginary spectator (E. Lecaldano, 2013). While Hume primarily associates sympathy with relationships among humans, he also extends it to animals—both in human-animal interactions and among animals themselves—thus presenting sympathy as a principle that transcends species boundaries. Another key figure in the ethics of sympathy is Max Scheler, especially in his work *The Nature of Sympathy*. Drawing on phenomenology, Scheler sees sympathy as a privileged pathway to apprehend the reality of the other as a center of value. This, in turn, allows us to perceive and develop the essential dimensions of life within the sphere of intersubjectivity.

Within environmental ethics, this theoretical lineage extends to philosophers such as Mary Midgley, whose anti-reductionism led her to develop a holistic moral philosophy aimed at overcoming the rigid distinction between humans and animals. In *Animals and Why They Matter* (1998), Midgley explores the complex relationships between humans and other species, with a particular focus on the emotional aspects of our moral choices—choices that are not purely rational, since we think first and foremost with our bodies and our feelings. In *Afectividad Ambiental* (2020), Omar Felipe Giraldo and Ingrid Toro invite us to cultivate sensitivity in order to nurture emotional awareness and rediscover the body's connection with the world—through direct contact with other forms of life and by reconfiguring our affective dimension in harmony with the planet's vital cycles. Sympathy, in this view, becomes the best way to participate in the whole and to rediscover an ethical dimension grounded in feeling, culminating in what has been called *affective ecologies* (von Mossner 2017).

Human behavior is inevitably driven by affective principles, even when it appears to be guided by a purely rational, deontological ethic devoid of feelings or passions. The human being cannot be split into affectivity and rationality, because every human expression is a synthesis of the two. To fully grasp the influence of affectivity on ethical behavior, however, we must reframe it: not as a trait to be controlled, but as something to be cultivated. While affective education certainly involves the ability to recognize and understand one's own emotions, it must go beyond this. Instead, it must foster the connective potential between the feeling, desiring subject and other beings—emphasizing the fundamental principle of sharing. The ethics of sympathy thus calls for the ability to cultivate harmony and complexity within the inner garden of one's own feelings and passions.

8.4 Cultural Shifts Toward an Ecological Ethics

The development of ecological and evolutionary thinking has profoundly transformed our concept of nature, as well as the attitude contemporary culture holds toward it. For a long time, nature was viewed as an unchanging entity, separate from human action—essentially a backdrop or stage for it. This perspective, as we have seen, can be traced back to Plato and Aristotle, who saw nature as a permanent and untouchable reality, distinct from human endeavor. This separation is also evident even in Kant's *Critique of Practical Reason*, where a clear divide remains between the moral law, which governs relations among human beings, and a nature governed by necessity. For Kant, ethics applies only to human interactions, defining what is just and permissible. Respect, in his view, is something owed solely to persons—that is, to human beings (Bartolommei, 1995). However, things changed with the Industrial Revolution and the rapid growth of urban centers, which became increasingly crowded and congested with the movement of goods and people. The rise of large factories, often located near cities, brought environmental issues to the forefront, as the widespread use of coal and other fossil fuels degraded air quality while industrial waste polluted rivers and groundwater. Cities faced an ever-growing accumulation of human-generated waste, while the streets became increasingly dirty due to the heavy use of horses for transportation.

The earliest critical reflections on the relationship between humans and nature can be traced back to late eighteenth- and early nineteenth-century thinkers. Among them were French researchers like Philibert Commerson and Bernardin de Saint-Pierre, who were concerned with preserving the natural beauty of Mauritius, and the Englishman Stephen Hales, who worked in Tobago in the Caribbean (Grove 1996). Early traces of what we might call proto-environmentalism—emphasizing the importance of nature and the need to live in harmony with it—can be found in the works of Jean-Jacques Rousseau, particularly in *Émile*. Concerns about the relationship between humanity and natural resources are also evident in the influential 1798 essay by Thomas Malthus, *An Essay on the Principle of Population as it Affects the Future Improvement of Society*. In it, Malthus warned that unchecked population growth would eventually lead to a shortage of resources, as population increased geometrically while resource production advanced only arithmetically. His argument significantly influenced Charles Darwin's theory of natural selection and, in some respects, foreshadowed the modern theme of limits to growth. This concern would later be echoed in the aforementioned Club of Rome, founded by Aurelio Peccei, especially in the 1972 report *The Limits to Growth* authored by Donella and Dennis Meadows, Jørgen Randers, and William Behrens III.

During the nineteenth century, literary and philosophical movements increasingly embraced the value of unspoiled nature. A notable example is the German movement known as *Romantische Naturphilosophie*, which flourished between 1790 and 1830. It placed renewed emphasis on beauty and an organic, holistic, and vitalistic view of nature, challenging the mechanistic and reductionist paradigms. Leading figures in this movement included Friedrich Schelling, Johann Gottfried Herder, Johann Wolfgang Goethe, and Friedrich Schlegel. These thinkers viewed nature as imbued with intrinsic purpose and hidden relationships, animated by an internal dynamism and shaped by ideal forms. Their revival of Spinoza's thought also suggests a kind of pantheism: nature is seen as a living, creative force marked by invisible connections—a world of ideal forms and constant metamorphosis, where microcosm and macrocosm mirror each other, and diversity exists alongside continuity among all living things.

England also had its own Romantic poetry that celebrated nature as a source of beauty and authenticity. This is especially evident in the works of William Wordsworth and Samuel Taylor Coleridge, who set their poems against the evocative landscape of the Lake District. As Coleridge wrote in a letter, "Nature has her proper interest; & he will know what it is, who believes & feels, that every Thing has a Life of its own, & that we are all *one Life*" (Griggs 1966 II, 864). Moving to the United States, a central figure among the forerunners of environmental thought is undoubtedly Henry David Thoreau. In his book *Walden, or Life in the Woods* (1899), he outlines a personal journey that, perhaps for the first time so explicitly, expresses the need to reconcile human life with the natural world in order to rediscover the very essence of existence. As Thoreau famously wrote: "I went to the woods because I wished to live deliberately, to front only the essential facts of life, [...] and not, when I came to die, discover that I had not lived." He continues: "I wanted to live deep and suck out all the marrow of life [...] to put to rout all that was not life" (1899, 93). Alongside Thoreau, the philosopher Ralph Waldo Emerson must also be mentioned. In his 1836 essay *Nature*, he presents the natural world as a creative and transcendental force, leaning toward pantheism and emphasizing the central, ontological importance of the relationship between the individual and nature.

Among other early works of environmental consciousness are George Perkins Marsh's *Man and Nature* (1864) and *A Wind-Storm in the Forest* (1894) by John Muir, in which he describes the beauty of nature in its wild, untamed state—unspoiled by human intervention—combining botanical observation with meditative solitude and awe for the landscape. In Italy, poet Giacomo Leopardi also reflected deeply on the relationship between humans and nature. His work evokes both fascination and tension between *natura naturans* (nature creating) and *natura naturata* (nature created). In his *Zibaldone di pensieri*, written between 1817 and 1832, Leopardi offers a biocentric view of nature, placing life—with all its fragility, and thus its strength—at the center. As philosopher Sergio Givone writes in *Storia del nulla*, Leopardi's worldview "is not nihilism, but an ontology of nothingness—an understanding that it is precisely because things are fragile, ephemeral, and mortal that they are worthy of being loved in their reality" (1995, 143).

Aside from the pioneering works of Aldo Leopold (*A Sand County Almanac* 1949b), Rachel Carson (*Silent Spring* 1964), and Lynn White Jr.'s essay *The Historical Roots of Our Ecological Crisis* (1967), environmental ethics, in the full sense of the term, began to develop more substantially in the 1970s through a series of influential publications. One key text was Barry Commoner's *The Closing Circle* (1971), followed by his later book *Making Peace with the Planet* (1990). A major moment came in 1973, during the World Congress of Philosophy held in Bulgaria. There, Australian philosopher Richard Routley delivered a pivotal presentation titled *Is There a Need for a New, an Environmental, Ethic?* That same year, the journal *Inquiry* published an article by Norwegian philosopher Arne Naess titled *The Shallow and the Deep: Long-Range Ecology Movement*, in which he distinguished between two kinds of environmentalism: a "shallow" form concerned with protecting natural resources insofar as they benefit humans, and a "deep" ecology rooted in biospheric egalitarianism, which aims to promote and preserve the full diversity of species and habitats (Attfield 2018). Also in 1973, Ernst Schumacher published his influential book *Small is Beautiful*, which focused on the problem of diminishing resources and humanity's tendency toward economic gigantism. He argued that this drive toward large-scale industrial systems leads, on the one hand, to uncontrolled exploitation of natural resources, and on the other, to a failure to recognize that small-scale systems are inherently more stable and sustainable.

The following year, in addition to John Passmore's already-mentioned *Man's Responsibility for Nature*, two other significant works were published: *Philosophy and Environmental Crisis* by philosopher William Blackstone, and *Should Trees Have Standing? Towards Legal Rights for Natural Objects* by legal scholar Christopher Stone. To round out this overview of 1970s environmental ethics, I should also mention Holmes Rolston III's 1975 essay *Is There an Ecological Ethic?*, which connects human-centered values with ecological dynamics, emphasizing the importance of resilient ecosystems. Finally, Kenneth Goodpaster's *On Being Morally Considerable* (1978) introduced a biocentric approach to moral standing, proposing various degrees of moral relevance and extending the scope of moral consideration beyond traditionally defined boundaries.

To summarize the main themes of environmental ethics—particularly over the past fifty years—we can identify the following key areas:

(i) The defense of the natural environment or wilderness, viewed as a space of authenticity, aesthetic pleasure, and personal growth;
(ii) The issue of development in terms of large-scale economic systems, energy-intensive production and consumption patterns, capitalist society, and the depletion of natural resources;
(iii) The technological question, specifically its impact on nature and its role in reinforcing a Promethean human attitude—one that fuels the idea of human domination over the planet;
(iv) The issue of animal rights, reflecting a profound cultural shift from a rural model centered on the utilitarian use of animals to an urban mindset based on companionship;
(v) The erosion of biodiversity, caused by the use of pesticides and herbicides in agriculture, deforestation followed by the progressive cementification of peri-urban areas, and increasing tropical deforestation leading to the destruction of vital biomes;

(vi) The issue of pollution, particularly of air and groundwater, along with fears related to radioactive contamination, excessive waste production, and, in particular, the proliferation of plastic and microplastics in seas and oceans;
(vii) The problem of climate change, linked to greenhouse gas emissions from fossil fuel use and large-scale livestock farming, with consequences such as glacier melt, rising sea levels, desertification of vast areas, and increasingly extreme weather events;
(viii) The epidemiological issue tied to disruptions in the virosphere, caused by ecosystem destruction, the migration of virus reservoirs, amplification effects from intensive farming, and spillover infections with pandemic potential—further exacerbated by globalization, increased mobility, and the presence of non-native animal species (Quammen 2012).

During the last two decades of the twentieth century, the discussion around these issues grew even more substantial, also thanks to the founding of important journals such as *Environmental Ethics*, established in 1979 by Eugene Hargrove and Holmes Rolston. The debate among the various positions outlined in the previous chapters became increasingly heated—on one side, the proponents of *shallow ecology*, such as Passmore or Jonas, who advocated for a mere extension of moral consideration to new subjects or a greater sense of human responsibility toward nature; on the other, those who, following the ideas of Leopold and Naess, argued for a profound paradigm shift in environmental ethics, favoring a *deep ecology* approach, as outlined by thinkers like Callicott and Rolston III. Among the essays that, though through different lenses, have addressed ecological issues over these two decades, it's worth mentioning: *In Defense of the Land Ethic* by John Baird Callicott (1989), *Respect for Nature: A Theory of Environmental Ethics* by Paul W. Taylor (1986), *Monocultures of the Mind* by Vandana Shiva (1993), *Remaking Society* by Murray Bookchin (1990), *Rebels Against the Future* by Kirkpatrick Sale (1996), and *The Way: An Ecological Worldview* by Edward Goldsmith (1992).

The development of a new understanding of the relationship between humans and non-human otherness—as well as biological processes in general—emerged more clearly by the end of the twentieth century, thanks to significant shifts in several disciplines. For instance, evolutionary theory was marked by the decline of reductionism and gene-centrism, and the emergence of *Extended Synthesis* (Pigliucci and Müller 2010); ethology saw the rise of cognitive science, which highlighted the existence of animal minds and moved beyond explanations based solely on reactive and instinctual automatisms (Griffin 1976). Likewise, epistemology saw the rise of complexity thinking, which led to a true paradigm shift in how complex systems are understood (Von Bertalanffy 1968; Simon 1975; Atlan 1979), along with developments in psychology (Maturana and Varela 1980), anthropology (Morin 1977), and biology (Gould 1990; Lewontin 1991). Finally, botany saw a shift in perspective on the plant world due to new knowledge about plant physiology and symbiotic processes.

8.5 Theoretical Issues in Environmental Ethics

Within ecological ethics, several theoretical and conceptual issues have emerged over forty years of debate. These have taken shape through the confrontation between different positions and, more broadly, through the cultural transformations that have particularly marked the early decades of the twenty-first century—for example, the rise of the posthumanist philosophical movement, which challenges the humanist vision of a self-sufficient human being detached from relationships with non-human entities.

Among the key theoretical issues that must be addressed are:

1. The concept of moral value—specifically, how to assign value to non-human entities in a way that is non-anthropocentric while avoiding the naturalistic fallacy.
2. The concept of the moral patient and its relationship to the moral agent, particularly concerning the foundations of moral recognition and the nature of the relationship established in moral action.
3. The problem of moral prescription, that is, on what principles and through which methods one can determine an appropriate course of action, especially in light of the plurality and irreducibility of competing interests when dealing with non-human entities.
4. The relationship between the evolution of human knowledge and operational capacities, and the resulting moral obligations—in other words, whether there is a correlation between the agent's level of knowledge or power and the extent of their moral duties.

I will address these four conceptual issues drawing on insights from several authors who, in my view, have offered important clarifications on these topics, while also referencing the ethical framework I developed in my own essay *Toward an Ethics of Empathy* (2025).

Let's start with the first point. The naturalistic fallacy, first described by George Edward Moore in *Principia Ethica* (1903), is often invoked to discourage extending moral value to non-human entities. According to this principle, there is a flawed reasoning in deriving prescriptions—that is, value judgments—from naturalistic descriptions. This view can be traced back to Hume's law, a metaethical principle which posits a fundamental gap between *is* and *ought*—that is, between descriptive and prescriptive propositions. Building on Hume's argument, Moore concludes that moral value cannot be identified through natural properties such as species diversity, ecosystem balance, the fact of being alive, or the capacity to feel pain. After all, the skeptic can always ask why species richness should constitute a moral value—that is, something capable of informing moral conduct (Peverelli 2005). A second issue related to the attribution of value concerns the fact that even an ecocentric perspective ultimately remains a human perspective (Tallacchini 1998).

In this regard, I believe several clarifications are in order. First, the definition of moral value arises from a series of intermediate considerations that cannot be reduced to the dichotomy proposed by Moore. Second, the concept of moral value does not coincide with that of prescriptive cogency, and therefore moral value attribution is not equivalent to Hume's law. Finally, moral prescription entails multiple levels of intervention and moral appeal, as demonstrated by distinctions between narrow and broad morality, various notions of justice (e.g., compensatory, distributive, retributive), and the difference between moral obligations or a lack of rights over someone. In countering the charge of the naturalistic fallacy, it is worth recalling the contributions of John McDowell (*Mind, Value, and Reality* 1998) and Hilary Putnam (*Beyond the Fact/Value Dichotomy* 1982), who argue that even if values are not simply found in the world independently of our judgment, this does not mean they are disconnected from our lived experience.

Regarding the fact that any perspective remains a human one, we have already seen that there is a profound difference between an *anthropocentric* approach—where only the human being is deemed worthy of moral value—and an *anthropocentered* perspective, which simply denotes a human standpoint on the world, one that evolves over time. As epistemology has shown (see G. Bachelard, 1984), each development in knowledge gradually reshapes the human viewpoint in an increasingly expansive and decentered direction. That said, I believe that beyond the contributions of McDowell and Putnam, the question of value itself still warrants deeper reflection. I will begin with the philosopher who has most consistently developed a theory of value: J. Baird Callicott. In particular, his essay *Intrinsic Value, Quantum Theory, and Environmental Ethics* (1985) defends the possibility of recognizing an autonomous, that is, non-anthropocentric, value. He does so by appealing to Hume's concept of sympathy and by critiquing the Cartesian dualism at the root of anthropocentric divisions. Starting from Aldo Leopold's land ethic, Callicott sees the attribution of value as a process of progressive extension: from parental behaviors to prosocial behaviors within groups, and from there to broader forms of inclusion—an idea that reprises Darwin's analysis in *The Descent of Man* (1871).

From my point of view, when analyzing the value of an entity, it is essential to distinguish four basic conditions:

(i) Extrinsic value: the value an entity has *for others*, which can be *instrumental* (used as a resource), *utilitarian* (value of use for other entities), or *ecological* (participatory or systemic value, inseparable from a whole, as exemplified by ecosystems—the holobiont being the most explicit manifestation);

(ii) Intrinsic value: the value an entity has *in itself*, by virtue of being a bearer of its own interests—interests that shape its behavior and orientation in the world. In this sense, a living being cannot be reified (even if it can be used), precisely because it possesses inalienable interests;

(iii) Inherent value: a value that is pertinent to an external entity, which may be *anthropo-referential* (its significance is linked to human interpretation), or *dialogical* (it confronts or questions another entity). The notion of *inherence* here refers to what concerns, relates to, or evokes a connection—a value that establishes an intimate or meaningful relationship;

(iv) Moral value: a quality or condition that prompts moral action, challenges one's conduct, can serve as a prescriptive guide, and falls within a specific ethical framework.

As for the extrinsic value (i) of an entity, it is clear that—despite Kant's principle of treating the human being always as an end and never merely as a means—there is always some degree of instrumental or ecological value in any entity. As Holmes Rolston III (1988) pointed out, there is nothing inherently wrong with this. On the other hand, the intrinsic value (ii) of an entity—its *for-itself* condition—is impossible to dismiss when we are dealing with a living being. That said, this condition does not automatically entail a moral prescription: I do not necessarily face an ethical dilemma when eating a salad or working in the garden. Nonetheless, I must acknowledge that when I encounter a living entity, I am in the presence of something that has its own interests. Robin Attfield, in his essay *The Good of the Trees* (1981), applied the notion of interests to the plant world. This essay opened the door to what would later become Plant Critical Studies and the ethics of plant life, linked also to research on plant perception (e.g., Mancuso and Viola 2015). When we shift our focus to the animal world, it becomes even clearer: not only do animals possess the capacity to feel pain, but their entire spectrum of conative behaviors (emotions and motivations) unequivocally expresses the concept of *for-self*, that is, the possession of interests that need to be protected or pursued.

The concept of inherent value (iii) is more complex. Though related to intrinsic value, it is not identical to it. Inherent value introduces the question of *pertinence*: the fact that an entity's intrinsic value may concern *my* perspective (anthropo-referential), preferences, or ethical reflection (dialogical). For example, the question arises when I choose whether to ignore an animal's sentience. In *inherence*, I recognize the other's value not merely through subjective attribution, but through informed recognition. This marks the difference between simple projective attribution (based on subjectivism) and intersubjective recognition (based on homology).

Moral value (iv), in turn, derives from inherence—specifically from the pertinence of the quality manifested by an entity with respect to its ethical relevance. In other words, it depends on whether the entity's condition can make an ethical demand on me. Not only the face of my fellow human being (Levinas 1969), but also the suffering of a non-human entity can do this, particularly within a utilitarian framework. Moral value, therefore, is tied to the moral criteria adopted within a given ethical system.

Now onto the second point. When discussing moral value, it's essential to consider the traditional dichotomy between the moral agent—the one who shapes their conduct based on prescriptive guidelines—and the moral patient, the one who receives the consequences of that conduct. Upon closer examination, however, this dualism creates a separation between the two entities and implies a kind of passive receptivity on the part of the moral patient. Once again, I turn to Lévinas, even though his conception of moral action remains rooted within an anthropocentric framework. From a metaethical perspective, however, Lévinas introduces a true paradigm shift by dissolving the agent–patient divide. In *Totality and Infinity* (1969), the French philosopher rejects the self-sufficient stance of the moral agent—as one who stands above and apart from the other—because the face of the other takes on an epiphanic role: it reveals itself and calls the agent to act. This is the foundation of what I refer to as the ***principle of interception*** in ethics—the idea that the moral agent is, in a sense, held hostage by their responsibility toward the other. The epiphany of the face, revisited in *Humanism of the Other* (2005), is an encounter, almost a form of captivation, where the agent cannot evade the call of the other and is thus bound to respond through moral action. In Lévinas' view, it is no longer possible to speak of a self-referential moral agent who follows only their own internal principles. Nor is it adequate to describe the recipient as a mere patient, a being who passively endures the consequences of a moral decision. The principle of interception—the active role played by the recipient in prompting the moral awareness of the agent—reshapes our understanding of moral relationships. Even the concept of ***moral extensionism***—a term coined by John Rodman in his essay *The Liberation of Nature?* (1977)—fails to capture this dynamic, because it still treats the recipient as a passive subject granted access to universal human rights from the outside.

It is equally essential to address a distinctive aspect of moral concern expressed by the so-called recipient. In my essay *Toward an Ethics of Empathy* (2025), I argued for the importance of viewing the recipient as a moral claimant. The ethics of empathy I propose can be seen as an evolution of the ethics of sympathy. However, it diverges from the latter by avoiding the principle of identification through analogy as a bridge for extending moral consideration. Instead, it emphasizes recognition through shared traits (homology) and the valorization of diversity. This empathetic approach involves an affective component, grounded in co-feeling—a shared experience rooted in homology, or a common metapredicate, e.g., the shared condition of being-a-body, as discussed by Ralph R. Acampora in *Corporal Compassion: Animal Ethics and Philosophy of Body* (2006). It also entails a cognitive component, based on understanding the specific characteristics of the other—essentially, the distinctive features that define their needs and capabilities. This dual structure allows us to avoid what John Rodman referred to as the neglect of particular traits of otherness—the unique needs and potentials of others. Centering these peculiarities as a moral compass offers a valuable framework for antispeciesist ethics. It encourages us to consider species-specific characteristics as morally relevant factors and to build moral reflection around a set of values understood as possibilities or opportunities to be safeguarded, which must then be tailored to each unique condition. Such an approach could help resolve the contradiction of attempting to overcome ethical anthropocentrism while still clinging to ontological anthropocentrism—a contradiction that risks producing not only ambiguities but also forms of cryptospeciesism.

Ultimately, the ethics of empathy is founded on two core principles. First, a redefinition of the moral agent, who—much like in the ethics of care—is called to develop their nature, bringing it to full fruition through both affective and cognitive empathetic capacities, in the spirit of virtue ethics. Second, a transformation in the status of the recipient, who is no longer a mere patient (a passive subject of moral action), but rather a moral claimant, capable of guiding moral orientation in terms of opportunities that should be granted based on their unique characteristics as an ontological other. At the same time, this claimant is able to intercept the moral agent through a direct act of assertion, by virtue of their very presence in the world.

Moving to the third aspect, the problem of prescription takes into consideration the debate on moral behavior. In my opinion, it is confronted with the obvious insufficiency of the different approaches and at the same time struggles to find negotiations between the different interests at stake. We know, for example, that if we want to defend a particular ecological reality we must defend it from disruptive entities such as allochthonous species by taking action to contain their invasiveness, but in doing so we would be in antithesis with Tom Regan's deontological view that considers each subject-of-a-life independent of everything else. Similarly, it is clear that the defense of a species from the risk of extinction may collide with the view of the preservation of the individual in terms of suffering, undermining Peter Singer's approach. As Elliott Sober points out in the essay *Philosophical Problems for Environmentalism* (1986) an animal conservationist faced with the alternative of whether to save an endangered species or the greatest number of individuals of a common species will choose the second option.

The tendency to absolutize the value of the capacity to suffer or the subject's autonomy, however, exposes itself to a large number of problems: in the first case, the inability to consider the biological value of suffering itself in the animal world and, at the same time, the fact that under anesthesia any action on the subject would become licit; in the second case, the evidence that there is no subject that can be defined as autonomous and absolute. This does not mean that Regan's and Singer's views have no validity, but that in order to find a course of action that takes into account this plurality of problems, it is necessary to adopt a kind of ethical pluralism that, despite the risk of contradiction, a pragmatic approach is preferable to remaining entrenched in absolute principles. Leopold's view, which summarizes the principle of justice in the ability of a course of action to preserve the integrity, stability and beauty of the biotic community, seems frankly too general, if ideally right, to deal with the concrete problems that arise in the field of ecological ethics.

When moving from abstract principles to real-world applications, ecocentric ethical theories face significant challenges. Arne Naess' ecosophy, for instance, highlights these difficulties. The Norwegian philosopher proposed four principles to guide eco-prescriptive practices: (i) the principle of intrinsic relationality between humans and nature, which should help us overcome the dualism of moral action by highlighting the interconnectedness of human presence into nature; (ii) the principle of biospheric egalitarianism, which holds that there are no hierarchies among living beings; which, however, makes it impossible to establish priorities or decisive criteria for resolving potential conflicts; (iii) the principle of diversity and symbiosis among living beings, according to which all species, by virtue of their diversity, contribute to the well-being of the whole—yet this raises challenges when the interests of individual species conflict with the interests of the whole; and (iv) the principle of the complexity of reality, which recognizes that ecological systems are characterized by a multiplicity of connections and exchanges, making it impossible to dissect them into isolated parts—though this complexity does not necessarily help when navigating the maze of possible choices.

Undoubtedly, these principles originate from the descriptive framework of ecology—not through direct derivation, but rather by posing challenges to philosophical ecology, as rightly emphasized by Naess (1989). An ecocentric ethical approach requires more than simply adopting basic ecological principles as the foundation for a coherent and ethically sound course of action. Rather, the analysis of the intrinsic and extrinsic characteristics of living organisms and their interconnections prompts us to reflect on the four dimensions of value discussed earlier, with particular attention to inherence—since it is from this concept that the moral agent's engagement becomes possible, and ethical guidelines can be drawn. This is not about strictly adhering to Hume's law, but rather about recognizing that ecology reveals the interconnections within the living world. However, we still need philosophical reflection to understand how these qualities are relevant—that is, how they raise ethical questions. Undoubtedly, there is a problematic circularity between description, philosophical reflection, and evaluation, since each of these stages poses challenges to the next. Still, the issue of translating values into prescriptive action requires grounding principles and values expressed in concrete terms.

This same eco-philosophical perspective is shared by J.B. Callicott in his essay *The Conceptual Foundations of the Land Ethic* (1987) and by H. Rolston in *Values Gone Wild* (1983). Callicott urges us to move beyond the ontological primacy of objects over relationships, since it is relationships that define the nature of organisms rather than the other way around. An ecocentric ethics must therefore expand to include these dimensions, through the recognition of the biotic community, while still paying attention to individual entities. According to Rolston, the recognition of values that go beyond the individual—values that refer to relational systems such as symbiosis, species, ecosystems, or even the biosphere as a whole—should not be interpreted subjectively or relativistically. As I suggested in discussing the relationship between intrinsic and inherent value, recognizing intrinsic value—a kind of inalienable "for-itself" that transcends being merely a resource—raises a question of relevance: how that quality engages the subject and addresses them in various ways, including morally. Inherence does not directly follow from description; rather, it arises from the challenge that the descriptive quality poses to the subject. In this sense, it can be discussed on an intersubjective level. A similar view is found in L. Lombardi Vallauri (1990), who argues that intrinsic value gives rise to relationality, without implying a relativistic attitude.

The fourth aspect I wish to consider is the relationship between the subject's knowledge and power, and the resulting moral obligations. As we have seen, both sentiocentric and biocentric theories of individual ethics, as well as ecocentric ones, regard objective knowledge of biological reality—such as the capacity to suffer, the possession of a self-directed *telos*, or the maintenance of ecosystem balance—as an essential guide for moral orientation. Although these theories constantly risk deriving value from fact, they maintain that scientific knowledge cannot be ignored in ethical reflection. This does not mean falling into the naturalistic fallacy, but rather recognizing that every form of knowledge has implications for moral behavior. Every acquisition of knowledge inevitably poses a challenge to the moral subject. If I know that an animal feels pain, I must ask myself whether it is right to ignore that fact. Ethics is not a discipline completely disconnected from other forms of knowledge—indeed, the idea of an uninformed ethics makes little sense. The "last man" thought experiment introduced by Richard Routley (1973)—in which we ask whether the last surviving human, on the verge of death, would act wrongly by destroying the environment—is an example of how ethics often begins with describable, conceivable scenarios.

People frequently ask why humans shouldn't pursue only their own interests, since all other species in nature do. But this kind of reasoning attempts to derive a value from a fact. By contrast, the principle of inherence asks whether, given certain factors that engage moral concern—namely, (i) in light of the moral values that humans aim to uphold (e.g., in virtue ethics); (ii) considering the operational capabilities humanity has achieved through technology; (iii) and in relation to the range of alternative actions available to us—the presence of intrinsic value in another being gives rise to a moral issue. As we can see, the principle of inherence serves as a kind of bridge between the recognition of intrinsic value and the development of moral value.

We have seen how Hans Jonas' principle of responsibility links human technical-operational capacity to moral responsibility: one becomes responsible for something only when, and to the extent that, one is capable of altering or compromising it. Alfred W. Crosby, in his essay *Ecological Imperialism* (1986), connects environmental issues to the history of Western culture, which has promoted a way of thinking and living grounded in an epistemology of domination and a global spread of technology. Similarly, the geographer Yi-Fu Tuan, in *Man and Nature* (1971), explores the relationship between technology and life, illustrating the ways humans have exerted force in domesticating plants and animals. According to Jonas—particularly in his essay *Toward a Philosophy of Technology* (1979)—technological action already entails a denial of the thing *as it is*, inevitably leading to its modification. In *Civilized Man's Eight Deadly Sins* (1974), Austrian ethologist Konrad Lorenz argues that humanity, having developed an immense destructive potential, must correspondingly strengthen its moral conduct. Taken together, these thinkers suggest that there is an implied relationship between the growth of human operational capacities and the development of greater ethical sensitivity. This view leads us to understand moral obligations not as absolutes, but as outcomes shaped by several variables:

(i) The particular conditions of life—Ethical prescriptions cannot be the same for all; what is expected of a European may differ from what applies to populations such as the Inuit or the San, who live by hunting.
(ii) Scientific advancement—If research in neurophysiology and ethology reveals particular qualities in non-human species, these insights cannot be ethically disregarded.
(iii) Technological capacity—Every increase in our ability to intervene in the world heightens our responsibility toward the biosphere.
(iv) Economic or power status of the moral agent—The more power a subject holds, the greater their moral obligations.
(v) The human condition as a whole—Humanity's decisive role in the balance of the planet places it on a different level than other members of the biosphere—not because of anthropocentrism, but because of the weight of its responsibilities.

Finally, another important aspect linking the principle of power with the development of ecological ethics concerns the intersectionality of ecological problems with other types of issues affecting vulnerable individuals, minorities, and disadvantaged human groups. Karen J. Warren, in her essay *The Power and the Promise of Ecological Feminism* (1990), emphasizes that human ethical orientations are shaped by a conceptual framework—an underlying set of beliefs, ways of thinking, and assumptions—that is androcentric in origin, meaning it reflects a masculine and patriarchal view of the world. According to this perspective, our approach is shaped by pre-conceptual structures that influence key philosophical concepts such as reason, rationality, and ethics. Warren argues that the patriarchal system is based on relations of domination, as it: (i) favors hierarchical thinking, organizing reality along a top–bottom axis; (ii) establishes value dualisms through oppositions rather than complementarity; and (iii) develops a logic of domination—an argumentative structure that legitimizes subordination. More broadly, Warren refers to these as *oppressive conceptual frameworks*.

In her 2025 essay *Politiche dell'antispecismo*, Elisa Ortolani explores the connections between animalized, feminized, and disabled bodies, showing how various forms of oppression share a common matrix and advocating for intersectionality among different struggles against oppression. Lastly, I want to mention Carol J. Adams' *The Sexual Politics of Meat* (1990), which investigates the relationship between patriarchal values, meat consumption, the exploitation of animals, and attitudes toward nature.

As these examples show, ethics must necessarily engage with other disciplines rather than remain self-referential, if it hopes to help humanity find viable, coherent, and practically applicable solutions to the ongoing crisis.

References

Acampora, Ralph A. 2006. *Corporal Compassion: Animal Ethics and Philosophy of Body*. Pittsburgh: Pittsburgh University of Pittsburgh Press.

Adams, Carol J. 1990. *The Sexual Politics of Meat: A Feminist-Vegetarian Critical Theory*. New York: Continuum.

Anscombe, Gertrude Elizabeth Margaret. 1958. Modern Moral Philosophy. *Philosophy: The Journal of the Royal Institute of Philosophy* 33 (124): 1–19.

Anscombe, Gertrude Elizabeth Margaret. 2000. *Intention*. Cambridge, MA: Harvard University Press.

Arendt, Hannah. 2018. *The Human Condition*. Chicago: University of Chicago Press.

Atlan, Henri. 1979. *Entre le cristal et la fumée: Essai sur l'organisation du vivant*. Paris: Seuil.

Attfield, Robin. 1981. The Good of Trees. *Journal of Value Inquiry* 15 (1): 35–54. https://doi.org/10.1007/BF00136626.

Attfield, Robin. 2018. *Environmental Ethics*. Oxford: Oxford University Press.

Bachelard, Gaston. 1984. *The New Scientific Spirit*. Boston: Beacon Press.

Bartolommei, Sergio. 1995. *Etica e natura: una rivoluzione copernicana in etica?* Roma: Laterza.

Bateson, Gregory. 1972. *Steps to an Ecology of Mind: Collected Essays in Anthropology, Psychiatry, Evolution, and Epistemology*. Chicago: University of Chicago Press.

Battaglia, Luisella. 2011. *Un'etica per il mondo vivente. Questioni di bioetica medica, ambientale, animale*. Roma: Carocci.

Benhabib, Seyla. 1992. *Situating the Self: Gender, Community, and Postmodernism in Contemporary Ethics*. Cambridge, UK e New York: Polity Press.

Bentham, Jeremy. 1948. *An Introduction to the Principles of Morals and Legislation*. New York: Hafner Publishing Company.

Birch, Charles, and John B. Cobb. 1991. *The Liberation of Life: from the Cell to the Community*. Denton: Environmental Ethics Books.
Blackstone, William T. 1974. *Philosophy and Environmental Crisis*. Athens: University of Georgia Press.
Bookchin, Murray. 1990. *Remaking Society: Pathways to a Green Future*. Boston: South End Press.
Boulding, Kenneth E. 1966. The Economics of the Coming Spaceship Earth. In *Environmental Quality in a Growing Economy*, ed. Henry Jarrett, 3–14. Baltimore: Resources for the Future/Johns Hopkins University Press.
Braidotti, Rosi. 2013. *The Posthuman*. Cambridge: Polity Press.
Bunyard, Peter, and Edward Goldsmith. 1988. *GAIA: The Thesis, the Mechanisms and the Implications*. Wadebridge: Wadebridge Ecological Centre.
Butler, Judith. 2004. *Undoing Gender*. New York: Routledge.
Callicott, J. Baird. 1985. Intrinsic Value, Quantum Theory, and Environmental Ethics. *Environmental Ethics* 7 (3): 257–275.
Callicott, J. Baird. 1987. The Conceptual Foundations of the Land Ethic. In *Companion to a Sand County Almanac: Interpretive and Critical Essays*, ed. J. Baird Callicott, 186–217. Madison: University of Wisconsin Press.
Callicott, J. Baiard. 1989. Animal Liberation: A Triangular Affair. In *In Defense of the Land Ethic: Essays in Environmental Philosophy*, 15–38. Albany: State University of New York Press.
Callicott, J. Baird. 2014. *Thinking Like a Planet: The Land Ethic and the Earth Ethic*. New York: Oxford University Press.
Campbell, Donald T. 1974. Evolutionary Epistemology. In *The Philosophy of Karl Popper*, ed. Paul Arthur Schilpp, 413–463. La Salle: Open Court.
Capra, Fritjof. 2004. *The Hidden Connections: A Science for Sustainable Living*. New York: Knopf Doubleday.
Carlson, Allen. 2010. Contemporary Environmental Aesthetics and the Requirements of Environmentalism. *Environmental Values* 19 (3): 289–314.
Carruthers, Peter. 1992. *The Animals Issue: Moral Theory in Practice*. Cambridge: Cambridge University Press.
Carson, Rachel. 1964. *Silent Spring*. New York: Fawcett.
Center for Countering Digital Hate. 2024. *The New Climate Denial*. London/ Washington, DC: Center for Countering Digital Hate.
Chakrabarty, Dipesh. 2014. *The Climate of History: Four Theses*. Chicago: The University of Chicago Press.
Clark, Stephen R. L. 1977. *The Moral Status of Animals*. Oxford: Clarendon Press.
Coll, Steve. 2013. *Private Empire: ExxonMobil and American Power*. New York: Penguin Books.
Commoner, Barry. 1971. *The Closing Circle: Nature, Man, and Technology*. New York: Alfred A. Knopf.
Commoner, Barry. 1990. *Making Peace with the Planet*. London: Victor Gollancz.

Cook, John, et al. 2016. Consensus on Consensus: a Synthesis of Consensus Estimates on Human-Caused Global Warming. *Environmental Research Letters* 11:4. https://doi.org/10.1088/1748-9326/11/4/048002.

Crosby, Alfred W. 1986. *Ecological Imperialism: The Biological Expansion of Europe, 900–1900.* Cambridge, MA: Cambridge University Press.

Darwin, Charles. 1871. *The Descent of Man, and Selection in Relation to Sex.* London: John Murray.

de Waal, Frans B. M. 2001. *Good Natured: The Origins of Right and Wrong in Humans and Other Animals.* Cambridge, MA: Harvard University Press.

de Waal, Frans B. M. 2019. *Mama's Last Hug: Animal Emotions and What They Tell Us about Ourselves.* New York: W. W. Norton & Company.

Derr, Thomas Sieger. 1974. *Ecologia e liberazione umana: critica teologica all'antropocentrismo.* Brescia: Queriniana.

Derrida, Jacques. 2008. *The Animals that Therefore I Am.* New York: Fordham University Press.

Dombrowski, Daniel A. 1984. *Vegetarianism: The Philosophy Behind the Ethical Diet.* Wellingborough: Thorsons Publishers Limited.

Donovan, Josephine, e Carol J. Adams, eds. 2007. *The Feminist Care Tradition in Animal Ethics: A Reader.* New York: Columbia University Press.

Dworkin, Ronald. 1977. *Taking Rights Seriously.* Cambridge, MA: Harvard University Press.

Ehrlich, Paul R., and Anne H. Ehrlich. 1968. *The Population Bomb.* New York: Ballantine Books.

Foot, Philippa. 2003. *Natural Goodness.* Oxford: Oxford University Press.

Gilligan, Carol. 1982. *In a Different Voice: Psychological Theory and Women's Development.* Cambridge, MA: Harvard University Press.

Giraldo, Omar Felipe, and Ingrid Fernanda Toro. 2020. *Afectividad ambiental: Sensibilidad, empatía, estéticas del habitar.* Chetumal: El Colegio de la Frontera Sur e Universidad Veracruzana.

Givone, Sergio. 1995. *Storia del nulla.* Roma-Bari: Laterza.

Goldsmith, Edward. 1992. *The Way: An Ecological World-View.* London: Rider.

Goodpaster, Kenneth E. 1978. On Being Morally Considerable. *The Journal of Philosophy* 75 (6): 308–325. https://doi.org/10.2307/2025709.

Gould, Stephen Jay. 1990. *Wonderful Life.* New York: W.W Norton & Company.

Griffin, Donald R. 1976. *The Question of Animal Awareness: Evolutionary Continuity of Mental Experience.* New York: Rockefeller University Press.

Griggs, Earl Leslie. 1966. *Collected Letters of Samuel Taylor Coleridge.* Vol. II, 1801–1806. London: Clarendon Press.

Grove, Richard H. 1996. *Green Imperialism: Colonial Expansion, Tropical Island Edens and the Origins of Environmentalism, 1600–1860.* Canberra: Australian National University.

Guattari, Félix. 2000. *The Three Ecologies.* London/New York: Continuum.

Hardin, Garrett. 1973. *Exploring New Ethics for Survival.* New York: Penguin Books.
Hargrove, Eugene C. 1989. *Foundations of Environmental Ethics.* Englewood Cliffs: Prentice Hall.
Heidegger, Martin. 1978. *Being and Time.* Oxford: Blackwell.
Hiskes, Richard P. 2008. *The Human Right to a Green Future: Environmental Rights and Intergenerational Justice.* Cambridge: Cambridge University Press.
Hursthouse, Rosalind. 2001. *On Virtue Ethics.* Oxford: Oxford University Press.
Hutcheson, Francis. 2004. *An Inquiry into the Original of Our Ideas of Beauty and Virtue in Two Treatises.* Indianapolis: Liberty Fund.
Jamieson, Dale. 2008. *Ethics and the Environment: An Introduction.* Cambridge: Cambridge University Press.
Johnson, Lawrence E. 1992. *A Morally Deep World. An Essay on Moral Significance and Environmental Ethics.* New York: Cambridge University Press.
Jonas, Hans. 1979. Toward a Philosophy of Technology. *The Hastings Center Report* 9 (1): 34–43. https://doi.org/10.2307/3561700.
Jonas, Hans. 1984. *The Imperative of Responsibility: In Search of an Ethics for the Technological Age.* Chicago: University of Chicago Press.
Kallhoff, Angela, Marcello Di Paola, and Maria Schörgenhumer, eds. 2019. *Plant Ethics: Concepts and Applications.* London: Routledge.
Kurzweil, Ray. 2005. *The Singularity Is Near: When Humans Transcend Biology.* New York: Viking.
Lasch, Christopher. 1979. *The Culture of Narcissism.* New York: W.W. Norton & Company.
Latouche, Serge. 2009. *Farewell to Growth.* Cambridge: Polity Press.
Latour, Bruno. 2017. *Facing Gaia: Eight Lectures on the New Climatic Regime.* Cambridge, UK: Polity Press.
Lecaldano, Eugenio. 2013. *Simpatia.* Milano: Raffaello Cortina Editore.
Leopold, Aldo. 1949a. *A Sand County Almanac and Sketches Here and There.* New York: Oxford University Press.
Leopold, Aldo. 1949b. *A Sand County Almanac.* New York: Oxford University Press.
Levinas, Emmanuel. 1969. *Totality and Infinity: An Essay on Exteriority.* Pittsburgh: Duquesne University Press.
Levinas, Emmanuel. 2005. *Humanism of the Other.* Urbana/Chicago: University of Illinois Press.
Lewontin, Richard C. 1991. *Biology as Ideology: The Doctrine of DNA.* New York: Harper Perennial.
Liu, Haoyang, e Jiangtao Zhai. 2025. Carbon Emission Modeling for High-Performance Computing-Based AI in New Power Systems with Large-Scale Renewable Energy Integration. *Processes* 13(2): 595. https://doi.org/10.3390/pr13020595.

Lombardi Vallauri, Luigi. 1990. Abitare pleromanticamente la terra. In *Il meritevole di tutela*, ed. Luigi Lombardi Vallauri. Milano: Giuffrè.
Lorenz, Konrad. 1974. *Civilized Man's Eight Deadly Sins*. New York: Harcourt Brace Jovanovich.
Lorenz, Konrad. 1977. *Behind the Mirror: A Search for a Natural History of Human Knowledge*. Londra: Methuen.
MacIntyre, Alasdair. 1985. *After Virtue: A Study in Moral Theory*. London: Duckworth.
Malthus, Thomas Robert. 1798. *An Essay on the Principle of Population as it Affects the Future Improvement of Society*. London: J. Johnson.
Mancuso, Stefano, and Alessandra Viola. 2015. *Brilliant Green: The Surprising History and Science of Plant Intelligence*. Washington, DC: Island Press.
Marchesini, Roberto. 2017. *Over the Human: Post-humanism and the Concept of Animal Epiphany*. Cham: Springer International Publishing.
Marchesini, Roberto. 2025. *Towards an Ethics of Empathy. Morality and the Recognition of Otherness*. Cambridge, UK: Ethics Press.
Marsh, George Perkins. 1864. *Man and Nature; or, Physical Geography as Modified by Human Action*. New York: Charles Scribner.
Maturana, Humberto R., and Francisco J. Varela. 1980. *Autopoiesis and Cognition: The Realization of the Living*. Dordrecht: D. Reidel Publishing Company.
McDowell, John. 1998. *Mind, Value, and Reality*. Cambridge, MA: Harvard University Press.
Meadows, Donella H., Dennis L. Meadows, Jørgen Randers, and William W. Behrens III. 1972. *The Limits to Growth: A Report for the Club of Rome's Project on the Predicament of Mankind*. New York: Universe Books.
Mercer, Philip. 1972. *Sympathy and Ethics: A Study of the Relationship Between Sympathy and Morality with Special Reference to Hume's Treatise*. Oxford: Clarendon Press.
Midgley, Mary. 1998. *Animals and Why They Matter: A Journey Around the Species Barrier*. Athens: University of Georgia Press.
Mihr, Anja. 2022. The Glocal Between the Local and the Global. In *Glocal Governance*, 11–24. Cham: Springer.
Mill, John Stuart. 1998. *Utilitarianism*. Oxford: Oxford University Press.
Moore, George Edward. 1903. *Principia Ethica*. Cambridge: Cambridge University Press.
Morin, Edgar. 1977. *La méthode (vol. 1: La nature de la nature)*. Paris: Seuil.
Mortari, Luigina. 2022. *The Philosophy of Care*. Wiesbaden: Springer VS.
Muir, John. 1894. A Wind-Storm in the Forests, Chapter 10. In *The Mountains of California*. Boston: Houghton, Mifflin and Company.
Murdoch, Iris. 1970. *The Sovereignty of Good*. London: Routledge Classics.

Næss, Arne Dekke Eide. 1973. The Shallow and the Deep, Long-Range Ecology Movement: A Summary. *Inquiry: An Interdisciplinary Journal of Philosophy and the Social Sciences* 16 (1–4): 95–100.

Næss, Arne. 1989. *Ecology, Community and Lifestyle: Outline of an Ecosophy.* Cambridge: Cambridge University Press.

Nancy, Jean-Luc. 2000. *Being Singular Plural.* Stanford: Stanford University Press.

Nicholson, Edward Max. 1970. *The Environmental Revolution: A Guide for the New Masters of the World.* London: Hodder & Stoughton.

Nobis, Nathan. 2002. Vegetarianism and Virtue: Does Consequentialism Demand Too Little? *Social Theory and Practice* 28 (1): 135–156.

Nordhaus, Ted, and Michael Shellenberger. 2009. *Break Through: From the Death of Environmentalism to the Politics of Possibility.* New York: Mariner Books.

Norton, Bryan G. 1988. *Why Preserve Natural Variety?* Princeton: Princeton University Press.

Noys, Benjamin. 2008. Accelerationism. https://leniency.blogspot.com/2008/10/accelerationism.html.

Nussbaum, Martha C. 2010. *Not for Profit: Why Democracy Needs the Humanities.* Princeton: Princeton University Press.

Nussbaum, Martha Craven, and Amartya Kumar Sen. 1993. *The Quality of Life.* Oxford: Clarendon Press.

Ortolani, Elisa. 2025. *Politiche dell'antispecismo: Superare lo specismo con un approccio intersezionale.* Milano: Ombre Corte.

Panikkar, Raimon. 1995. *Invisible Harmony: Essays on Contemplation and Responsibility.* Minneapolis: Fortress Press.

Passmore, John. 1974. *Man's Responsibility for Nature.* London: Duckworth.

Peverelli, Raffaella, ed. 2005. *Valori selvaggi. L'etica ambientale nella filosofia americana e australiana.* Milano: Medusa.

Pigliucci, Massimo, and Gerd B. Müller. 2010. *Evolution. The Extended Synthesis.* Cambridge: MIT Press.

Popper, Karl R. 1972. *Objective Knowledge: An Evolutionary Approach.* Oxford: Oxford University Press.

Potter, Van Rensselaer. 1971. *Bioethics: Bridge to the Future.* Englewood Cliffs: Prentice-Hall.

Potter, Van Rensselaer. 1988. *Global Bioethics: Building on the Leopold Legacy.* East Lansing: Michigan State University Press.

Pouteau, Sylvie. 2018. Plants as Open Beings: From Aesthetic to Plant-Human Ethics. In *Plant Ethics: Concepts and Applications*, ed. Angela Kallhoff, Marcello Di Paola, and Maria Schörgenhumer, 82–97. London: Routledge.

Pulcini, Elena. 2013. *Care of the World: Fear, Responsibility and Justice in the Global Age.* Dordrecht: Springer.

Putnam, Hilary. 1982. Beyond the Fact–Value Dichotomy. *Critica* 14 (41): 3–12.

Quammen, David. 2012. *Spillover, Animal Infections and the Next Human Pandemic*. New York: W.W Norton & Company.
Rawls, John. 1971. *A Theory of Justice*. Cambridge, MA: The Belknap Press of Harvard University Press.
Rawls, John. 1993. *Political Liberalism*. New York: Columbia University Press.
Reece, Rogers. 2024. AI's Energy Demands Are Out of Control. Welcome to the Internet's Hyper-Consumption Era. *Wired US*, July 11.
Regan, Tom. 2004. *The Case for Animal Rights*. Berkeley: University of California.
Rodman, John. 1977. The Liberation of Nature? *Inquiry* 20 (1–4): 83–145.
Rollin, Bernard E. 1981. *Animal Rights and Human Morality*. Buffalo: Prometheus Books.
Rolston, Holmes, III. 1983. Values Gone Wild. *Inquiry: An Interdisciplinary Journal of Philosophy* 26 (2): 181–207.
Rolston, Holmes, III. 1988. *Environmental Ethics: Duties to and Values in Nature*. Philadelphia: Temple University Press.
Ross, William D. 1930. *The Right and the Good*. Oxford: Clarendon Press.
Routley, Richard. 1973. Is There a Need for a New, an Environmental, Ethic? *Proceedings of the XVth World Congress of Philosophy* 1:205–210.
Rowlands, Mark. 2009. *Animal Rights: moral Theory and Practice*. Basingstoke: Palgrave Macmillan.
Rozzoni, Stefano, and Plinio Limata, eds. 2022. *The Economy of Francesco. Un glossario per riparare il linguaggio dell'economia*. Roma: Città Nuova.
Rushkoff, Douglas. 2013. *Present Shock: When Everything Happens Now*. New York: Current.
Ryder, Richard D. 1971. Experiments on Animals. In *Animals, Men and Morals*, ed. Stanley Godlovitch, Roslind Godlovitch, and John Harris, 81–100. London: Grove Press.
Sagoff, Mark. 1985. *Risk-Benefit Analysis in Decision Concerning Public Safety and Health*. Dubuque: Kendall/Hunt.
Saito, Kohei. 2020. *Hitoshinsei no "Shihonron" [Capital in the Anthropocene]*. Tokyo: Shueisha.
Sale, Kirkpatrick. 1996. *Rebels Against the Future: The Luddites and Their War on the Industrial Revolution – Lessons for the Computer Age*. London: Quartet Books.
Scheler, Max. 2008. *The Nature of Sympathy*. London: Routledge.
Schneider, François, Giorgos Kallis, and Joan Martínez-Alier. 2010. Crisis or Opportunity? Economic Degrowth for Social Equity and Ecological Sustainability. *Journal of Cleaner Production* 18 (6): 511–518. https://doi.org/10.1016/j.jclepro.2010.01.014.
Schweitzer, Albert. 1953. *The Philosophy of Civilization*. New York: Macmillan.
Shiva, Vandana. 1993. *Monocultures of the Mind: Perspectives on Biodiversity and Biotechnology*. London: Bloomsbury Publishing PLC.

Simberloff, Daniel. 1980. A Succession of Paradigms in Ecology: Essentialism to Materialism and Probabilism. *Synthese* 43:3–39. https://doi.org/10.1007/BF00413854.

Simon, Herbert A. 1975. *The Sciences of the Artificial*. Cambridge, MA: MIT Press.

Singer, Peter. 1975. *Animal Liberation*. New York: New York Review/Random House.

Slote, Michael. 2007. *The Ethics of Care and Empathy*. New York: Routledge.

Smith, Adam. 2010. *The Theory of Moral Sentiments*. London: Penguin Classics.

Sober, Elliott. 1986. Philosophical Problems for Environmentalism. In *The Preservation of Species. The Value of Biological Diversity*, ed. Bryan G. Norton, 173–194. Princeton: Princeton University Press.

Stengers, Isabelle. 2015. *In Catastrophic Times: Resisting the Coming Barbarism*. Lüneburg: Open Humanities Press and Meson Press.

Stone, Christopher D. 1974. *Should Trees Have Standing? Toward Legal Rights for Natural Objects*. Palo Alto: William Kaufmann.

Tallacchini, Mariachiara, ed. 1998. *Etiche della terra: Antologia di filosofia dell'ambiente*. Milano: Edizioni FrancoAngeli.

Taylor, Paul W. 1986. *Respect for Nature: A Theory of Environmental Ethics*. Princeton: Princeton University Press.

Thoreau, Henry David. 1899. *Walden; or, Life in the Woods*. New York: Thomas Y. Crowell Company.

Tronto, Joan C. 1993. *Moral Boundaries: A Political Argument for an Ethic of Care*. New York: Routledge.

Tuan, Yi-Fu. 1971. *Man and Nature*, Resource Paper no. 10, Commission on College Geography. Washington, DC: Association of American Geographers.

Udall, Stewart L. 1963. *The Quiet Crisis*. New York: Holt, Rinehart & Winston.

United Nations. 1992. Rio Declaration on Environment and Development, Principle 15. In *United Nations Conference on Environment and Development, Rio de Janeiro, Brazil*, June 3–14. https://www.un.org/en/development/desa/population/migration/generalassembly/docs/globalcompact/A_CONF.151_26_Vol.I_Declaration.pdf.

Vanderheiden, Steve. 2001. Environmental Ethics and the Globalization of Environmental Justice. In *Ethics, Environmental Justice and Climate Change*, ed. Paul G. Harris, 41–61. Cheltenham: Edward Elgar Publishing.

VanDeVeer, Donald. 1979. Of Beasts, Persons, and the Original Position. *The Monist* 62 (3): 368–377.

Vernadsky, Vladimir I. 1998. *The Biosphere*. New York: Copernicus.

von Bertalanffy, Ludwig. 1968. *General System Theory: Foundations, Development, Applications*. New York: George Braziller.

von Mossner, Alexa Weik. 2017. *Affective Ecologies: Empathy, Emotion, and Environmental Narrative*. Columbus: The Ohio State University Press.

Warren, Karen J. 1990. The Power and the Promise of Ecological Feminism. *Environmental Ethics* 12 (2): 125–146. https://doi.org/10.5840/enviroethics199012221.

Weiss, Edith Brown. 1989. Climate Change, Intergenerational Equity and International Law: An Introductory Note. *Climate Change* 15 (1–2): 327–335. https://doi.org/10.1007/BF00138858.

Westra, Laura. 1993. *An Environmental Proposal for Ethics: The Principle of Integrity.* Boston: Rowman and Littlefield.

White, Lynn, Jr. 1967. The Historical Roots of Our Ecological Crisis. *Science* 155 (3767): 1203–1207.

Wilson, Edward O. 1986. *Biophilia.* Cambridge, MA: Harvard University Press.

CHAPTER 9

The Fundamentals of Philosophical Ecology

9.1 Premise

Philosophical ecology is not concerned with studying the environment as a space of shelter and sustenance for various species, nor is it focused on mapping the relationships between living beings—that is the role of *descriptive* ecology. Instead, philosophical ecology aims to spark a deep paradigmatic shift, which might be framed as follows: *Can ecology help us think in a non-anthropocentric and more effective way about the urgent challenges of the twenty-first century?* As a starting point, we could say that philosophical ecology questions the essentialist and disjunctive paradigm that still underpins much of our thinking.

There is a close relationship between the epistemological foundation of descriptive ecology—including the practical outcomes of its studies—and the broader principles of ecological thinking (Iofrida 2019). We can imagine these two domains as connected by communicating vessels: ideas and reasoning flow between them, even as philosophical ecology calls for a more radical shift in mindset. Now, while there are continuities and mutual influences between descriptive and philosophical ecology, the transformation of thought that we urgently need demands a new phenomenological approach to reality—one in which entities are no longer seen as expressions of an inner essence. Ecological thinking, in this sense, is grounded in a *relational ontology*, made not of barriers but of thresholds. In this view, an inherent quality cannot manifest through simple emanation; instead, it

R. Marchesini, *Philosophical Ecology*,
https://doi.org/10.1007/978-3-032-04987-2_9

always emerges through encounter. The philosophical impasse that philosophical ecology seeks to overcome lies in the lingering essentialist view of predicates—a view in which *otherness* can only surround, contain, provide resources, or, worse, contaminate. Philosophical ecology invites us to abandon this logic and embrace a more interconnected, reciprocal understanding of being.

Humanism bears undeniable responsibility for this disjunctive conception of ontology. Therefore, my research stems from a reexamination of the humanistic paradigm, questioning its assumptions (Marchesini 2002; Braidotti 2013; Ferrando 2019), and highlighting the connection between the human being and the biosphere (Karpouzou and Zampaki 2023). The necessity for philosophical ecology arises precisely from the need to overcome predicative essentialism, so as to integrate into ontology something that biological description has long recognized: *there is no such thing as separation and autonomy in the domain of life.* Each entity relies on others to exist and manifest itself; no quality can be considered the outcome of autarchic emanation and pure self-expression, and no configuration is realized through simple autopoiesis. This awareness transcends the descriptive prospectus, urging us toward philosophical reflection. What needs to change is our very perspective on the predicate, as it becomes evident that no quality is ever realized autonomously.

This line of inquiry facilitates an expanded perspective for formulating a meta-theory of predicates that shifts from an essentialist stance to a relational conception. In other words, instead of regarding predicates as the pure expressions of entities themselves, it views them as expressions of the entities' relations (Bateson 1991; Frigerio 2023). Our conventional way of thinking tends to isolate beings from the connections that sustain them, uprooting them in an attempt to understand their characteristics through internal analysis. This approach, marked by a dichotomous-oppositional stance, is fundamentally anti-ecological, as it separates entities from their relations and endeavors to distill them into their supposed purity.

Ecological thinking calls for profound changes both in ontology, redefining existence as relational, and in epistemology, considering knowledge not as a linear process but as an emergence resulting from contamination with external reality. Furthermore, each entity is, in turn, the outcome of a prior (phylogenetic) constitutive relation, implying that nothing is realized *ex nihilo*. Embracing these two principles, which alone mark a departure from essentialism, we come to understand that the eco-ontological approach is fractal. Indeed, even the features that may seem intrinsic and inherent, because they are enclosed within the entity, are the product of a preceding relationship. Thus, the pursuit of an essential core becomes entirely futile, much like peeling through the layers of an onion hoping to find its "true core." The debate on the internal and external relationships of an entity—consider thinkers like Bradley, Deleuze, and Wahl—should thus be revisited through: (i) an evolutionary perspective, where internal relations are the phylogenetic outcomes of external pressures, such as selective forces on a lineage; and (ii) an ecological perspective, where environmental factors influence the expression of internal characteristics. Ecological principles reveal the limitations of essentialist approaches to existence, where qualities are seen as intrinsic and entities are studied in isolation. This inward-looking view restricts our understanding and fosters separation between beings. In this chapter, I translate foundational ideas from descriptive ecology into ecological thinking, emphasizing key concepts for evaluating the predicates shown by an entity and the dynamics governing an event.

Ecological thinking requires challenging a philosophical mindset that is deeply ingrained in cultural tradition and that, for this reason, appears intuitive and very difficult to question. Kant would term it a *synthetic* a priori *judgment*. This perspective, bolstered by humanism, prompts us to seek the predicates of entities within their essence rather than through their interconnected relations. It involves buying into the illusion of inherent purity, attributing specific essential qualities to beings rather than focusing on their connections (Candiotto and Pezzano 2019). This non-ecological thinking hampers the understanding that an entity's phenomenology relies only partially on its internal qualities. What a being reveals of itself results from the interplay between its relational potential and the specific situation it finds itself in. Its manifestation is always co-defined by the relationships it enters into.

In other words, we must understand the entity as implicitly predicatively open—that is, as a being inherently constituted to relate to the world, forming its qualities not by simply expressing internal traits, but through the relations it contracts. At the same time, it is important to recognize that a biological entity is not immersed in a generic web of potential relations, but is always situated within a specific range of possible, probable, or necessary interactions. Not every being is under the relational influence of every other being. Moreover, when we speak of living beings, we must acknowledge that they are not passive within these relational dynamics. They possess agency—the capacity to actively shape the interactive context in which they are embedded. For this reason, philosophical ecology cannot be reduced to a mere philosophy of relations.

One of the key contributions of philosophical ecology is the redefinition of nature from a posthumanist perspective—that is, a critical questioning of the humanist dualisms that sharply divide the human from the non-human. However, this revision of the concept of nature does not seek to eliminate its conceptual value. As Merleau-Ponty (1995) insightfully notes, nature remains the irreducible Other: that which resists human intervention while also serving as the reservoir from which human activity emerges—though never fully exhausting it. In this sense, a philosophy that seeks to embrace the meta-principles of ecology must be grounded in a different understanding of the human being. It must follow a post-Promethean logic—based on a clarified definition of human nature, a unified re-evaluation of bodily phenomenology that overcomes the mind–body dualism, an acceptance of finitude, and a recognition of the human being's ethological tendencies. This allows us to understand our inclinations and vulnerabilities without falling into determinism. To redefine nature, then, is to accept its ontological plurality, which does not erase human specificity, but rather overcomes the dichotomous view of nature that places the human in opposition to it.

9.2 The Concept of Nature

The word *nature* has accumulated so many meanings over time that it has lost its status as a reliable guide or "polar star" for orientation (Daston 2019). It's therefore understandable that some contemporary thinkers—such as Bruno Latour (2017) and Timothy Morton (2012)—have called for abandoning the concept altogether. I find their critique of the term *nature* as a vague catch-all invoked to support wildly different—and often contradictory—claims to be legitimate. Indeed, when *nature* is used as a broad container of tautological principles, selectively interpreted to confirm one's biases (e.g., *it is against nature*, *it is a natural right*, *it aligns with nature*), such appeals to "nature" become not only unhelpful, but misleading and prescriptive in problematic ways.

Still, I question whether discarding the term entirely is the best path forward. Might it be more productive to critically revisit the concept, overcoming the interpretive frameworks that have distorted its meaning—such as dualism, in its many shapes? In my view, the concept of nature has been misused within the humanist project, which sought to elevate humans by defining them in opposition to everything else (hence the nature vs. culture dichotomy). This misinterpretation is especially evident when nature is understood as a fixed, inviolable principle. Such a view blurs the distinction between domains that are indeed governed by necessity—like physical and chemical laws—and those that allow for variation and agency, such as species identity. The latter, as Sean Carroll (2016) reminds us, often includes a wide range of behavioral and developmental possibilities—not only in humans, but across the living world.

Since the term *nature* refers to multiple layers of the biosphere, as Helmuth Plessner suggests in *Levels of Organic Life and the Human* (2019), it is always necessary to specify what kind of natural phenomenon we are talking about—whether it's a physical law, such as the boiling point of water, a chemical reaction, or the behavior of a particular animal. One can only agree with Bryan Magee's observation that a law of nature is descriptive—but this does not mean it is necessary or inviolable (1973). Still, to treat the concept of nature as purely descriptive, limited to *what-is* rather than a domain of virtuality, of *what-could-be*, is to revive the humanist dualism that casts the non-human realm as passive and governed solely by chance and necessity (Monod 1971), in contrast to the human sphere, which is thought to be defined by free will. Instead, we should recognize that different levels of organic life involve varying degrees of necessity, chance, probability, and possibility. From this perspective, descriptions may sometimes have strong predictive power, but in other cases they should be treated more like field functions; any prescriptions we might draw from these descriptions do not follow directly, but must be treated as problematic.

Similarly, what has been said about the concept of the *animal*—used as a counter-term to define the human—can also help unravel the false dualisms at the core of humanist thought. In the case of the animal, the dualism can be overcome relatively simply by expanding the concept to include humans, treating our species as another expression of animality, even if this leaves the complex task of identifying the ontological specificity of the human within that broader category. The human vs. non-human dichotomy—most famously expressed in the Cartesian split between *res extensa* and *res cogitans*—is one of the key foundations of modern anthropocentrism. Dismantling this edifice is no easy task, as it is built upon a maze of ad hoc distinctions and conceptual constructs.

Let us return, then, to the concept of nature, and look at one of the most common counter-terms used to draw a dividing line between the human and the non-human: *culture*. This gives rise to the first major opposition—nature vs. culture. If the term *nature* evokes the Epimethean principle of necessity—of the inability or impossibility to say no, to refuse the ineluctability of a law or predetermination—then culture, by contrast, opens the space of possibility: the Promethean realm of negation and transformation, the domain of free will and self-determination that casts humans as world-builders. In this view, culture becomes an index of autopoietic plurality, the triumph of the principle of negation, an expression of creativity—and at the same time, a marker of distance from nature.

In *Beyond Nature and Culture* (2013), anthropologist Philippe Descola argues that the nature/culture divide is not a universal principle, as we are often led to believe, but a construct of the modern era—and one that emerged relatively late even within Western thought. While it is true that the rigid form of this opposition must be attributed to modernity, still the seeds of the dichotomy were already evident in Greek philosophy, such as Sophism. Descola shows how other cultures—such as those rooted in animism or totemism—tend to emphasize a continuity between the human and the non-human. Yet one might ask whether these assimilative frameworks truly account for the diversity found in nature, or whether they simply reflect a form of projective thinking. The mechanism of *analogy* invoked by Descola—understood as the recomposition of differences through a network of similarities that connects intrinsic properties—fails, in my view, to fully capture the continuum of life. This continuum arises not through analogy, but through homology (see my critique of Nagel).

It remains a fact that the non-human world is not governed by a deterministic nature, as species continually invent new physiological and expressive styles—an idea I explored in *The Creative Animal* (2022). That said, human beings undoubtedly display a particular predisposition for the development of technical and cultural artifacts—but they are by no means the sole possessors of such capabilities. There is ample evidence of cultural and technical traditions in non-human species (Bonner 1980; Mainardi 1974). We might describe the cultural dimension of our species as simply one among many forms of niche construction, so as to erase even the distinction between the natural and the artificial. In this view, techno-culture becomes one of the possible natural expressions of a species (Beck 1980).

As already mentioned, this is not to diminish the peculiarities of human beings. I would regret deeply if that were the impression I gave, as I am fully aware of how extraordinary our species is in many respects. That, however, is not the issue. The mistake of humanism, in my view, was its attempt to explain the uniqueness of the human by constructing counter-terms—such as *nature* or *animal*—to which it denied any agency, reducing them to passive, deterministic mechanisms. In doing so, it turned nature into a kind of inert backdrop, a static tarpaulin or stage on which only one protagonist—the human being—could act. This paradigm cast the human as an exile in a foreign land, compelled to struggle against nature in order to unleash its own ingenuity. Once we accept that nature contains multiple ontological possibilities, this does not diminish the human; rather, it allows us to understand—beyond the Promethean narrative—how our species has been able, for better or worse, to accomplish what it has. And it has done so not by cutting itself off from other living beings, but by drawing deeply—not only in terms of material resources but also of inspirations, ideas, and models—from the vast reservoir of biodiversity.

A posthumanist redefinition of the human thus begins with a philosophy that shifts from a vertical ontology of disjunction and transcendence to a horizontal vision of participation. It is therefore essential to avoid the risk of naturalistic determinism—that is, the danger of retaining within the concept of nature the connotations that humanism has historically imposed on it. I want to stress that this is not my intention. To illustrate this point, I will refer to Giacomo Leopardi's poem *The Solitary Sparrow*, in which the poet reflects on the solitary habits of *Monticola solitarius*, a passerine bird roughly the size of a blackbird, known for its cobalt-blue plumage. In describing the bird's behavior, Leopardi writes: "all thy wishes are/ Obedient to Nature's law" (1887, 55).[1] At first glance, we might think Leopardi is drawing a clear distinction between human customs—seen as *not* the outcome of nature—and those of the lonely sparrow, which would thus appear dictated and necessary by its nature. However, the entire poem hinges on a comparison between the social tendencies Leopardi sees as inherent to him—tendencies he laments having forsaken—and the solitude of the sparrow, addressed with a generic plural "you" that appears to refer to the species. If Leopardi had meant the possessive "your" (*vostre*) to refer to the entire non-human context, the entire emotional and philosophical structure of the poem would collapse. There is, however, one word that dispels any such ambiguity: *vaghezza* (translated by Townsend as "wishes," but whose meaning is closer to *vagueness*). Leopardi attributes this trait to the solitary sparrow, and in his lexicon, *vaghezza* expresses

[1] The original reads "di natura è frutto ogni vostra vaghezza."

the indefinite longing—something that *does*, not simply something that *is*. This single word is enough to dissolve the specter of natural determinism.

The idea that even what we call artificial should be considered part of the natural world was developed by Donna Haraway in her *Cyborg Manifesto* (2016), where she introduced the concept of *natureculture*. In *The End of Nature* (1989), Bill McKibben argues that nature—understood as a realm untouched by human influence—no longer exists. This view underpins the concept of the Anthropocene, a supposed geological epoch in which human activity has become the dominant force shaping the planet (Crutzen and Stoermer 2000). Here we encounter the second major oppositional dualism: natural vs. artificial. This dichotomy, like that of human vs. non-human, has been central to the humanist project of separating humanity from its natural background. Including human activity within the domain of nature, and refusing to oppose it to the artificial, can risk—as Stephen Vogel warns in *Against Nature: The Concept of Nature in Critical Theory* (1996)—rendering the concept of nature so vague that it loses its usefulness. I believe Vogel's concern is valid, but only if one continues to operate within humanist assumptions. That is not necessarily the case, especially if we adopt a perspective—such as philosophical ecology—that aims to move beyond those assumptions. This approach seeks a nuanced view that avoids the pitfalls of vague holism ("everything" without gradients), rigid dualisms (opposed entities), and a-specific interconnectedness (everything connected to everything else). At the same time, it aspires to a more precise understanding of relations, seen as systemic and transformative rather than static or given.

Reintegrating human activity into the natural context is a necessary step if we are to overcome the dualism between human and non-human. Human techno-cultural behavior should be seen as one of many expressions already observable in other species—such as the use of medicinal plants, the building of termite mounds, orca dialects, and the diverse customs of chimpanzees. In this light, artifice is not an exclusively human trait. Rather, humanity's particular tendency toward the artificial is a marker of *specificity*, not *specialness*—and thus, the artificial falls back within the realm of the natural. This shift in perspective is important—but it must not be used to justify unchecked techno-cultural expansion as a mere extension of nature. When we speak of biodiversity, we must not lose sight of our relationship with otherness. Nature is not simply a blank canvas for the expression of human potential; it is, first and foremost, an encounter with the *irreducible Other*—an Other that, however repressed or dominated, never leaves us (Iofrida 2019).

A philosophical ecology must therefore rest on an understanding of finitude, a respect for irreducibility, and a vision grounded in reciprocity and limits (Merleau-Ponty 1995). Instead of opposing nature and artifice, we can assess the activity of any species—including our own—based on its compatibility with this Other. Does it harmonize with the whole of which it is a part, or does it undermine it? Recognizing the naturalness of the so-called artificial does not grant license for any and all human activity. Every species—not just humans—can threaten its own ecosystem, driving other species to extinction and ultimately jeopardizing its own survival. This is not hypothetical; it has already happened several times in the history of our planet. A key example is the Great Oxidation Event, about 2.45 billion years ago, when cyanobacteria dramatically altered Earth's atmosphere through oxygen production, triggering a mass extinction of anaerobic life. So, yes—it was natural. But it was also suicidal.

An essential figure in the development of philosophical ecology—particularly for his exploration of nature as an element of irreducibility—is undoubtedly Maurice Merleau-Ponty. His focus on the somatic dimension of experience restores the principle of existence to the body. For Merleau-Ponty, being-there is, above all, a rootedness in the world. He invokes the Earth as a fundamental dimension of life, criticizing the Western model of disembodied, detached rationality (1995). Nature, in his view, is not only the expression of what we are—it is also the encounter with that which resists us, the irreducible other that both opposes and grounds us. Beginning in the late 1950s, this confrontation with the Other became increasingly urgent, emerging in the shadow of the atomic age and the real possibility of an event capable of erasing humanity entirely. In response to this crisis, Merleau-Ponty felt the pressing need for a cultural renewal—one that would place nature once again at the center of philosophical thought.

To counter a rigid and impermeable ontology, he introduced two fundamental concepts into philosophical ecology. The first is *chiasm*, which describes a porous ontology—an interweaving of entities that are open to conjunction and mutual influence. The second is *inherence*, understood as both pertinence and participation: a form of intrinsic being that is always already in relation. According to this view, entities refer to and imply each other. Sharp distinctions must give way to recognition of their intersections, their deep correspondences. This leads to the necessity of a

Stiftung—a conceptual re-founding grounded in cultural transformation. It also calls for a rethinking of the divide between the descriptive and the prescriptive: not to collapse the two into a vague indistinction, but rather to understand their chiasmatic relation.

So let's address the third dualism: descriptive vs. prescriptive. This refers to the idea that natural laws can dictate prescriptive norms of behavior—that something is reprehensible because it is "against nature," or justified because it is "natural." Bruno Latour highlights and critiques this assumption in *Facing Gaia* (2017), and rightly so—it demands our full attention. To address this issue, I would like to emphasize two points. First, in the living world, there are no universal behavioral laws. Nature does not prescribe one fixed path over another—whether selfish or altruistic behavior, heterosexuality or homosexuality, group solidarity or complete disregard. Life probes the field of possibilities, and every viable path is explored. Hence, biological diversity results from a wide array of choices and adaptations. Second, each species does exhibit specific tendencies—morphological, developmental, physiological, and behavioral—so it is possible to sketch a species-specific description. However, this is not a matter of strict determinism but rather of a spectrum of possibilities. Individuals navigate this range over the course of phylogeny, often taking novel or unexpected paths. In the realm of life, there are always margins of indeterminacy—spaces for choice and innovation (Carroll 2016).

Given the absence of universal rules, it makes little sense to treat nature as prescriptive, or to critique the concept of nature on the grounds that it implies normative justification (Magee 2004; Harré and Harris 1993). If the term *nature* may serve to indicate not a universal character—except in the case of physical-chemical laws, which are inviolable by necessity—but rather a species-specific tendency, Robin Attfield reminds us in *Environmental Thought* (2021) that even human nature does not entail a rigid set of prescriptions. Instead, it consists of inclinations and a broad range of possibilities. One might argue that such a plastic and open-ended definition of nature weakens the concept or renders it useless, but this would only be true if we insisted on forcing nature back into the humanist framework of necessity. Under an eco-ontological lens, however, nature is not a set of mandates but a horizon of possibility—a relational and onto-poietic field of becoming.

Let us now consider the critique advanced by Timothy Morton in *Ecology without Nature*, where he argues that nature is an arbitrary rhetorical construct, lacking any independent or genuine existence outside the texts we produce about it (2007, 21–22). For a long time, we have viewed nature as something pure and untouched by human influence—a legacy drawn from the thought of Rousseau, Emerson, Thoreau, Goethe, and other proponents of *Naturphilosophie*. This has fostered a kind of oleographic, almost nostalgic vision of nature, often leaning toward primitivism. Keith Thomas, in *Man and the Natural World* (1983), traces this shift from an exploitative view of nature to a purely aesthetic one between the fourteenth and nineteenth centuries—a transformation that still echoes in our own time. Yet, this aestheticization continues to support the human–non-human dichotomy. Consider, for instance, the concept of openness, typically attributed to animals. It can be framed positively, as in Rainer Maria Rilke's *Eighth Duino Elegy* (1961), or negatively, as in Heidegger's *The Fundamental Concepts of Metaphysics*. In either case, the concept is instrumental in reinforcing the distinction between humans and other species, furthering an anthropocentric perspective.

The same is true of the idealized image of pristine nature, which implicitly places humans outside of it, as though inhabiting a separate realm. While it is undeniable that human activity can endanger biodiversity, it has also historically contributed to it. Certain rural practices—such as hemp retting, manuring, or interplanting vines with broadleaf trees—have enhanced ecological diversity, unlike the extractive methods of industrial agro-livestock farming. Today's rewilding projects, for instance, are not rooted in the outdated ideal of an untouched nature. Instead, they are informed by a more ecological understanding of the relationships among living organisms. These approaches recognize the roles played by different species within broader ecosystems—such as that of *umbrella species*—and are shaped by a dynamic view of biodiversity.

Another noteworthy text by Timothy Morton that seeks to move beyond the idea of nature is *Dark Ecology: For a Logic of Future Coexistence* (2016). In this work, Morton—rightly, in my view—critiques the notion of nature conceived as an absolute Other, as wild, pristine, and untouched. Such a conception, in fact, merely repackages the human vs. non-human dichotomy, which inevitably fuels both the view of nature as a resource to be exploited and as an entity to be preserved in its purity—two sides of the same anthropocentric coin. Morton thus targets what he calls *heteronormative environmentalism*, which he sees as rooted in paternalism and the stewardship model of the biosphere. He traces this attitude back to the Neolithic revolution, a historical shift he labels *agrilogistics*, grounded in practices of management and domination. According to Morton, this shift gave rise to a *forma mentis* shaped by violent and arbitrary distinctions, which in turn fostered phenomena such as patriarchy, inequality, substantivalism, and even the principle of non-contradiction—"poisonous fruits" later reinforced by the agricultural and industrial revolutions.

Among these arbitrary constructs, Morton also includes the very concept of nature. Here, however, I believe his argument falters. There is a clear distinction between critiquing how the concept of nature is interpreted and dismissing it altogether as a hollow idea. This is particularly evident in *Dark Ecology* itself, where Morton consistently refers to natural entities—demonstrating that the category still holds functional and conceptual value. In this sense, one cannot help but agree with Robin Attfield's (2018) observation that Morton, by repeatedly referencing nature, falls into contradiction. Morton also encourages us to reject the syndrome of facts, which he believes only produces anxiety—generating guilt in some and denialism in others. Yet it's not entirely clear what is being denied here. Moreover, by casting facts as naïve and obsolete, this stance risks playing into the hands of those who already question the legitimacy of scientific evidence. Finally, we arrive at a further paradox: Morton's claim that consumption should not be demonized. Already in *Ecology without Nature* (2007), the theorist of hyperobjects defended consumption as a mode of encountering things, interpreting desire as openness to otherness—two concepts in blatant contradiction.

As we have seen, the theme of finiteness is central to philosophical ecology. In this framework, relationality is not static or stable but unfolds through evolutionary dynamics—*sub specie transformationis*. Entities, by virtue of being in relation, are bound to one another in reciprocal ways. As Laura Candiotto and Giacomo Pezzano emphasize in *Filosofia delle relazioni*, it is essential "to highlight the processual and tensional character of the relation, whereby the relation represents the tension between terms (…), which is equivalent to connoting the relation as a limit" (2019, 66). Each entity possesses an internal organization that shapes its selective relational openness to the external world. However, this does not translate into totipotence; rather, it defines a relational position that marks the entity's horizon of possibility.

In the essay *Is the Concept of Nature Dispensable?* (2007), Robin Attfield argues that the term *nature* is indispensable for several reasons: (i) the flourishing or decline of an organism depends on its nature; (ii) every being holds intrinsic value regardless of external considerations; (iii) without the concept of nature, the needs of any organism—including human needs—would be entirely flexible and subject to manipulation; (iv) the expressions of an entity arise from elements internal or proper to that entity; and (v) without the concept of nature, the very notion of alienation would be meaningless. Thus, the concept of nature, reinterpreted through the lens of philosophical ecology, offers a meaningful anchoring point—particularly in relation to the concept of *difference*, which the traditional view of nature as a passive backdrop for human self-assertion failed to provide. If ecology studies the interrelationships among organisms within an environment, then philosophical ecology can be assigned a more radical and expansive role: that of *thinking ecologically* (Amoroso 2023).

There are, however, aspects of *Dark Ecology* that deserve attention—first and foremost the concept of *ecognosis*: a form of sensitivity that fosters deep engagement with otherness through a sense of wonder and fascination with the world. On the other hand, to see in the deep ecology of philosophers like Aldo Leopold and Arne Naess a naïve tendency to idealize nature as pristine purity strikes me as overly simplistic and forced. Deep ecology, in fact, is grounded in a radically non-anthropocentric impulse—not of detachment, but of complete immersion and openness to otherness, even to the point of identification.

Morton's approach, rooted in a purely humanistic tradition, raises questions about his ability to overcome the *agrilogistics* he critiques—a framework whose most obvious product was, in fact, humanism itself. To frame alienation from the world as alienation from the self is emblematic of a humanist, Vitruvian conception of the world. But setting aside this point, which pertains more to the legacy of the anti-ecological modern paradigm than to *dark ecology* itself, one thing remains clear: to speak of *alienation* already presupposes the existence of nature, since without it, there would be nothing from which to be alienated (Attfield 2018).

In *The Ecological Thought* (2010), Morton speaks of a kind of universal interconnectivity—summarized in the phrase "All life forms are interconnected"—but this vision risks negating ecological thought altogether. Ecology, by definition, involves differentiation: the identification of biomes, ecological niches, and varying degrees of connectedness. If everything were uniformly interconnected, there would be no meaningful distinctions, no ecological geography, and none of the historically specific interaction trajectories described, for example, by Anna L. Tsing in *The Mushroom at the End of the World* (2015). Philosophical ecology is not a trivial extension of descriptive ecology. Rather, it draws from some of its key principles or meta-principles to develop a philosophical reflection—but it cannot be reduced to a form of humanist philosophy simply applied to ecological issues.

9.3 Philosophy of Relations

In *The Principles of Mathematics* (1903), Bertrand Russell argued that Western thought has historically shown a distinct antipathy toward relations—an idiosyncrasy epitomized by Aristotle's concept of substance as a predicatively autonomous entity. By contrast, a connective or dialectical understanding of being-in-relation—expressing the inherent union between entities—can be found in thinkers such as Spinoza and Hegel. However, a sustained ontological investigation of relations can be considered a development of twentieth-century philosophy (Horstmann 1999).

Among the cultural transformations that have helped to elevate relation to the status of ontological fundamental—marking a shift from mere connectionism to genuine relationalism—we must count the rise of Darwinism and ecology, the advent of quantum physics, the emergence of information theory, and the spread of complexity thinking. This broader intellectual climate of the twentieth century fostered a paradigmatic shift in ontology: the collapse of essentialist frameworks in favor of relational approaches. As a result, reality came to be viewed increasingly as composed of things that always exist *in relation* (Frigerio 2023).

Russell himself laid the groundwork for the classic debate between internal relations, understood as constitutive of an entity's intrinsic nature, and external relations, which occur between entities (Russell 2008). From the outset, this distinction has fueled a fundamental philosophical divide: on one side, those who view an entity's qualities as expressions of intrinsic features; on the other, those who see them as products of external relational networks—a debate that continues to this day (DeLanda 2002; Harman 2002). These contrasting perspectives have developed into two radical and opposing models: one that absolutizes intrinsic predicates, as in Object-Oriented Ontology (Kleinherenbrink 2019), and another that pushes relationalism to its extreme, positing *relations without relata*, as in Ontic Structural Realism (Ladyman and Ross 2007).

My intention is to move beyond this antithesis and arrive at an eco-ontological view of living organisms—one that treats internal and external relations as part of a continuous flow of conversion, whereby external relations become internalized and, conversely, internal relations influence the external. The first reflection I wish to offer concerns the internal character of an entity. Whereas in a substantivalist (non-relational) view, the characteristics of an entity are defined by the being-in-itself, in an eco-ontological framework, what is intrinsic: (i) does not directly produce the predicate, because the entity possesses a plurality—albeit finite—of possible manifestations (intrinsic virtuality); and (ii) defines the field of relational possibilities the entity can entertain with the external world (relational inherence). In other words, the entity *is* insofar as it is in relation. It cannot be meaningfully qualified outside of relation—not because it is empty or "liquid" (i.e., lacking internal features), but because its fullness lies in both the field of possible forms it can assume (intrinsic virtuality) and the space of relations it can enter into (relational inherence).

Living beings exhibit predicative multipotency, observable across various domains of biology—from the phenotypic translation of the genome to the differential development of systems during ontogeny. Outside of relationality, the entity is phenomenologically absent, as in a state of ontological autonomy it would express no predicate—only potentiality. In a relational conception of the entity, the interactive field in which it is situated—that is, the web of external relations it is part of—actualizes its predicative potentialities. The term *actualize* here refers not to the unveiling of a hidden truth (as if removing a veil), but to the translation of a virtuality into a form. It is a matter not of revealing but of bringing out. To be clear, this approach does not deny the intrinsic; rather, it rethinks it as a range of possibilities that are actualized through relation and a set of structures that bind the entity to the world, making it inherent. Thus, different predicative outcomes are attributable to a given entity depending on the relational condition in which it is situated.

In this conception, the relation brings out a predicate potentially present within the entity's range of virtuality. Yet this alone is not sufficient. A second intrinsic factor intervenes in the process of actualization: the entity's specific susceptibility to different types of relation. This susceptibility can be articulated through: (i) intrinsic electivity—the entity's particular receptivity to certain external relations, for example via specific receptors; (ii) intrinsic positionality—the entity's embeddedness within a relational field, such as a specific ecosystem or climatic/meterological condition, which makes certain relations probable or necessary; (iii) heteronomy—the entity's being a priori constituted in such a way as to enter certain relations, as in the case of fish gills presupposing the presence of dissolved oxygen in water. Because of this, external relations in the phenomenology of life may appear contingent, but are often already presupposed by the entity's constitution. Hence the notion of ***relational inherence***: a pertinence of relation that the entity exhibits due to its patterns-that-connect (Bateson 1979).

In this framework, the inherent characteristics of an entity determine which relations are:

(i) possible, as defined by its relational electivity—such as a cell's membrane receptivity or permeability;
(ii) probable, based on the likelihood of encountering certain external entities due to its ecological condition;
(iii) necessary, referring to conditions without which the entity cannot exist—for instance, a cell's dependence on others for survival;
(iv) expected, that is, relations already inscribed in the entity's intrinsic nature—such as its relationship with a microbiome.

The entity, therefore, cannot assume just any predicate, because actualization depends on two boundaries of possibility: (i) the kind of virtuality the entity possesses—whether a given actualization is possible—determined by the internal relations (i.e., the organization of relata) that shape its configurability, or what outcomes can be actualized within it; and (ii) the kind of inherence the entity presents—whether a relation with a particular external entity is compatible with its own structure, based on parameters of relational sensitivity. An entity may resist actualization in two ways: (i) by refusing conversion into a particular form, i.e., a constraint or resistance to configuration; (ii) by resisting a particular external relation, i.e., interactional refractoriness or lack of susceptibility.

The internal condition, as a field of possible predications and of relational electivity, gives us a picture of the entity characterized by constraints and actualization possibilities, meaning a defined perimeter within which the entity can be configured by external relations, and relational probabilities, determined by its selective ability to relate to specific external entities—its inherence. Without falling into essentialism, which equates an entity's internal constitution with the predicates it expresses, an eco-ontological view instead acknowledges the intrinsically relational nature of the entity, as both its virtuality and its inherence depend on internal relations. Indeed, these relations define the range of organizability of its constituent parts, in the case of virtuality, and the structures of connection and susceptibility to external relations, in the case of inherence. Having established that there is no overlap between the core of an entity and the predicates it manifests—that is, between its internal constitution and the qualities it presents—we can say that the two are linked only through influence, in terms of possibilities of actualization and internal electivity.

Let us now turn to external relations. Given the internal content of an entity, it follows that actualization is not determined solely by what lies within. It also depends on the nature of the external relations the entity is engaged in at a given moment. In this sense, every ontology is the result of an ontopoietic process—one that always requires otherness in order to be realized. The relation, then, does not merely reveal the core of a being; it plays an active role in shaping it. The entity possesses multiple predicative possibilities, yet this does not make it "liquid." Its internal relations define a specific range of virtuality and relational pertinence. At the same time, these possibilities need relation to be brought out, so the predicates that emerge are not "unveiled" in the traditional sense, nor are they hypothetically unveilable—let alone created *ex nihilo*. They are actualizable.

The internal characteristics of an entity, shaped by its internal relations, do not correspond to any specific predicate. Rather, they define a field of predicative potential: qualities not yet actualized, but capable of being actualized in various—though not limitless—ways. To assume otherwise would be to fall into the opposite error of essentialism: the idea that predicates are invented entirely through relation. We cannot think of an entity's predicate as a trait that directly stems from it or is expressed autonomously—something that could be discovered through internal analysis or deduced from its own principles alone. But neither can we treat the predicate as entirely dependent on relation, like a liquid conforming to the shape of its container. External relations neither reveal nor invent predicates. Instead, they bring them out. They are co-factors in the genesis of the predicate and participate in its formation. It is therefore misguided to either believe that predicates can be extracted from an entity in some "alchemical" way—through the pursuit of purity—or to claim, as ultra-relationalism does, that external relations alone generate predicates.

The essentialist conception views predicates as autonomous expressions of the entity's essence, treating external factors as accidental or even corrupting. In this view, external relations merely obscure or alter the entity's authentic character, which could only reveal itself in a condition of purity. The environmental conception takes the opposite stance: it denies the entity's internal qualities, specificity, and relational protagonism, reducing it to a fluid construct entirely shaped by external relations. An eco-ontological conception, instead, overturns both these views. It recognizes that an entity can only manifest itself within a relational condition—but crucially, this manifestation involves selecting one of many predicative forms from among its internal possibilities. This does not mean denying the core of the entity; it only means rethinking what that core is. If the defining features of an entity are a field of predicative possibilities to be actualized and a relational electivity toward the world, then the entity does possess its own consistency—rooted in its internal relations. But precisely because of this internal structure, the entity cannot manifest itself except through relation. In other words, what we call predicates of the entity are expressions, or unfoldings, of the entity. The relationship between the entity's internal or constitutive relations—which define its virtuality, or its field of possible actualizations—and its inherence, that is, its relational susceptibility toward the world, is not incidental. Likewise, the interaction with external or declining relations is not optional or secondary. It is indispensable to predication itself.

To summarize, the philosophy of relations rests on two main factors: (1) external relations—the stable or occasional interactions the entity engages in, which drive the process of actualization, that is, the emergence of a specific form from the multiplicity inherent within the entity; and (2) internal relations—the specific organization of the entity that determines its range of virtualities, or possible actualizations, and that forms the connective structures linking the entity to the world, shaping its interactive electivity. Eco-ontology seeks to move beyond relational dualism—not by denying that the entity relates simultaneously to an internal and an external dimension, but by framing these two forms of relation as part of a continuous flow of transformation. In this view, what is external can become internal, and what is internal can be projected outward.

In this sense, from an eco-ontological standpoint we speak of external relations as *expressive relations*—that is, relations that bring about the actualization of certain predicates. These predicates emerge from the interaction between the entity—as something internally organized through a specific systemic configuration and particular susceptibilities to external influences—and the external elements it encounters. External relations can take various forms:

(i) *occasional*—possible but with low probability or even random, so the resulting predicate appears unusual;
(ii) *probable*—likely to occur because the entity is already equipped to handle that specific interaction, as in prey-predator dynamics;

(iii) *expected*—anticipated by the internal constitution of the entity, as with the ruminal microbiome of a bovine, which cannot function without such symbiosis;
(iv) *produced by the entity itself*—as in niche construction, where the entity reorganizes its environment to shape possible external relations;
(v) *endogenous*—such as endosymbioses, which are not only functionally relevant but also constitutive of the entity itself;
(vi) *persistent*—stable and enduring relations marked by close, ongoing interactivity with the entity.

Similarly, internal relations are *constitutive relations*—those responsible for shaping the entity's field of virtuality and inherence. These relations define a set of qualities specific to the entity, determining how it may be actualized. This includes:

(i) the *field of virtuality*, shaped by the degree of organizational constraint among internal parts—whether they are rigidly structured or allow for configurational plurality;
(ii) the *internal material* of the entity, which sets both the quantitative and qualitative parameters for its possible configurations;
(iii) *heteronomy*, referring to internal structures that are already predisposed to interact with external inputs to produce morpho-physiopoiesis;
(iv) *probability gradients* of different actualizations—while external conditions play a role, the entity itself defines varying levels of likelihood for each outcome;
(v) *inherence*, the entity's capacity to engage specific external relations (possible, probable, or obligatory) through connecting patterns;
(vi) *sensibility*, the presence of particular receptors attuned to specific external stimuli, such as those found on a cell membrane.

In an eco-ontological view, relations do two things: (i) they conjugate the entity by the way it interacts with the world, turning its virtual potential into an actual state with specific predicates; and (ii) they constitute the entity by shaping its internal organization (configurability), its intrinsic resources (relata), and its particular sensitivity to the external world (inherence). This last aspect determines the degree and type of openness the entity has toward its environment. For example, we now know that an ovum is not simply passive to sperm penetration—it actively releases chemical signals to attract certain sperm and slow down or block others. Therefore, the internal relations that define the entity's inherent character establish both the constraints and possibilities of actualization, as well as the selectivity of its connections to the world, which together determine how open the entity is to predicative relations. As a result, both the possible configurations and the relational interactions of the entity are finite. This leads to two key questions about the link between internal and external relations: (1) What kind of relationship exists between them—are they additional, juxtaposed, complementary, or something else? (2) Do external relations influence internal ones and vice versa, or are they completely separate realms?

Regarding the first point, it is important to stress that external relations, when they actualize a virtual condition, act like a sculptor that brings out a statue from a block of marble. In living systems, all selectionist actions follow this model—as explained by Gerald M. Edelman's *Neural Darwinism* (1988), for example. External relations should not be understood as merely additive or juxtapositional. Similarly, actualization is not about filling a void or compensating for a lack in the entity's configuration—there is no inverse relationship between inside and outside. Instead, virtuality is characterized not by absence, but by redundancy. The relationship is directly proportional: the internal and external terms support each other and depend on the resources each provides. In other words, the more complex an entity's internal relations, the wider its range of possibilities and its inherence. This means that (i) its possible actualizations will be more numerous, and (ii) its capacity to relate to the external world will be greater. Therefore, virtuality and inherence are not caused by a lack of internal content, but by an abundance of possibilities and receptivity.

Regarding the second point, while it is clear that an entity, despite certain resistances, remains open to possible relations, its external relations are never merely accidental. The essentialist view cannot explain the entity's natural orientation toward external relations or how its internal constitution is already predisposed for them. A living organism is actually the product of the external relations experienced by its species through selective pressures. It should not be seen as an autonomous essence generating its intrinsic characteristics independently, but rather as an entity whose constitution is shaped by the external relations it has previously encountered. Therefore, when discussing living beings, it is clear that internal and external relations are not sharply separate but mutually influence each other. A good example is adaptation, where an entity's internal constitution results from selective pressures by external factors, so that the entity's core—in its configurability and inherence—is shaped by external relations experienced throughout its phylogeny. At the same time, during development, the entity's internal relations affect which external relations become part of its ontogeny.

On the other hand, the world in which the entity exists is not a generic whole or an undifferentiated mass of interactive possibilities. Every living being has its own specific ecology of insertion, so, as already noted, external relations cannot be seen as a generic condition where everything connects with everything else. One of the basic principles of philosophical ecology is the concept of ecological position: the entity is always embedded within a specific field of interactivity. Thus, on the one hand, internal relations define the entity's intrinsic organization—that is, the actualizing conjugations it can assume and the external relations to which it is susceptible—and therefore the predicates the entity can potentially express; on the other hand, the external world sets the possible and probable external relations through its own ecological organization—its extrinsic virtuality.

Extrinsic virtuality, in turn, should not be understood as a random collection of relations but is best analyzed according to clear parameters inferred from descriptive ecology, such as: (i) the proximity of external factors the entity may encounter, since if a factor is absent in the entity's environment, it is unlikely to affect it; (ii) the relevance of the external element to the specific morphopoietic or physiological dynamics of the entity—whether it is essential to the entity's survival or constitution; (iii) the chain of effects or indirect influence, where an external factor, even if distant, influences another factor that directly affects the entity; and (iv) the variability level of the environment the entity inhabits.

I use the term extrinsic virtuality to avoid determinism, because even external relations have gradients of possibility, sometimes approaching certainty. Moreover, there is no strict separation between the entity's intrinsic characteristics and its external relations—not only because the entity has relational inherence but also because it helps shape its field of interactivity, for example, by modifying its environment, building a niche, and moving through its territory. Eco-ontology thus departs sharply from both essentialism and aspecific relationalism—not merely by acknowledging that an entity has internal and external relations, but by emphasizing that the predicate manifested by the entity expresses the totality of internal and external relations, meaning relations are not accidental or secondary factors but the direct result of relational concurrence. Likewise, eco-ontology posits that internal and external relations are not separate but act as co-factors, linked by a logic of direct proportionality and reciprocal influence. Living organisms are the product of a continuous flow of relations constantly shifting between external and internal, gradually revealing a multiplicity of predicates that do not belong to the thing-in-itself but to the entire relational organization the entity takes on at a given time. Dividing relations into internal and external only makes sense if we consider the entity at a single moment; viewed diachronically, what is external at one time becomes internal at another.

9.4 The Foundations of Eco-Ontology

If we follow evolutionary thinking, it becomes clear that living beings have specific relational characteristics. They cannot be treated as amorphous or passive entities governed solely by the laws that apply to interactions between abiotic elements—whether those laws are linear or based on complex systems theory. First, living beings are the result of relational processes that have developed over time and become embedded through heredity. These inherited structures are then projected forward through development, metabolism, and reproduction—processes defined by the internalization of external influences, and which must be understood within a diachronic framework. Second, these processes unfold across multiple layers and stages. The structure of a living being thus emerges as a multilayered system of introjected relational events, where each internal layer is the product of a prior interaction with the external world. For this reason, we cannot fully grasp the morphological, physiological, or expressive features of an organism without viewing it through the lens of this internalized relationality.

Evolutionary theory clearly shows that a species' interactions with its environment shape its development, progressively aligning organisms with the specific features and demands of their habitat. This relationship between a life form's mode of being and the risks and opportunities of the environment drives internal transformation. In this sense, we can speak of an indirect internalization of external relations: selective pressures carry environmental information into the biological organization of the organism. Differential replication—where traits suited to the environment are passed on—is a powerful mechanism for converting external (exo-) relations into internal (endo-) ones. What once governed a population's relationship with its environment now regulates the ontogenetic processes and internal dynamics of the organism itself. Phylogeny is just one example of this exo-endo conversion. Others include the relationship between genome and epigenetics, embryogenesis, early-life experiences, attachment and socialization, and the physiological activation of body systems.

Ecology, too, reveals specificities in the relationships between entities—peculiarities that cannot be reduced to a simplistic explanatory holism, where the whole overrides individual differences and distinct relational mechanisms. Nor, as said, can it be adequately captured by a vision of hyper-connectionism, which claims that everything is connected to everything else, while neglecting the fact that, for each organism, relationships fall into distinct categories: (i) expected relationships, which are an integral part of the organism's constitution; (ii) necessary relationships, essential to the life of that particular organism; (iii) influencing events, which vary in probability and impact; (iv) factors that are present but have no effect on that organism; and (v) elements that are not only uninfluential but also have virtually no chance of interacting with the organism.

Ecology shows us that living beings are not only shaped by relationships, but also actively construct their existence through them. They organize relational systems, alter their environment, establish symbiotic or competitive ties, use otherness as a resource, and build life-niches that allow them to regulate external influences—dampening some, amplifying others. The agency of living organisms is revealed in this *relational protagonism*, especially in their creation of protective shells of existence: true ecological abodes equipped with buffering systems that filter and stabilize their interactions with the outside world. In this light, ecology is not simply the science of relationships: it seeks to organize the dynamics between organisms and their environment within an evolutionary framework. Therefore, a philosophical interpretation that seeks to engage with ecology cannot merely be a philosophy of relationships, even while recognizing their importance. Philosophical ecology takes on the task of building a philosophical reflection rooted in the meta-principles of ecology. Its aim is to go beyond scientific description and fully address the relationship between human beings and the world—ethically, epistemologically, and ontologically—in order to help us *think* ecologically (Amoroso 2023).

When dealing with the phenomenology of life, relations are never just interactions between entities—they are the conditions of life itself. This is what led me to speak of an eco-ontology: a perspective that takes into account the deep dependencies among organisms, their co-evolutionary dynamics, their structural adaptation to anticipated external inputs (heteronomy), the internalization of external factors, and their metastable organization—which makes ontology, in this view, ontopoiesis. I propose six basic points as the foundations of eco-ontology:

1. Internal relations are potential configurations among components (intrinsic configurability). External relations shape the internal characteristics of an entity by influencing how its phenotypic possibilities are realized in specific configurations.
2. The internal relations of an entity arise not only from its potential attributes (predicative virtuality) but also from its specific interactive electivity (relational inherence), which determines what kinds of external relations are possible, likely, influential, necessary, or expected by the entity.
3. Internal relations are the introjection of prior external relations, seen in both phylogeny and ontogeny. What we call intrinsic qualities—whether of configurability or relational selectivity—mirror the ontopoietic process through which they were shaped.
4. External relations may appear accidental, but in living organisms, they typically occur with specific degrees of probability and necessity, depending on the organism's ecology. These relations are always implicated in the constitution of the organism and are often prefigured by it (heteronomy), since internal structure emerges only in response to external inputs.
5. Calling them "external" or "environmental" relations presumes a fixed boundary between inside and outside, but the fractal nature of living entities undermines this distinction. Every "inside" is itself the result of prior relational conditions. We should instead speak of ontological relational planes—interwoven layers where internal and external co-constitute one another in a kind of ontopoietic yin-yang.
6. External relations are often thought of as imposed on the organism, but in reality, they are frequently sought out, or inherently woven into its way of being. Hence, we can speak of a relational ontology, in which the organism's predicates are not distorted by its relations but actualized through them. The being is made-to-relate: relations do not obscure the essence of the entity—they reveal it.

Before addressing these six foundational points through a discussion of the theses that shape relational models—see, for instance, Christian

Frigerio's *Ricomporre un cosmo in frammenti* (2023)—I want to introduce a few key concepts I consider fundamental to eco-ontology: (i) relational inherence, (ii) metastable condition, (iii) diachronic dimension, and (iv) field of virtuality.

Starting with *relational inherence* (i), I use this term differently from its traditional meaning as "pertinence-to-self." Instead, I define it as relational pertinence: the entity's inherent capacity to form elective affinities with specific external relations. Inherence, in this sense, is an intrinsic feature of the entity, but does not define its predicates; it produces a specific affinity or sensitivity toward external entities; it establishes the entity's openness to certain forms of otherness; and it influences how the entity can relate—what kinds of relations are possible, likely, necessary, or expected. The second concept is the *metastable condition* (ii) of the entity. This refers to the fact that its internal relations are not fixed by structural determinism but remain within a range of configurability. Some aspects of the entity are more stable, while others remain flexible and open to reconfiguration, giving the entity varying degrees of plasticity. The third concept is the *diachronic dimension* (iii). Even inherence and metastability change over time, meaning that the qualities of the entity cannot be captured at a single moment. A living being, therefore, cannot be fully understood in a synchronic framework—it must be seen in its temporal unfolding.

Finally, the fourth concept is the *field of virtuality* (iv). From an eco-ontological perspective, this includes both intrinsic virtuality, the potential configurations the entity itself can assume, and extrinsic virtuality—the potential external factors with which it might interact. This virtuality, however, is not infinite—it is situated within a field that includes a) a clearly defined range of possibilities, organized by degrees of possibility, probability, influence, necessity, and heteronomy; b) zones with high constraining value, which are particularly sensitive to change, and other zones with high degeneracy, which are more robust and less affected by variation; and c) dimensions of relative homeostasis, such as thresholds that confer a certain stability, maintained through homeodynamic processes and compensatory feedback mechanisms. When these stability mechanisms fail—when a homeostatic threshold is exceeded—the entity can shift from one state to an entirely different one. Thus, while the transformation of the entity may be continuous in its process, from a predicative standpoint, it often follows abrupt transitions—as illustrated by the theory of punctuated equilibrium (Gould and Vrba 1982).

Having made these propaedeutic considerations, we can now turn to the different relational theses, examining both their content and their connection to the six foundational points of eco-ontology. From an eco-ontological perspective, one begins with the idea that predicates cannot be reduced to the intrinsic properties of an entity. It is the external relation that gives the entity its configuration—and thus, its predication. The internal relational properties of the entity make these external relations predicatively impactful, in the sense that they convert the virtual into the actual, producing ontopoiesis. We could also say that external relations are often (I would say always) implicated in the intrinsic predicates of the entity. In living beings, the entities with which one comes into contact are, generally speaking, expected or foreseeable—contrary to the distinction drawn by David Lewis in *On the Plurality of Worlds* (1986). In fact, in an eco-ontological view the internal constitution of the entity reflects external relations previously developed by that entity; these relations are presupposed in the entity's predicative expressions and thus are never entirely unforeseen; furthermore, the predicate is the result of the relations the entity establishes with the external world—not because the entity lacks content or is passive in the face of interaction, but because it is redundant with possibilities.

It is also important to clarify that there are different types of relations, and not all of them carry ontopoietic significance—hence the need to distinguish between them. For this reason, when speaking of eco-ontology, we do *not* refer to the following types of relations:

(i) *comparative relations*, in which two entities are compared to assess their manifest qualities (e.g., color, weight, shape);
(ii) *differentiative relations*, which help us navigate reality by formulating distinctive, categorical, correlative, or causal coordinates;
(iii) *additive or juxtapositive relations*, which do not alter the internal configuration of the entity but merely add predicates;
(iv) *integrative relations*, which position the entity within a higher-level element, acting as a container or new plane of reality;
(v) *conjunctive relations* (e.g., being the child of), which merely indicate proximity or connection;
(vi) *dialectical or oppositional relations*, in which the relation is resolved through the overcoming of the relation itself.

These types of relations are, so to speak, ontologically costless, as they do not contribute to or modify the ontology of the entity (Armstrong 2010).

So let's go back to *ontopoietic* relations—those that matter ontologically for the entity—starting from the relevance thesis, which holds that the relation is fundamental to defining the identity of the terms: if the relation changes, the relata change as well (Frigerio 2023). In an eco-ontological view, this proposition is valid, but not absolute: the entity possesses *multipotent*, not *totipotent*, configurability. Moreover, through inherence, it exhibits relational selectivity and thus presents possible, probable, necessary, and expected indices of actualization. Added to this is the fact that change is never absolute—metastability does not imply liquidity, but rather plasticity; therefore, there are limits to change. We can say, then, that in an eco-ontological framework, internal and external relations stand in a multivariate relationship: each changes as the other changes, though always within a flow matrix. If we consider certain relations as ontopoietic, then both external and internal relations are necessary for the predication and ontological evolution of the entity, defining points of possibility and points of necessity (Schaffer 2010).

In an ontopoietic conception, there is no strict divide between the predicative manifestation of the entity and its ontological transfiguration. Rather, there is a close—albeit selective—implication, whereby external relations tend to be internalized (Rojek 2020). As mentioned, I refer to this process as *ontopoietic mirroring*. In phylogeny, it leads a species to incorporate into its specific traits the external relations it has experienced—selective pressures that, through adaptation, are mirrored in the habitat; or, as Lorenz put it, the a priori elements are phylogenetically a posteriori. In ontogeny, mirroring causes the individual to reflect the environment in which they grew—consider, for example, the development of a tree's crown or roots depending on its access to light or water. At the same time, while it is true that the canopy reflects the tree's growth environment, the way an oak reflects that environment differs significantly from an aspen. In other words, an eco-ontological approach does *not* equate to an externalism that treats the relations between entities as freely interchangeable or indifferent to intrinsic characteristics (Lewis 2020; Meillassoux 2006). In an eco-ontological conception, one should not speak of a generic interconnection between entities, but rather of degrees of dependence, whereby the external relations available to an entity are never absolute—let alone occasional—but always defined by an external field of virtuality that determines which relations the entity can entertain.

Let us now consider the *generativity thesis*, which holds that relations are not only fundamental in defining the characteristics of the relata but also generate the relata themselves (Frigerio 2023). According to my eco-ontological approach, this thesis is acceptable only from a *diachronic* perspective: the external relation in the here and now does not generate the intrinsic contents of an entity, as these depend on three factors: (i) the *ante-rem*, that is, the relations that preceded the evolution of the entity (e.g., phylogeny); (ii) the *in-re*, the specific configuration that occurs during the external relation; (iii) the *post-rem*, what follows after the external relation has taken place. In an organism, for instance, external relations modify the structure of various systems through the law of effect (what yields an advantage tends to grow) and the law of exercise—what is used tends to grow.

Let us turn now to the *structural relations thesis*, which is fundamental in eco-ontology. This thesis holds that relations are responsible for the entire structure of reality, such that predicative terms are nodes of subsistent relations, and predicative utterances should be reformulated in relational terms (Ladyman and Ross 2007). However, I do not fully agree with the hypothesis of radical deflationism regarding the essentialism of entities proposed by these authors. As they describe it, their view entails a synchronic relational structuralism, whereas an eco-ontological perspective is necessarily *evolutionary*—structures are diachronic entities. In this view, an entity exhibits its own a priori internal structure, and for this reason it is open to, but can also resist, further configuration.

Alongside these aspects of inherent multipotent—but not liquid—configurability, we must also consider the basic principles of ecology, which require us to speak of *relation geographies*, where influences are embedded within a networked matrix. Ecology rests on the assumption that no entity can relate to every other in a kind of relational generalism or totality of influences. Instead, ecology is based on the definition of shells of influence, whereby external relations must be evaluated according to systemic rules such as: proximity, interdependence, sensitivity (or conversely, refractoriness), recursiveness, and emergence. In other words, in an eco-ontological conception, the structural relations model holds true—but only insofar as it respects: (i) the evo-devo nature of the entity, that is, the diachronic dimension that defines the entity's virtuality and inherence; and (ii) its *ecological positioning*, which establishes the geographies of the relations relevant to that specific entity.

The importance of the relationship between relations and time—whereby the entity is indeed in relation, but in a diachronic dimension—leads to the view that time itself may be a function of relations, or may even emerge through connections (James 2014; Connolly 2011; Madelrieux 2006). The diachronic character of an entity's intrinsic properties and its specific ecological location reveal another key aspect of the external relations the entity receives in time: namely, their vectorial nature. These relations convey to the entity a position that, on the one hand, situates it within a precise matrix of possible/necessary relations, and on the other hand, defines an asymmetry between its past and future (Paci 1956).

It is evident, for example, that a developing mammal possesses greater developmental potential—i.e., ontogenetic virtuality—than an adult or elderly subject. In such cases, the conversion of the virtual into the actual, while extracting a form, also reduces potentiality by pruning away other previously possible avenues (Edelman 1988). A similar point can be made regarding the concept of *evolvability* in evolutionary biology, in relation to the gradient of genetic variability within a species (Lehman and Stanley 2013). In other words, extrinsic relations do not act upon a condition of interactive genericism because there is a field of virtuality intrinsic to the process of actualization, defined by well-established constraints and possibilities—among which is inherence, i.e., relational selectivity. Likewise, there is a field of extrinsic virtuality, whereby the entity does not relate to all possible others but is situated within a matrix of possible/necessary relations.

An ecological perspective does not consider the interconnectedness of entities in a generic or undifferentiated way, because a living being is always situated within gradients of external relationality—possible, probable, necessary, and expected. Therefore, eco-ontology always specifies the field of external relations in which a subject is embedded. From its perspective, in fact, relation is not simply connection (Forster 1910), nor does it mean that everything is connected to everything else (Commoner 1971), or that a butterfly's flutter might alter distant weather patterns as in the so-called *butterfly effect* (Lorenz 1963). In reality, the connections between most entities and events are relatively tenuous—often to the point of insignificance (Attfield 2018). Ecology investigates precisely the field of extrinsic relationality in living beings, which can at times be extremely complex, as in the mentioned case of the multiple factors compromising leopard frogs (*Rana pipiens*) in North America—but which nonetheless operate through specific cycles of interaction (Kiesecker 2002).

Another key factor influencing an entity's susceptibility to external relations is its specific inherence—that is, its selective receptivity in interacting with external elements. In living organisms, particularly animals, this relational electivity depends on three main aspects: accessibility, susceptibility, and orientation toward the external world. In terms of ***accessibility***, we can refer to von Uexküll's concept of the *Umwelt*, alongside subsequent studies in the ethology of perception—ranging from the definition of sensory apparatuses (Hughes 1999), to sign stimuli (Tinbergen 1951), to the concept of ***affordance*** (Gibson 1966). ***Susceptibility***, instead, indicates how permeable or exposed an entity is to external influences. For instance, the skin of an amphibian makes it particularly sensitive to pollutants, turning it into an effective environmental indicator (Cain et al. 2011). Likewise, dopaminergic pathways are networks of neurons specifically susceptible to activation by dopamine (Carlson 2013). The third aspect is ***relational orientation***, shaped by the affective states of the organism—its emotions and motivations—which define the type of copula, or coupling, it forms with the external world (Deleuze and Guattari 1987).

I agree, then, that everything we encounter—or that surrounds us—has a form of relational consistency. However, I disagree with the view that intrinsic properties do not exist (Ladyman 2017). In an eco-ontological framework, an entity has a distinct relational position shaped by its internal characteristics—its phylogenetic and ontogenetic history, as well as its inherent evolvability. An organism is always situated within the space-time of ecology and evolution, and this defines both the kinds of actualization available and the kinds of external relations possible. Moreover, every entity is embedded in a diachronic flow in which the status of relations constantly shifts—what is external can become internal, and vice versa. It is therefore essential to embed this ontopoietic principle of relations within a broader evo-devo conception—one that accounts for the being's transformation over time—and an ecological one, which considers the entity's positioning.

9.5 Philosophical Ecology and Epistemology

The world perceived by human beings is undoubtedly multifaceted, yet it possesses a certain solidity and stability—even amid continuous change and transformation. This is why we speak of things like concreteness, objectivity, being down-to-earth, and reality. What surrounds us is not only revealed through sensory cues, but also as discrete entities, making discrimination a fundamental activity for navigating the labyrinth of reality. Experience—and thus knowledge—rests on this discreteness, which shapes the kind of relationship between the self and otherness, and more broadly, with the world. Even in animals with relatively simple nervous systems, we see the first traces of subjectivity emerging from the function of reafference (Godfrey-Smith 2020; Vallortigara 2021). Experience arises from a relationship with the world—not only through sensory input but also through other forms of information acquisition, such as the immune response. This allows individuals to internalize the formal and functional properties of external elements, enabling them not only to assess, e.g., use or predictive value, but also the entity's meaning, its emergence from the context, and its specific qualities. Clearly, this relationship is shaped by the receiver's particular position within the web of reality, making perception of the world inherently perspectival (Giere 2006).

At this point, building on this notion of perspectival positionality, we can ask whether ecological thinking might offer epistemological insights consistent with the ontological reflections developed above. There is no doubt that cognitive functions are among the traits most subject to selective pressures, and that a living being's ecological position is the primary factor shaping its interaction with the world. While the perspectival principle is well established—demonstrating that data is always processed through intrinsic filters, such that different instruments or organisms transpose and represent the same phenomena in different ways—we may ask whether this leads us beyond mere perspectival relativism. As we have seen, the entity—our primary object of inquiry—possesses an intrinsic range of virtuality that allows it to assume different forms and functions depending on how it is actualized through external relations. If perspectivism brings us into the realm of evolutionary epistemology (Vollmer 1983; Stanzione 1984), we can now add that, depending on the ecological position of the inquirer—that is, the nature of the interaction they initiate—it is reality itself that becomes reconfigured, giving rise to a new plane of reality (Marchesini and Celentano 2021). This calls for a further consideration: if we combine the evolutionary principle of introjection with the ecological principle of positionality, we must conclude that the entity is already pre-formed for certain types of relations with the world. This is true (i) because its internal qualities already relate it to a particular world-context; and (ii) because it is embedded in a specific niche of interaction with external entities. Knowledge—in both its perceptual and cognitive dimensions—is therefore inseparable from the entity's inherence and ecological positioning.

To know means to come into contact with certain qualities of the world—a set of experiential references that allow us to evaluate, discriminate, correlate, and establish causality, among other things. Yet we know that the qualities an entity displays do not exhaust its being; its phenomenal consistency depends on three factors: (i) the entity's intrinsic characteristics—its range of virtuality and degree of inherence; (ii) the relational field in which it is embedded, and thus the way it is shaped by external factors at a given moment, since an entity is always in relation; (iii) the observational plane adopted during inquiry, largely determined by the apparatuses of knowledge—so that the same landscape appears differently depending on the species or the instruments employed. As Konrad Lorenz argued in *Behind the Mirror* (1977), the representational apparatuses we use to perceive the world are themselves shaped by the world. In this sense, what appear to be a priori categories are, once more, a posteriori outcomes of the phylogenetic process. The observer, too, is a relational entity. Evolutionary epistemology builds on this principle, viewing human knowledge as a partial and situated perspective—shaped by our species' evolutionary history. What we perceive is not a universal and objective image of reality, but a version already configured to meet specific adaptive needs (Campbell 1974; Popper 1972).

Moreover, perception is not merely a matter of receptivity. The whole body, in its ecological positionality, is in relation with the external world. Each species, as Jakob von Uexküll (2010) insightfully noted, constructs its own context-world. Still, it is important to move beyond the image of the *Umwelt* as a sealed bubble that encloses each living being within its own isolated context. From an evolutionary perspective—grounded in phylogenetic homologies—it becomes clear that different *Umwelten* share overlapping margins. These overlaps are shaped by two main factors: (i) taxonomic proximity, or shared evolutionary traits, which means that humans and chimpanzees, for example, share substantial experiential overlap; and (ii) adaptive convergence, where species inhabiting the same environment and facing similar challenges tend to develop comparable perspectives. The idea that humans—and other species—are doomed to remain locked within their own phylogenetic perspective is a misunderstanding of the *Umwelt* concept. Ontogenetic processes—growth, maturation, and learning—continuously reshape our image-apparatuses of the world. They alter the subject's inherence, that is, its capacity to be affected by and respond to external stimuli.

Knowledge itself can transform how a subject perceives the environment, through mechanisms of salience or habituation. In a park, for instance, an entomologist may see primarily insects, a botanist will notice the diversity of plants, and an ornithologist will be attuned to birdsong. Each observer highlights different phenomena, bringing forward a distinct slice of the whole. Science has radically reshaped how we perceive the world, while technology has extended our perceptual reach—allowing us to both transcend and reflect critically on our anthropocentric perspective (Wolpert 1994; Bachelard 1938). At the same time, a perspectivist approach shows that by shifting the parameters of observation—especially scale—we encounter entirely different phenomenologies of reality. Reality, in this sense, can be expressed in many ways, depending on the lens we choose to adopt (Callebaut 2012).

That said, a key question raised by philosophical ecology is whether there exists a final and definitive configuration of reality—however perspectival and, according to some, ultimately unknowable—or whether there is nothing beyond perspectivism itself. In other words, is reality devoid of any inherent form outside our processes of investigation, or does the very act of knowing—precisely because it is an external relation involving a specific plane of interaction with the entity under investigation—serve to reveal a distinct phenomenological layer of the real? In this sense, it is important to clarify that I am not suggesting an ontological nihilism or a relativism of perspectives. By speaking of a range of intrinsic virtuality, I acknowledge a bounded field of possible configurations—but no totipotency. There are limits. I fully agree that the real has its own texture and is not simply the product of our investigative apparatuses; there is indeed a distinction between the epistemological and the ontological. And yet, what exactly constitutes that texture—and on what ontological plane it can be located—remains an open and contested question (Austin 1962).

If we assume that the core of an entity is not a fixed essence but something configurable, then each entity can manifest across multiple planes of reality—not only based on the perspectival lens applied, but also because it is ontologically predisposed to assume different conformations depending on the particular web of external relations activated in a given inquiry. But, I repeat, this is not because knowing *creates* the predicates of the entity; rather, it is because, by placing itself in relation with the entity, it actualizes them—that is, it brings them forth on a particular ontopoietic plane already contained within the entity's own range of possibilities. What I said in the previous section regarding the range of predicative possibilities—deriving from the entity's inherent virtuality, its relational salience, and the plane of external relations in which it is embedded—applies equally on the epistemological level. The investigative act is, in every respect, a relational event, and since there exist multiple planes of investigation, we will encounter a plurality of predicative planes for any given entity—each revealing what that entity is, when observed from that specific standpoint. As Hilary Putnam suggests, "a chair can be usefully and truthfully described in the language of physics, of carpentry, of furniture design, or of etiquette," to which we might add the kind of wood used, the species of tree from which it came, and so on (2012, 15). Thus, the problem of knowledge must be situated within a matrix defined by the relationship that arises at a given moment between the internal relations of the entity—which delineate its openness (what is possible) and its interactive electivity (what is probable)—and the external relations that become available, constituting an equally virtual and elective field of interactions for the entity to access.

The world as it appears to us is the result of an organizational arbitrariness—but it is not a mere illusion, nor an invention. Rather, it is the outcome of the particular matrix of organization we have imposed on the virtuality of reality. By virtuality, I do not mean an empty or amorphous substrate, malleable to any imposed form, but a content that can be arranged according to multiple patterns—yet one that has its own resistance (Ferraris 2014), that cannot take on all thinkable forms. To investigate an entity is to interrogate it—that is, to position oneself on a specific ecological plane of interaction, which inevitably generates a structure of relations, an epistemic fabric comparable to niche construction. It is the plane of questioning that defines epistemic ecology, and so its partiality should not surprise us: the plane reflects the subjectivity of the question, the specific selection of what is to be observed, the kind of information being sought. But this partiality does not equate relativism.

Dialogue, in this sense, is a way of organizing reality. Epistemics is the act of placing oneself on a given plane of reality to extract useful information from it. This is not an approximation, but an emergence—an epistemological niche construction that reveals not only a perspective, but possibilities of the real. In investigating and explaining phenomena, we confront planes of reality, each of which gives rise to different predicates of the entity depending on the kind of relation established. These external relations are structured by possibility and probability, as is readily observable in the ecological analysis of a living organism's interactions with its world-context (Marchesini 2018, 7–19). A reality plane is thus a particular snapshot of the real: not invented, not illusory, not purely perspectival, but concretely drawn from the possibilities of reality. In this sense, we can say that, insofar as it is assimilated to external relations, investigation is always ontopoietic: it gives rise not merely to a point of view, but to a concrete configuration of the real.

By assuming a scale of positioning, the questioner brings forth a version of reality shaped by the ecological context in which they are situated. Because both the questioner and what is being questioned exist in relation, the process of inquiry becomes ontopoietic: it brings into focus one particular configuration of the real, selected from among its many virtual possibilities. *Knowledge, then, is not about uncovering fixed truths but about actualizing certain aspects of an entity*, depending on the nature of the inquiry. It is an act that allows the entity to take on a particular form—it prompts the entity to show itself in one way rather than another. This leads to what I see as a key idea in ecological epistemology: we don't access the internal contents of an entity through what it shows us, but rather through all the situations when our theories or questions exceed its field of possibility. In other words, the being's resistance reveals the limits of its possible expressions. Hence two things: first, the existence of multiple ways an entity can be manifested; second, a boundary that defines the limits of those manifestations. As Maurizio Ferraris (2012, 155), put it "The salient feature of sentience is its not confirming, not realizing, but refuting our expectations and our knowledge, revealing with clarity that there is something distinct and separate (…) this resistance seems intrinsically connected with reality in the ontological sense." He concludes: "The best way to qualify 'real' […] is always negative." If an entity can only be expressed in a finite number of ways, then its true nature—its core—emerges precisely where it resists any further variations.

To conclude, the ecological epistemology I've outlined here—and explored in more detail elsewhere (Marchesini 2018, 2022)—can be summarized in the following tenets:

1. Reality has no final, fixed configuration. It is characterized instead by a "virtual field": finite, yet rich in possibilities. While not everything is possible—there are definite points of resistance—within those limits, many configurations can emerge.

2. Every act of knowledge is a way of organizing that virtuality. It brings about a "plane of reality," much like a statue is shaped from a block of marble: not created out of nothing, but actualized from a pre-existing potential.
3. Each reality plane emerges from a specific dialogue between subject and world. It is shaped by a particular mode of inquiry—a question asked to reality—that determines what aspect of the real becomes visible. This is not merely a matter of perspective.
4. Every living being, based on its own physiology, does not interact with the full virtual field of reality, but with an "ecological dimension"—a more limited range of virtual possibilities shaped by its bodily relation to the world. Each epistemic condition is thus specific to the organism's way of being in its environment.
5. Yet different organisms do not inhabit entirely separate or disconnected planes of reality. Because life must respond to shared constraints—both internal needs and external conditions—there is a transpecific dimension to knowledge.
6. Each species develops a species-specific reality plane, but these are never sealed bubbles. Due to inherited homologies and adaptive convergence, there are overlaps and shared structures across species.
7. Each individual is born with a particular perspective, but can go beyond it—hence the notion of *intus-legere*, reading-into reality, beyond appearances. Thus, the *Umwelten* are in constant transformation.
8. When an entity shapes its own intersection with the world—through singular, lived experience—it gives rise to a plane of reality that does not precede the act of knowing. This can be called "predicative emergence." In this process, the individual is an active protagonist, much like a sculptor shaping marble.
9. What the individual perceives as immutable laws are not the given possibilities, but the limits—what cannot be actualized. Knowledge is thus shaped by resistance; finality exists only in what denies realization.
10. Finally, knowing is not separate from other forms of interaction—such as dispositions or orientations. Change the ecological context of the questioner, the questioned, or the structure of their interaction, and the plane of reality that emerges will change too.

References

Amoroso, Prisca. 2023. *Ecologia*. Roma: Castelvecchi.

Armstrong, David M. 2010. *Sketch for a Systematic Metaphysics*. Oxford: Oxford University Press.

Attfield, Robin. 2007. Is the Concept of Nature Dispensable? *Proceedings of the Twenty-First World Congress of Philosophy* 5 (25): 59–63. https://doi.org/10.5840/wcp2120075210.

Attfield, Robin. 2018. *Environmental Ethics*. Oxford: Oxford University Press.

Attfield, Robin. 2021. *Environmental Thought: A Short Introduction*. 2nd ed. London: Polity Press.

Austin, John L. 1962. *Sense and Sensibilia*. Oxford: Oxford University Press.

Bachelard, Gaston. 1938. *La formation de l'esprit scientifique: Contribution à une psychanalyse de la connaissance objective*. Paris: Librairie Philosophique J. Vrin.

Bateson, Gregory. 1979. *Mind and Nature: A Necessary Unity*. New York: Dutton.

Bateson, Gregory. 1991. *A Sacred Unity: Further Steps to an Ecology of Mind*. New York: William Morrow.

Beck, Benjamin B. 1980. *Animal Tool Behavior: The Use and Manufacture of Tools by Animals*. New York: Garland STPM Press.

Bonner, John Tyler. 1980. *The Evolution of Culture in Animals*. Princeton: Princeton University Press.

Braidotti, Rosi. 2013. *The Posthuman*. Cambridge: Polity Press.

Cain, Michael L., William D. Bowman, and Sally D. Hacker. 2011. *Ecology*. Sunderland: Sinauer Associates.

Callebaut, Werner. 2012. Scientific Perspectivism: A Philosopher of Science Response to the Challenge of Big Data Biology. *Studies in History and Philosophy of Biological and Biomedical Sciences* 43 (1): 69–80.

Campbell, Donald T. 1974. Evolutionary Epistemology. In *The Philosophy of Karl Popper*, ed. Paul Arthur Schilpp, 413–463. La Salle: Open Court.

Candiotto, Laura, and Giacomo Pezzano. 2019. *Filosofia delle Relazioni*. Genoa: Il Nuovo Melangolo.

Carlson, Neil R. 2013. *Physiology of Behavior*. Boston: Pearson.

Carroll, Sean M. 2016. *The Big Picture: On the Origins of Life, Meaning, and the Universe Itself*. New York: Dutton.

Commoner, Barry. 1971. *The Closing Circle: Nature, Man, and Technology*. New York: Alfred A. Knopf.

Connolly, William E. 2011. *A World of Becoming*. Durham: Duke University.

Crutzen, Paul J., e Eugene F. Stoermer. 2000. The 'Anthropocene'. *Global Change Newsletter* 41 (May): 17–18.

Daston, Lorraine. 2019. *Against Nature*. Cambridge, MA: The MIT Press.

DeLanda, Manuel. 2002. *Intensive Science and Virtual Philosophy*. London: Continuum.

Deleuze, Gilles, and Félix Guattari. 1987. *A Thousand Plateaus: Capitalism and Schizophrenia*. Minneapolis: University of Minnesota Press.

Descola, Philippe. 2013. *Beyond Nature and Culture*. Chicago: University of Chicago Press.

Edelman, Gerald M. 1988. *Neural Darwinism: The Theory of Neuronal Group Selection*. New York: Basic Books.

Ferrando, Francesca. 2019. *Philosophical Posthumanism*. London: Bloomsbury Academic.

Ferraris, Maurizio. 2012. Esistere è resistere. In *Bentornata realtà: Il nuovo realismo in discussione*, ed. Mario De Caro and Maurizio Ferraris, 139–145. Torino: Einaudi.

Forster, E. M. 1910. *Howards End*. London: Edward Arnold.

Ferraris, Maurizio. 2014. *Manifesto of New Realism*. Albany, NY: SUNY Press.

Frigerio, Christian. 2023. *Ricomporre un Cosmo in Frammenti: Filosofia, Ecologia e Relazioni*. Sesto San Giovanni: Mimesis.

Giere, Ronald N. 2006. *Scientific Perspectivism*. Chicago: University of Chicago Press.

Godfrey-Smith, Peter. 2020. *Metazoa. Animal Minds and the Birth of Consciousness*. Glasgow: William Collins.

Gould, Stephen Jay, and Elisabeth Vrba. 1982. Exaptation – A Missing Term in the Science of Form. *Paleobiology* 8 (1): 4–15.

Haraway, Donna Jeanne. 2016. A Cyborg Manifesto. Minneapolis: University of Minnesota Press.

Harman, Graham. 2002. *Guerrilla Metaphysics: Phenomenology and the Carpentry of Things*. Chicago: Open Court.

Harré, Rom, and Roy Harris. 1993. *Linguistics and Philosophy: The Controversial Interface*, Language & Communication Library. Vol. 13. New York/Oxford: Pergamon Press.

Horstmann, Rolf-Peter. 1999. What Is Hegel's Legacy and What Should we Do with it? *European Journal of Philosophy* 7 (2): 275–287. https://doi.org/10.1111/1468-0378.00087.

Hughes, Howard C. 1999. *Sensory Exotica*. Bradford: Bradford Book.

Iofrida, Manlio. 2019. *Per un paradigma del corpo: una rifondazione filosofica dell'ecologia*. Macerata: Quodlibet.

James, William. 2014. *The Will to Believe*. Cambridge: Cambridge University Press.

Karpouzou, Peggy, and Nikoleta Zampaki, eds. 2023. *Symbiotic Posthumanist Ecologies in Western Literature, Philosophy and Art: Towards Theory and Practice*. Berlin: Peter Lang.

Kiesecker, Joseph Michael. 2002. Synergism between Trematode Infection and Pesticide Exposure: A Link to Amphibian Limb Deformities in Nature? *Proceedings of the National Academy of Sciences of the United States of America* 99 (15): 9900–9904. https://doi.org/10.1073/pnas.152098899.

Kleinherenbrink, Arjen. 2019. *Against Continuity: Deleuze's Speculative Realism.* Edinburgh: University of Edinburgh.

Ladyman, James. 2017. An Apology for Naturalized Metaphysics. In *Metaphysics and the Philosophy of Science: New Essays*, ed. Heather Matthew, John Slater, and Z. Yudell, 1–18. Oxford: Oxford University Press.

Ladyman, James, and Don Ross. 2007. *Every Thing Must Go: Metaphysics Naturalized.* Oxford: Oxford University Press.

Latour, Bruno. 2017. *Facing Gaia: Lectures on the New Climatic Regime.* Cambridge, UK: Polity.

Lehman, Joel, e Kenneth O. Stanley. 2013. Evolvability Is Inevitable: Increasing Evolvability without the Pressure to Adapt. *PLoS ONE* 8 (4): e62186. https://doi.org/10.1371/journal.pone.006218.

Leopardi, Giacomo. 1887. *The Poems of Giacomo Leopardi.* Translated by Frederick Townsend. New York: G. P. Putnam's Sons.

Lewis, David. 1986. *On the Plurality of Worlds.* Oxford: Blackwell.

Lewis, David. 2020. *Sulla pluralità dei mondi.* Sesto San Giovanni: Mimesis.

Lorenz, Edward N. 1963. Deterministic Nonperiodic Flow. *Journal of the Atmospheric Sciences* 20 (2): 130–141. https://doi.org/10.1175/1520-0469(1963)020<0130:DNF>2.0.CO;2.

Lorenz, Konrad. 1977. *Behind the Mirror: A Search for a Natural History of Human Knowledge.* Londra: Methuen.

Madelrieux, S. 2006. Pluralisme anglaise et pluralisme américain: Bertrand Russell et William James. *Archives de Philosophie* 69 (3): 375–393.

Magee, Bryan. 1973. *Karl Popper.* London: Collins (Fontana Modern Masters).

Magee, Bryan. 2004. *Confessions of a Philosopher.* London: Phoenix/Orion.

Mainardi, Danilo. 1974. *L'animale culturale.* Milano: Rizzoli.

Marchesini, Roberto. 2002. *Post-Human. Verso nuovi modelli esistenziali.* Torino: Bollati Boringhieri.

Marchesini, Roberto. 2018. *Beyond Anthropocentrism. Thoughts for a Post-Human Philosophy.* Sesto San Giovanni: Mimesis International.

Marchesini, Roberto. 2022. *The Creative Animal: How Every Animal Builds Its Own Existence.* Cham: Palgrave Macmillan.

Marchesini, Roberto, and Marco Celentano. 2021. *Critical Ethology and Post-Anthropocentric Ethics: Beyond the Separation between Humanities and Life Sciences*, Numanities – Arts and Humanities in Progress. Vol. 16. Cham: Springer.

McKibben, Bill. 1989. *The End of Nature.* New York: Random House.

Meillassoux, Quentin. 2006. *Après la finitude: essai sur la nécessité de la contingence.* Paris: Seuil.

Merleau-Ponty, Maurice. 1995. *La nature: Notes, cours du Collège de France. Suivi de Résumés de cours correspondants.* Paris: Éditions du Seuil.

Monod, Jacques. 1971. *Chance and Necessity: An Essay on the Natural Philosophy of Modern Biology.* New York: Alfred A. Knopf.

Morton, Timothy. 2007. *Ecology Without Nature: Rethinking Environmental Aesthetics*. Cambridge, MA: Harvard University Press.
Morton, Timothy. 2010. *The Ecological Thought*. Cambridge, MA: Harvard University Press.
Morton, Timothy. 2012. *Ecological Thought*. Cambridge, MA: Harvard University Press.
Morton, Timothy. 2016. *Dark Ecology. For a Logic of Future Coexistence*. New York: Columbia University Press.
Paci, Enzo. 1956. *Ancora sull'esistenzialismo*. Torino: Edizioni Radio Italiana.
Plessner, Helmuth. 2019. *Levels of Organic Life and the Human*. New York: Fordham University Press.
Popper, Karl R. 1972. *Objective Knowledge: An Evolutionary Approach*. Oxford: Oxford University Press.
Putnam, Hilary. 2012. *Philosophy in an Age of Science: Physics, Mathematics, and Skepticism*. Cambridge, MA: Harvard University Press.
Rilke, Rainer Maria. 1961. *Duino Elegies*. Berkeley: University of California Press.
Rojek, Pawel. 2020. Internalisation of Realism. *Philosophia* 48:1575–1593.
Russell, Bertrand. 1903. *The Principles of Mathematics*. Cambridge: Cambridge University Press.
Russell, Bertrand. 2008. *A Critical Exposition of the Philosophy of Leibniz*. New York: Cosimo Inc.
Schaffer, Jonathan M. 2010. Monism: The Priority of the Whole. *Philosophical Review* 199 (1): 67–87.
Stanzione, Massimo. 1984. Epistemologia evoluzionistica: confronti e critiche. In *Evoluzione e modelli*, ed. Barbara Continenza et al. Roma: Editori Riuniti.
Thomas, Keith. 1983. *Man and the Natural World: Changing Attitudes in England 1500–1800*. London: Allen Lane.
Tinbergen, Niko. 1951. *The Study of Instinct*. Oxford: Clarendon Press.
Tsing, Anna Lowenhaupt. 2015. *The Mushroom at the End of the World: On the Possibility of Life in Capitalist Ruins*. Princeton: Princeton University Press.
Vallortigara, Giorgio. 2021. *Pensieri della mosca con la testa storta*. Milano: Adelphi.
Vogel, Steven. 1996. *Against Nature: The Concept of Nature in Critical Theory*. Albany: State University of New York Press.
Vollmer, Gerard. 1983. Mesokosmos und objektive Erkenntnis. In *Die Evolution des Denkens*, ed. Konrad Lorenz and Frans Wuketits. München: Piper.
von Uexküll, Jakob. 2010. *A Foray into the Worlds of Animals and Humans: With a Theory of Meaning*. Minneapolis: University of Minnesota Press.
Wolpert, Lewis. 1994. *The Triumph of the Embryo*. New York: Oxford University Press.

CHAPTER 10

Conclusions

Philosophical ecology is a broad, interdisciplinary inquiry that spans anthropology, ethics, ontology, and epistemology. Drawing on key metaprinciples from ecological thought, it aims to establish a more integrated framework for understanding reality. Even though it moves beyond ecology understood strictly as a descriptive science, philosophical ecology is inspired by some of its core ecological concepts—such as symbiosis and niche—that help illuminate patterns of connection and mutual interdependence (McIntosh and Feltrin 2024). These concepts allow us to move beyond the limitations of both traditional humanistic assumptions and vague generalizations like "everything is connected to everything else."

At the heart of philosophical ecology lies the principle of relation. However, this principle is not developed in abstract or generic terms, but rather through the specific lens of the living world. It does not conceive of the entity as a passive "being-in-relation"—a notion that implies an undifferentiated relational openness—but as a *protagonist of relation* (Macelloni 2023). This means that a living being is already shaped by its phylogeny to be predisposed toward certain kinds of relations (relational inherence), and it occupies a particular position within an interactive environment (relational ecology). To fully grasp the relational condition of a living being, then, we must consider the unique phenomenology of life itself.

Philosophical ecology seeks to reframe our relationship with nonhuman otherness, moving beyond both the humanist logic of

R. Marchesini, *Philosophical Ecology*,
https://doi.org/10.1007/978-3-032-04987-2_10

disjunction—which casts the human as a separate and exceptional entity—and the anthropocentric view that reduces the non-human to a mere resource. From this perspective, it becomes evident that the challenges humanity faces cannot be dismissed as purely technical or operational issues; rather, they are deeply rooted in the dominant philosophical paradigm. Philosophical ecology, then, is not simply a response to today's ecological crises. Its ambition is broader: to foster a cultural metamorphosis in how problems are framed and addressed. It proposes an eco-logical metric—one that recognizes non-human entities not as resources to be used, but as integral components of human ontology. In this context, philosophical ecology promotes specific areas of research that highlight the foundational role of otherness in the ontopoietic development of the human condition and its cultural expressions. These include eco-criticism, which explores the relationship between literature and the environment—both as a mirror of and influence on our connection to nature—and zoo-anthropology, which emphasizes the profound inspiration human beings have drawn from natural entities, particularly other animals, in the shaping of culture.

Philosophical ecology also aims to establish a shared foundation for dialogue among different ethical approaches to our relationship with non-human otherness. One of its central concerns is the inclusion of non-human entities within the sphere of moral consideration. As long as these entities are seen solely in terms of their usefulness—as "available goods"—even the most trivial human needs will take precedence, and every development policy will inevitably aim to convert biodiversity into a human product.

This ethical dimension is perhaps the most familiar aspect of ecological thought—so much so that philosophical ecology is often mistaken for a variant of environmental ethics. However, I would argue that ethics cannot be the starting point. Before we can engage in moral reflection, we must first examine the ontological foundations—particularly the phenomenology of life and the concept of subjectivity—since all ethical positions ultimately stem from underlying value attributions. Only through a rigorous critique of anthropocentric ontology can we begin to face the ecological challenges that now press upon us with increasing urgency. At the same time, the need for a discipline capable of helping us think ecologically—even if only by telling new stories (Berry 1990) or dispelling old myths (Harari 2011), serving as a stimulus for the imagination and enabling a shift in perspective and a change in our epochal narrative (Servigne et al. 2021)—is clear and urgent. It is a challenge we can no longer avoid.

Yet while the pressing nature of our current situation makes a timely change of course compelling, I believe the need for philosophical ecology is rooted in more than just the crises we face today. The critique of anthropocentrism arises from a deeper metamorphosis within Western thought—one that began to take shape even before the twentieth century, marked by the gradual decline of humanism. That intellectual edifice, which stood for five centuries and undoubtedly yielded important achievements, regarded the human being as separate from the non-human and hierarchically superior to nature. It began to falter philosophically with the legacy of Nietzsche, and ontologically with the implications of Darwinian evolutionary theory. Likewise, the technological revolutions of the past fifty years have also delivered a powerful blow to the presumptions of uniqueness and superiority of the human species. The result has been a deeply diminished view of the human, particularly in terms of ontological distinctiveness, uniqueness, and universality. This shift has been shaped by, among other things: (i) paleoanthropology's revelation of multiple human species; (ii) cognitive ethology's discoveries about animal intelligence and cultural behaviors; and (iii) insights from artificial intelligence that expose the partiality of human cognition and the pervasiveness of epistemic biases.

The ecological crisis—already evident by the 1980s and made dramatically worse by climate change—has finally caught up with us. More than the crisis itself, however, it is the evident inability of global policies to reverse or even slow the trend of ecological degradation that has forced a reckoning. As Konrad Lorenz anticipated in *Civilized Man's Eight Deadly Sins* (1974), this reckoning concerns not only our responsibilities but also our limitations. The intensification of global capitalism in the decades since has only exacerbated the issues Lorenz identified, effectively legitimizing the idea that the non-human world is little more than a low-value resource to be freely exploited (Moore 2016).

A shift in philosophical perspective is urgently needed—but such a change is unlikely within a cultural framework still entrenched in anthropocentrism. This framework clings to *techne* either as the weapon of a reawakened Prometheus or as a protective shell—a technological exoskeleton—offering refuge, as in Peter Sloterdijk's (2016) revival of the nature-versus-culture dualism. Thinking ecologically, by contrast, requires us to move beyond the methodological assumptions rooted in the humanist tradition—assumptions that have long underpinned the ontological separation between humans and the non-human world. These include:

1. Essentialism—the belief that an entity's properties stem innately from within, so that true understanding of a being requires cleansing it of all external influences;
2. Autonomy—the idea that entities are self-contained, independent, autarchic, and can be fully understood in isolation from their relationships;
3. Dualism—in its various forms of complementarity, estrangement, or outright opposition, all of which imply a fundamental detachment between an entity and the external elements that help constitute it;
4. Passivity and relational genericism—the notion that entities are passive recipients of external relations, rather than active participants in shaping the world they inhabit.

In opposition to all the above, philosophical ecology emphasized the relational quality of ontology. However, it is not merely a philosophy of relation. What sets it apart are certain foundational principles. Philosophical ecology embraces:

(i) the evolutionary nature of the entity, meaning its development is shaped by its interactions with the external world—through phylogeny, ontogeny, and physiogeny—so that intrinsic traits, understood as internal potential and relational inherence, are in fact reflections of past external relations;
(ii) the diachronic character of the entity, which resists evaluation in synchronic terms. A subject is insofar as it has-been and becomes—hence, ontology is understood as a dynamic flow; and
(iii) the ecological nature of the entity, meaning it is situated in a specific context within the world, where everything is indeed connected—but not indiscriminately. Rather, connections occur through structured layers, patterns, and relational sequences, so the precise matrix of connections must be defined.

In this sense, philosophical ecology speaks of ontopoiesis as the entity's inevitable mode of being—rooted in the phenomenology of life—because any manifestation of an entity is always an actualization that occurs through otherness. It is the relation that brings the predicate into being. As a result, the idea of separating an entity from the world becomes untenable: external relations never remain such, but are always internalized through various mechanisms of assimilation. No living being can be treated as a self-contained entity, but only as a sympoietic element—co-creating and giving rise to hybrid forms.

The theories of relation developed over the past century have rightly emphasized the importance of relation in understanding the ontology of entities. However, because they remain rooted in a humanist paradigm, they tend to present relations in a dualistic way—such as in the debate between internal and external relations (Frigerio 2023). This is the original sin of humanism: it inevitably replicates, almost like a virus, explanatory models that, though differing in form and terminology, remain centered on dualism and opposition. The underlying goal is always to separate the human being from everything else, which leads to dividing knowledge into humanities versus natural sciences, as Charles Percy Snow famously described in *The Two Cultures* (2013). Furthermore, humanism cannot escape essentialism, as shown in theories of human incompleteness (Gehlen 1988) or poignancy (Heidegger 1995), where external elements may influence how an entity appears but not its intrinsic nature. Over time, dualism and opposition have taken many forms, but ultimately they rest on the same foundational idea: the separation of subject and world. Whether this separation is framed as a device of salience—like Agamben's anthropological machine (2003)—or as an interference or concealment of the entity—such as Heidegger's principle of *alétheia* (1978)—it assumes mutual estrangement. This assumption makes it impossible to adopt a relational logic that avoids all forms of dichotomous attribution.

As soon as one abandons the idea of an entity's ontological autonomy, one is often accused of denying its distinctiveness—of reducing the entity to something formless. But this is a misunderstanding. To attribute an ontopoietic dimension to the entity—where internal and external relations are seen only as phenomenological moments, not as absolute distinctions—is not to deny its core. When Morton (2007) claims that a new ecology requires stepping away from science and returning to the humanist tradition, he ends up reinstating the very disciplinary divisions his conceptual renewal claims to overcome. His move is, in fact, a step back from the opportunity to fully move beyond—consistently with an ecological perspective—the split between the human and the world. Indeed, the real obstacle to ecological thinking is precisely dualism—not because we must fall into an undifferentiated monism, but because we need to recognize that the reality of living beings unfolds across multiple planes of ontopoiesis, not through binary opposition (Deleuze and Guattari 1987). These planes are so deeply entangled that the external relations of one become internal relations for the next. Any attempt to frame them in a dichotomous logic is ultimately artificial.

Dualism does not only manifest as outright opposition; it also appears as juxtaposition, where a relation is acknowledged while maintaining the impermeability of the two terms. Such a dualistic, disjunctive reading is inherently non-ecological: it leads inevitably to a model of complementation, where the relationship is reduced to the need to fill a lack with an external element. Philosophical ecology reveals the fallacy of this reasoning: the capacity of an entity to assume different forms and to internalize external relations is not a sign of deficiency, but of redundancy. The relations an entity establishes with the external world are actualizing processes—mechanisms that convert its internal virtuality (the latent range of possible configurations) into realized forms. In this view, there is a directly proportional relationship between a system's complexity and the breadth or variability of its potential manifestations, as shown by the wiring of the human central nervous system.

This stands in contrast to the logic of complementation, which is based on inverse proportionality. The dualist, disjunctive framework of humanism can only support a complementary model, one that sees, for instance, the development of human culture as a response to a natural deficiency. Starting from such a model—where the more important external relations become, the weaker the intrinsic nature of the entity appears—inevitably leads to the devaluation, if not outright negation, of the entity's own being. But to argue that an entity can only manifest itself through relations is not to deny or diminish its distinctiveness; rather, it is to affirm its endogenous complexity and redundancy. In this sense, philosophical ecology draws deeply from complexity theory and from thinkers who have explored chaos dynamics, threshold-based systems, actualization processes, and nonlinear causality.

The philosophical debate on internal versus external relations continues to rely on the assumption of the thing-in-itself as stable and separate from the world—a residue of the fixist and pre-established conception of entities that Darwinian evolution and ecological thought have already dismantled. These disciplines have shown that living beings are not autonomous or self-contained, but fundamentally shaped by and situated within their environments. Their morpho-functional traits emerge through adaptation, revealing an inherent relationality; and through niche construction, they actively influence and are influenced by their ecological context. In both cases, life is not passive but elective and participatory. Humanist thought has largely ignored these insights. It continues to treat the intrinsic nature of entities as isolated from the world, failing to recognize that the intrinsic reflects the world and the extrinsic defines the entity's position within it. This persistent dualism—alternating between favoring internal or external relations depending on the thinker—ultimately relies on tautological reasoning. Philosophical ecology aims to move beyond this flawed way of thinking by acknowledging that entities evolve through increasing integration with their environments, internalizing external elements, and actively shaping the relational contexts in which they are embedded through protagonism and agency (Rignani 2018).

In my view, the difficulty in genuinely engaging with philosophical ecology lies in philosophy's enduring attachment to a form of anthropocentric immunity—an attachment that leads it to overlook the ontological upheaval introduced by evolutionary theory, which dismantled essentialism (Mayr 1982), and by ecology, which introduced the notion of the being's predicative heteronomy (Gilbert and Epel 2015). As a result, even in the face of evolutionism, ecology, and complexity thinking, philosophical discourse remains prone to falling back into a disjunctive conception of agency. To think ecologically, therefore, means adopting a different perspective. The essentialist distortion—widely recognized as a major obstacle to scientific understanding—stems from a non-ecological way of interpreting phenomena (Bachelard 2002). This distortion has affected not only our relationship with the biosphere but also our view of human destiny. It has led us to see our condition as separate from the world, fostering a sense of distance and alienation from everything non-human (Steiner 1975). To fully grasp why ecological concerns are often ignored and nature undervalued, we need to question a deeper ontological assumption that goes beyond the impact of human activity on the environment. In other words, we must rethink our underlying mindset, because we are still stuck on the idea of fixed essences, which prevents us from thinking in truly eco-ontological terms (Marchesini 2024).

Indeed, the greatest mistake threatening our ability to understand the current crisis is the tendency to treat the biosphere as merely a place where we live—an environment that simply hosts us and supplies the resources we need—while seeing it as something separate, not a co-creator of our existence. When we talk about an environmental crisis, we often emphasize precisely the biosphere's role as the source of planetary resources and its role as a mere backdrop, set against the presumed ontological impermeability of humans. This is like imagining the brain as something lodged inside the body but fundamentally foreign to it, or seeing the genome as a detached control room for shaping the phenotype. This is what I call predicate essentialism—where the connotation is reduced to an emanation of the entity itself (Janssen-Lauret 2021), severing it from its relations. This is not only a clear epistemological obstacle that prevents us from understanding phenomena like the polyphenism of living beings and their capacity to acquire different forms through their intrinsic virtuality of configurations; it is also a heavy burden on the ontological understanding of entities themselves, and ultimately on human ontology.

In this book, I have proposed an eco-ontological approach as a different way of understanding the predicative dimension of the living, one that can never be grasped by separating the entity from its relational field—its specific being-with. This being-with goes beyond mere being-there because it presupposes the entity's ontological permeability to changes brought by the external world, transforming ontology into ontopoiesis. Beyond my own eco-ontological proposal, based on an evolutionary and ecological vision of being-in-relation, I believe philosophical ecology offers a fertile ground for developing a new cultural paradigm—one that can fully incorporate the meta-principles and significance that evolutionism and ecology have contributed to ontology. Only by adopting a new, eco-ontological perspective, can we hope to truly change our way of relating with otherness. Only then will we be able to truly address the current environmental crisis.

References

Agamben, Giorgio. 2003. *The Open: Man and Animal.* Stanford: Stanford University Press.

Bachelard, Gaston. 2002. *The Formation of the Scientific Mind.* Bolton/Manchester: Clinamen Press.

Berry, Thomas. 1990. *The Dream of the Earth.* San Francisco: Sierra Club Books.

Deleuze, Gilles, and Félix Guattari. 1987. *A Thousand Plateaus: Capitalism and Schizophrenia.* Minneapolis: University of Minnesota Press.

Frigerio, Christian. 2023. *Ricomporre un Cosmo in Frammenti: Filosofia, Ecologia e Relazioni.* Sesto San Giovanni: Mimesis.

Gehlen, Arnold. 1988. *Man: His Nature and Place in the World.* New York: Columbia University Press.

Gilbert, Scott, and David Epel. 2015. *Ecological Developmental Biology: The Environmental Regulation of Development, Health and Evolution.* Oxford: Oxford University Press.

Harari, Yuval N. 2011. *From Animal into Gods.* Israel: Kinneret Zmora-Bitan Dvir.

Heidegger, Martin. 1978. *Being and Time.* Oxford: Blackwell.

Heidegger, Martin. 1995. *The Fundamental Concepts of Metaphysics: World, Finitude, Solitude.* Bloomington: Indiana University Press.

Janssen-Lauret, Frederique. 2021. Anti-Essentialism, Modal Relativity, and Alternative Material-Origin Counterfactuals. *Synthese* 199 (3–4): 8379–8398.

Lorenz, Konrad. 1974. *Civilized Man's Eight Deadly Sins.* New York: Harcourt Brace Jovanovich Inc.

Macelloni, Manuela. 2023. *La filosofia del cane: Orme per un futuro post-umanista.* Milano: Mimesis.

Marchesini, Roberto. 2024. *Posthumanist Manifesto.* New York: Lexington Books.

Mayr, Ernst. 1982. *The Growth of Biological Thought: Diversity, Evolution, and Inheritance.* Cambridge, MA: Belknap Press of Harvard University Press.

McIntosh, Shoshana, and Andrea Natan Feltrin. 2024. Dismantling Human Supremacy: Ecopedagogy and Self-Rewilding as Pathways to Embodied Ethics and Cross-Species Solidarity. *The Relations* 12 (2). https://doi.org/10.7358/rela-2024-02-mcfe.

Moore, Jason W. 2016. *Anthropocene or Capitalocene, Nature, History and the Crisis of Capitalism.* Oakland: PM Press.

Morton, Timothy. 2007. *Ecology Without Nature: Rethinking Environmental Aesthetics.* Cambridge, MA: Harvard University Press.

Rignani, Orsola. 2018. *Metafore del corpo post-umanista: Michel Serres.* Sesto San Giovanni: Mimesis.

Servigne, Pablo, Raphaël Stevens, and Gauthier Chapelle. 2021. *Another End of the World Is Possible: Living the Collapse (and Not Merely Surviving It).* Cambridge: Polity Press.

Sloterdijk, Peter. 2016. *Not Saved: Essays After Heidegger.* Malden: Polity Press.

Snow, Charles Percy. 2013. *The Two Cultures and the Scientific Revolution.* Mansfield Centre: Martino Fine Books.

Steiner, George. 1975. *After Babel: Aspects of Language and Translation.* Oxford: Oxford University Press.

Index

R. Marchesini, *Philosophical Ecology*,
https://doi.org/10.1007/978-3-032-04987-2

F

G

H

I

K

L

M

N

O

P

FSC
www.fsc.org
MIX
Papier | Fördert
gute Waldnutzung
FSC® C083411

Kolibri 360 GmbH
Ferdinand-Jühlke-Straße 7
99095 Erfurt, Deutschland
produktsicherheit@kolibri360.de